CURRENT TOPICS IN RADIATION RESEARCH

ADVISORY BOARD

H.S. Kaplan, *Department of Radiology at Stanford University School of Medicine, Palo Alto, California, U.S.A.*

R. Latarjet, *Institut de Radium, Paris, France*

K.G. Zimmer, *Institut für Strahlenbiologie, Kernforschungszentrum Karlsruhe, Germany*

CURRENT TOPICS IN RADIATION RESEARCH

EDITED BY

MICHAEL EBERT

AND

ALMA HOWARD

Paterson Laboratories, Christie Hospital & Holt Radium Institute, Manchester, U.K.

Volume 13 – 1978

NORTH-HOLLAND PUBLISHING COMPANY
AMSTERDAM

© *North-Holland Publishing Company* – *1978*

All rights reserved. No part of this publication may be reproduced, stored in a retrieval system, or transmitted, in any form or by any means, electronic, mechanical, photocopying, recording or otherwise, without the prior permission of the copyright owner.

ISBN: 0 444 85164 x

Library of Congress Catalog Card Number: 65-7977

Publishers:

NORTH-HOLLAND PUBLISHING COMPANY – AMSTERDAM • NEW YORK • OXFORD

Sole distributors for the U.S.A. and Canada:

ELSEVIER NORTH-HOLLAND, INC.
52 Vanderbilt Avenue
New York, N.Y. 10017

PRINTED IN THE NETHERLANDS

PREFACE

This volume will be the last in the series Current Topics in Radiation Research. In the fourteen years since its inception more than eighty papers from authors in fourteen countries have been included, as well as selected contributions from three international meetings. The range of subject matter of the papers reflects the diversity of fronts along which radiation research has been advancing and the several disciplines in which it has its roots.

In choosing authors and material for this series we have been guided by three main principles. Firstly, real progress in science depends on basic research and it is the fundamental approach which in the long run proves to be the quickest and most effective route to the solution of practical problems. Secondly, no mechanical aid to information collection and synthesis can replace the individual human mind, so that unconstrained articles by research workers themselves are likely to be the most informative and stimulating sources of progress in science. And thirdly, radiation research, in spite of the great diversity of systems and purposes covered by it, has throughout a common focus in the events immediately following the absorption of energy, and it is well worthwhile to try to keep this focus in mind and help radiation workers to be informed about progress in fields neighbouring their own.

With the ever-increasing industrial and technical use of radiation, its importance and expansion in radiotherapy and diagnostic medicine, the use of nuclear reactions to bolster dwindling energy supplies throughout the world, and current concern and apprehension about the development of nuclear weapons, it is obvious that radiation research continues to be a field of prime importance. We are grateful to our publishers for giving us, through this series, the opportunity of contributing towards the progress and understanding of this field, and we regret that we shall no longer be continuing this contribution.

THE EDITORS

CONTENTS

Preface v
C.S. Potten, *The cellular and tissue response of skin to single doses of ionising radiation* 1
S. Apelgot and J.P. Adloff, *Transmutation effects of ^{32}P and ^{33}P incorporated in DNA* 61
W.E. Liversage and R.G. Dale, *Dose-time relationships in irradiated weevils and their relevance to mammalian systems* 97
E. Polig, *The localised dosimetry of internally deposited alpha-emitters* 189
Subject Index 329
Author Index 332

THE CELLULAR AND TISSUE RESPONSE OF SKIN TO SINGLE DOSES OF IONISING RADIATION

Christopher S. POTTEN

Paterson Laboratories, Christie Hospital and Holt Radium Institute, Withington, Manchester M20 9BX, England

Radiation interaction with an organised tissue such as skin results in many alterations at the cellular, subcellular and even biochemical level. Undoubtedly the interaction with DNA is the most important. This can result in the alteration of the genetic programme, damage at the chromosomal level and even loss of DNA or chromosomes. A significant consequence of this can be loss of reproductive capacity. The skin consists of many intimately interrelated cell populations all of which are replacing lost cells by division although the replacement rates may vary considerably. The effects of single doses of ionising radiation on these various cellular constituents are reviewed. The predominant cell population is the epithelial population. However, even this one cell type has several subpopulations of cells. These are not only evident in the skin appendages but also in the morphologically uniform interfollicular basal layer. The clonogenic cell survival assay, the light and electron microscopic behaviour of the cells after irradiation and the frequency of morphologically distinct dead cells (pycnotic or apoptotic cells) all suggest that there are different radioresponsive cell populations in the epidermis. Most radiobiological studies have considered only one cell type, usually the epidermal cells, in total exclusion of the other cell populations. However, it seems likely that the behaviour of both damaged and undamaged epithelial cells is affected by the wellbeing of neighbouring non-epithelial cells. Capillaries, for example, are sensitive to low doses and show a rapid response. Epithelial cells are likely to be influenced by these changes in the oxygen, nutrient and chemical message transporting elements. The role of the other cell types in determining the skin reaction is less clear or totally unknown. Fibroblasts and Langerhans cells are two cell populations that may well influence epidermal cell behaviour considerably. The radiobiological behaviour of these cells is relatively unknown. These points are reviewed and discussed.

CONTENTS

1. INTRODUCTION 3
   1.1. Objectives 3
   1.2. The skin as a composite tissue 3
   1.3. Epidermal organisation 6
   1.4. Epidermal cell kinetics 7

2. GENERAL EFFECTS OF IRRADIATION 9
   2.1. Historical perspective 9
   2.2. Erythema and skin reaction 10
   2.3. Epilation 11
   2.4. Effects on sebaceous glands 14
   2.5. Effects on epidermal vasculature 14
   2.6. Effects on skin pigment cells (melanocytes) 15
   2.7. Effects on the dermis 16

3. RADIATION SURVIVAL CURVE DATA FOR EPIDERMAL CELLS 17
   3.1. Review 17
   3.2. Microcolonies or macrocolonies? 21
   3.3. Regenerating cell doubling time 23
   3.4. Cell migration during epithelialisation 23
   3.5. Epithelial cell numbers 24
   3.6. Clonogenic cell numbers 24
   3.7. Variability in survival curves 29
   3.8. Epithelial cell D_0 values 29
   3.9. Summary of epidermal cell survival data 31

4. RADIATION INDUCED CELL DEATH IN EPIDERMIS 31
   4.1. Pycnotic or apoptotic cells 31
   4.2. Changes in basal density 34
   4.3. Behaviour of non-pycnotic/apoptotic cells 38
   4.4. Summary 44

5. LANGERHANS CELLS 45
   5.1. Description 45
   5.2. Effects of irradiation 49
   5.3. Possible functions 51

6. GENERAL CONCLUSIONS AND FUTURE OBJECTIVES 53

ACKNOWLEDGMENTS 54

REFERENCES 54

1. Introduction

1.1. Objectives

This report summarises and reviews some of the literature on epidermal response to irradiation with particular emphasis on the fact that skin is a tissue composed of many intimately interrelated cell populations and subpopulations each of which may respond in its own way to irradiation. The literature review is not necessarily comprehensive but representative and the objectives are to emphasise the biological side of the radiobiology of skin and to suggest that though epidermal basal cell killing may be the predominant damaging process it is not the only one and it is not clear that all epidermal cells respond similarly to radiation. The epidermal response may be influenced by damage to other cellular constituents.

1.2. The skin as a composite tissue

The skin is a complex organ composed of many cell types all intimately interrelated to ensure an effective functional tissue. The structural arrangement (thickness, basal layer folding, hair follicle density etc.) varies slightly from region to region but the basic cellular constituents are constant (fig. 1 and table 1). I present these not as an introduction to somewhat basic skin

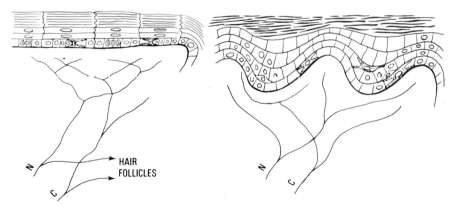

Fig. 1. Schematic representation of two types of skin histology. Left hand diagram a typical thin, flat slowly proliferating epidermis with a single basal layer and columns of differentiating and cornified cells (EPUs). N, nerve fibres; and C, capillaries are abundant in the dermis; dendritic melanocytes stippled. Right hand diagram a more rapidly proliferating epidermis with undulating basal layer (rete pegs), some suprabasal cell proliferation, and an apparent random arrangement of cornified cells.

Table 1
Skin cellular constituent [a]

Epidermis:	Keratinocytes;	Proliferative;	Stem?
			Committed?
		Differentiated	
	Melanocytes;	Melanoblasts?	
		Melanotic (melanocytes)	
	Langerhans cells;	various appearances	
		(different functional states?)	
	Infiltrating cells;	migrating blood cells?	
	Merkel cells;	(nerve endings?)	
Dermis:	Fibroblasts		
	Macrophages		
	Histiocytes		
	Mast cells		
	Chromaffin cells		
Hair follicles:	Keratinocytes;	Proliferative;	Stem?
			Committed?
		Differentiated	
	Melanocytes;	Melanoblasts	
		Melanocytes (melanotic)	
Sebaceous glands:		Proliferative	
		Differentiated	
Sweat glands (eccrine and apocrine)			
Nerves and nerve endings:			
Vascular supply:		Endothelial cells	
		Circulating blood cells	
		Lymphatic system	
Muscle cells			
Adipose cells			

[a] Based on Montagna [1962]; Jarrett [1974].

biology but to show the number and complexity of the cellular constituents each one of which will respond to the damaging effects of ionising radiation. Their individual response will be determined by their individual radiosensitivities which will be dependent on their proliferative activity or degree of differentiation, the level of oxygenation, the proximity of other dead or damaged cells, their functional relationship with neighbouring cells, etc.

The effects of irradiating the skin are clearly the composite consequence of the radiation response of the individual cellular constituents and it may be misleading to consider the response of one cell type to the exclusion of all

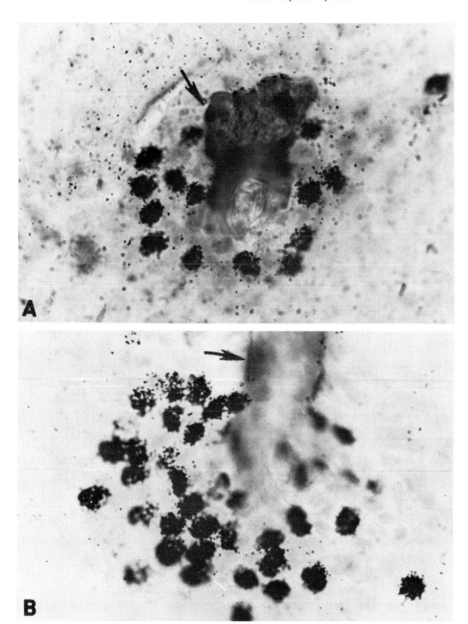

Fig. 2. Mouse epidermal sheet autoradiograph prepared 3–6 days after irradiation showing many heavily labelled cells in and around a hair follicle (arrow). A. Showing a small focus of labelled cells appearing late (6 days). B. Showing a large focus appearing earlier (3 days). A. 1,200 rad day 6; B. 1,000 rad day 3. A, B × 900.

others. The radioresponse of only a few of the cellular constituents has been studied (epidermis, melanocytes, hair follicles, with some information on sebaceous glands and vascular tissue). The information obtained has in some cases been somewhat limited.

The hair follicles and their associated sebaceous glands are formed from invaginations of the epidermal basal layer. This process occurs in rodents during the later stages of foetal development and continues for the first few days post partum. Once this invagination process is over no new follicles are formed [Claxton, 1966].

It is possible that after severe radiation depletion new follicles can be derived from regenerating epithelium which could account for the presence of occasional hairs in regenerating foci [Withers, 1967a]. It is, however, well documented that hair follicles or sebaceous glands can provide cells for epidermal regeneration (fig. 2) [Bishop, 1945; Eisen, Holyoke and Lobitz, 1956; Kligman, 1959; Roberts, 1965; Krawczyck, 1972; Oduye, 1975; Argyris, 1976; Al-Barwari and Potten, 1976]. Sebaceous glands can also reform from follicle canal cells [Montagna and Chase, 1950].

Some indication of the intimate interrelationship between the various cell populations is given by the concomitant change in epidermis, hair follicles, sebaceous glands, vasculature, biochemistry of the skin (e.g. glycogen), adipose changes etc. during cyclic hair growth (spontaneous and experimentally induced), [Hamilton, Howard and Potten, 1974; Potten, 1971; Johnson and Ebling, 1964; Chase, Montagna and Malone, 1953; Moffat, 1968; Moretti, Baccaredda-Boy and Rebora, 1969].

1.3. Epidermal organisation

A relatively new development in skin biology has been the recognition that in some epidermal regons from many mammalian species (including man) the cornified cells, which are large, thin flattened, roughly hexagonal plates of keratin, are arranged in precise columns [MacKenzie, 1969, 1970; Christophers, 1971a,b,c]. The demonstration of these columns initially involved some complex histological techniques but they have now been seen in several more routine types of preparation such as epidermal sheets separated at the basement membrane [MacKenzie, 1969, 1970; Potten, 1974] and 1 micron epon-embedded material or the electron microscope [MacKenzie, 1972; Allen and Potten, 1974]. In the latter case the column boundaries can be traced to the basal layer and can be seen to have a minimum overlap with neighbouring columns and to show a regular alternating pattern at the column boundaries [Allen and Potten, 1974; Potten and Allen, 1975a; Potten, 1976a]. The columnar arrangement exists in many regions but a more random arrangement is found on the palmar and plantar surfaces and in any region

Table 2

Cellular constituents of mouse Epidermal Proliferative Units		Possible function of the EPU basal cells	
Cornified cells (dead)	6+	Differentiated proliferating cells	~6 ($T_c \sim 100$ h)
Maturing cornified cells (metabolically active)	3(2–4)	Post-mitotic maturing cells	2–3
Spinous and granular layer cells		Stem cells (clonogenic)	1 (T_c possibly 7–10 days)
		Langerhans cells	1 (slowly cycling)
Basal cells	9–12 (mean for dorsum 10.6)		
Langerhans cells	1 (included in the 9–12)		
Melanocytes	~0.3		

Data based on Potten [1974, 1975a]; Potten and Hendry [1973]; Allen and Potten [1976, 1974]; MacKenzie [1975a, b, c]; Potten [1975b]; Christophers et al. [1974]; Iversen et al. [1968].

that has been stimulated into increased cell production (fig. 1) [Allen and Potten, 1976; Potten and Allen, 1975a]. There is a slightly different regular arrangement in tail epidermis [Allen and Potten, 1976] and a more complex columnar arrangement in the tongue filiform papilla [Hume and Potten, 1976]. The presence of these highly organised colunms of differentiated, differentiating and proliferating cells has led to the suggestion that the epidermis can be regarded as being composed of many discrete somewhat autonomous functional groupings of cells; i.e. the epidermal proliferative unit (EPU) concept [Potten, 1974]. The composition of these EPU can be estimated and is summarised in table 2. The validity of some of the functional roles suggested for the cells is still in dispute and the degree of overlap between some categories is not clear but these points are discussed elsewhere [Potten, 1975a, 1976a].

1.4. Epidermal cell kinetics

There are several review papers which include data on epidermal cell kinetics [Cleaver, 1967; Cameron, 1971; Weinstein and Frost, 1969; Potten,

1975a] and although the picture is not completely clear it appears that the epidermal cell cycle time (T_c) [Howard and Pelc, 1953] for most body regions and mammalian species lies between 90 and 120 h [Iversen, Bjerknes and Devik, 1968; Hegazy and Fowler, 1973a; Potten, 1975a]. However, there are several reports where the estimates for T_c are longer; a notable case being that of mouse ear epidermis. In some cases there is clearly confusion between turnover and cell cycle times, i.e. the functional epidermal basal cell growth fraction (GF) [Mendelsohn, 1962] is unknown or not considered. A review of the literature often reveals contradictory results when T_c values based on mitotic or labelling studies or percent labelled mitosis data are compared. There is also some conflict about the growth fraction data and the significance of continuous labelling studies for GF determination. The GF probably lies between 0.6 and 0.85 [Potten, 1975a, 1976a]. The duration of the DNA synthetic (S) [Howard and Pelc, 1953] phase (T_s) is about 11 h while the duration of mitosis (T_m) is between 1 and 2 h [Potten 1975a]. Some epidermal cells may be blocked in G_2 [~5%, Gelfant, 1966; Gelfant and Candelas, 1972] while others might be blocked in G_1 [i.e. in G_0, Lajtha, 1964] as suggested by the very long G_1 and wounding experiments where there is a fairly synchronous burst of DNA synthetic activity starting about 10 h after stimulation [Potten, 1972; Potten and Allen, 1975b]. These two blocks are also implied from the extensive work on skin chalones [Bullough, 1973a,b; 1976; Houck, 1976; Elgjo, 1973; Elgjo, Laerum and Edgehill, 1971, 1972].

The major controversy in epidermal cell kinetics is whether or not cell proliferation proceeds via a probabilistic uniform basal layer of stem cells [Osgood, 1957; Lajtha, 1963; Leblond, Clermond and Nadler, 1967], a model where all basal cells have the same proliferative capacity, are self-maintaining, and produce cells that migrate or terminally differentiate at random along the basal layer. Alternatively a more deterministic model could be envisaged where cell production or migration is not random but precisely programmed possibly according to the age (in the cell cycle or in terms of number of cycles) of the cell. Within this scheme the basal layer could either consist of uniform stem cells or of stem cells and differentiated proliferative cells which are not self-maintaining and have a limited division potential [Potten, 1975a, 1976a]. Each model has different implications for the interpretation of the cell kinetic and radiobiologic data.

As will be discussed later, some of the radiobiological results can be explained by a model where the basal layer is composed of at least two types of proliferating keratinocytes. The data also appear to suggest that the fraction of basal cells that are capable of extensive regeneration is less than 0.1. A more deterministic two-compartment model is now believed to exist in the intestinal mucosa [Leblond and Cheng, 1976; Cheng and Leblond, 1974;

Potten and Hendry, 1975; Potten, 1976b, c] and may exist in the tongue epithelium [Hume and Potten, 1976] and has for a considerable time been accepted for the haematopoietic system [Lajtha, 1964; Bruce and McCulloch, 1964]. It seems improbable that the epidermis would operate on a completely different type of scheme.

The high degree of structural organisation now known to exist in the epidermis, with precise columns of differentiated cells, would seem inconsistent with a random sequence of cell migration. This organisation suggests that both the basal layer and suprabasal layers (columns) are divided into functional units of proliferation and differentiation: the EPU concept [Potten, 1974; Allen and Potten, 1974]. The cell kinetic data suggest that each unit produces a little more than one cell a day, which when one considers the significant circadian rhythms [Tvermyr, 1972; Hegazy and Fowler, 1973a; Grube, Auerbach and Brues, 1970] is probably precisely one cell a day. The cells towards the centre of the group of 10 basal cells in the EPU have a slower turnover than the more peripheral cells [Potten, 1974, 1976a; MacKenzie, 1972, 1975a,b] and yet respond quickly and effectively to stimulation [Potten, 1974]

2. General effects of irradiation

2.1. Historical perspective

Roentgen described his results on the production of X-rays in December 1895 and very soon demonstrated their potential for radiology by producing and X-ray image of his hand. This somewhat rash demonstration was soon commonly used amongst roentgenologists and within a few years it took its toll in the form of moderate to severe skin injuries on the hand. The damaging effects of Roentgen's rays were further illustrated 3 months after Roentgen's discovery when an attempted radiological localisation of a metallic foreign body in the head was conducted. The result was a fairly complete scalp epilation [reported by Spear, 1953 and Ellinger, 1951]. The first use of radiation to cure cancer was in 1899 [reported by Ellinger, 1941]. Soon after Roentgen's report Becquerel described experiments on uranium and the emission of natural radiations in a paper to the Paris Academy of Science (November, 1896), followed within a few years by the work of the Curies. In 1901, Becquerel ill-advisably carried some radium in his pocket. This incident, like that with Roentgen, resulted in skin damage that was obviously related to the invisible radiations being generated naturally or artificially. This was further demonstrated by Curie using his own skin as experimental material. Even though his burns took several months to heal the damage was not considered serious and these burns were commonly inflicted deliberately. By 1902, six years after Roentgen's report, there were 167 recorded cases of accidental radiation burns.

It was not long before the carcinogenic action of these radiations was recognised by the early radium and roentgen-ray workers. By 1911 54 cases of malignant change had been recorded.

With this sort of beginning it is hardly surprising that the biological effects of ionising radiations very quickly became a field of great research activity. The skin and its associated structures being one of the first tissues to show an effect, and also being readily accessible were extensively investigated. The work in the period to the mid-nineteen fifties was concerned mainly with erythema reaction and epilation.

2.2. Erythema and skin reaction

The early radiation-induced burns were characterised by an initial three phase transient reddening of the skin (erythema) and if sufficient doses were used an ultimate desquamation and ulceration. It was soon evident that the second phase of erythema was, under standard physical and biological conditions, remarkably dose dependent and reproducible so that for many years this was the common way of expressing a biologically effective dose (skin erythema dose). The overall erythema is characterised by at least three phases: an initial somewhat variable phase (within hours or a few days); a second more reproducible phase (10–28 days) and a final rather variable phase which may develop much later.

It is clear that much of the first and last phases could be attributed to early (dilatation) and late effects (telangiectases) on blood vessels. The second phase may also be due in part to some vascular effects but it also involves other factors including direct epidermal cell killing and dermal damage.

The second phase in erythema in fact turned out to be so sensitive to variations in radiation quality, field size, dose rate, skin colour, thickness, hairiness, stage of hair growth, moistness, etc. that it became difficult to compare results from one hospital to another because it was rarely possible to standardise all these factors. This type of experimental approach has been improved and expanded into the extensive work currently being done using the end point defined as Skin Reaction.

The gross reaction of skin, which in mouse usually has a threshold of more than 1000 rad, does not begin until the 6–8th day after irradiation, usually reaches a peak value (which at the highest dose is ulceration) on the 12–25th day and, providing the dose is not too high, usually heals within 1–2 months [Denekamp et al., 1966; Denekamp, Ball and Fowler, 1969; Denekamp and Fowler, 1966; Hegazy and Fowler, 1973b; Field et al., 1975;

Fowler et al., 1974; Leith et al., 1975]. The reaction appears slightly earlier (4–5 days) if the skin is plucked, or is from mice homozygous for the hairless gene [Hegazy and Fowler, 1973b; Devik, 1962]. A skin reaction appears much more rapidly (1–2 days) if ultraviolet (UV) irradiation is used [Forbes, 1966; Forbes and Urbach, 1969]. The reaction appears within a few hours after UV, and may reach a peak within a few days; pigmentation changes usually occur on about the 5th day [Holti, 1955; Breit and Kligman, 1969]. Some of the UV reaction may be more readily attributed to vascular changes (dilation).

Irradiated field contraction is a dose dependent reaction that has a threshold and time course very similar to skin reaction. Contraction is assumed to be the consequence of dermal cell population damage [Hayashi and Suit, 1972].

It is not clear exactly what role basal cell depletion, suprabasal cell loss (thinning of epidermis in terms of strata) and vascular and dermal damage each play in the development and expression of skin reaction in view of the points already outlined.

2.3. Epilation

It was obvious from the early days of radiobiology that the hair follicles were sensitive to exposure to ionising radiation. However, at the time, it was not realised how this sensitivity and response was dependent on the hair growth cycle. Hair growth is cyclic with a 21 day period of growth in small rodents [Dry, 1926; Chase, Rauch and Smith, 1951]. The cycles are separated by periods of inactivity of varying length during which the few persisting follicle germ cells do not divide. When growth is initiated (either spontaneously or experimentally by plucking out the mature hair) there is a burst of mitotic activity and the follicle increases rapidly in size. In rodents the cycles are well synchronised and waves of synchronous growth pass down the animal's body while in humans each follicle is more autonomous (with the exception of the early post partum period). During the growth phase the follicle cells divide as rapidly as any in vivo mammalian cells (cell cycles of about 12 h). The resting follicle in mouse skin may have a depth in the dermis of about 250 μ while a growing follicle can be 2–4 times this depth and produces about 1 mm of hair a day. Both growing and resting follicles are formed by infolds of the epidermis and lie at an angle of about 30° in the skin (fig. 3). It is not clear how many precursor (stem) cells the follicles possess but a rough estimate might be 20 for a resting follicle while there may be 1–3 $\times$ 10^3 proliferative cells in a growing follicle. It is probable that only some of these are capable of follicle regeneration (i.e. are stem cells). Part of the variability in the depth, size and cellularity of follicles is due to the fact that

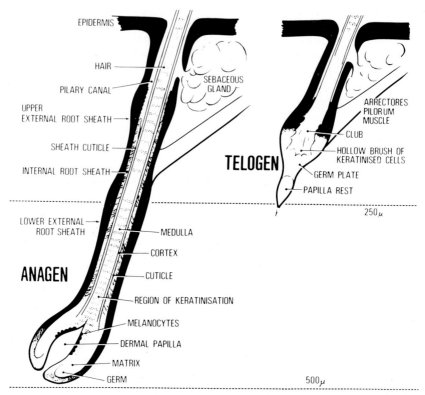

Fig. 3. Schematic representation of mouse hair follicles during the growth phase (anagen) and the resting phase (telogen) showing the various regions of the follicle and the differences in depth.

there are four very distinct types of hair and thus also of hair follicles in small rodents [Dry, 1926].

Since the total number of cells and their cycle activity varies during the hair cycle the radioresponse will also vary. Unfortunately the early workers did not appreciate the significance of the hair growth cycle and this accounts for the variability in the reported sensitivities. The second problem is that the end point studied was not well defined; frequently the distinctions between total hair loss, partial hair loss, temporary and permanent loss were not defined.

A dose of radiation to a resting hair follicle will kill follicle precursor cells according to their individual radiosensitivities (sensitivity of non-cycling (G_0) [Lajtha, 1963] cells) and the overall degree of oxygenation of the follicle. This cell killing activity may result in a loosening of the hair which may even fall out. Whether or not this epilation is permanent will depend on whether all the vital stem cells are killed, whether those that survive succeed in regenerating a functional follicle, (which may be very dependent on their position, i.e. whether they retain contact with the dermal papilla which has a follicle organising function [Cohen, 1969; Oliver, 1969]) and when after irradiation the end point is assayed since it will clearly take time to reform a follicle from one or a few surviving cells.

Irradiation of a growing follicle has more complex results. As with the resting follicle its ultimate survival depends on the sensitivity and level of cell killing in the stem cell compartment. This is unlikely to be the same size as the dividing cell compartment [Dubravsky, Hunter and Withers, 1976]. The rapidly dividing cells will, however, be sensitive to blockage in progression through the cell cycle (mitotic delay), to direct killing, and also to a cessation of new cell formation from any stem cell compartment. The consequence of this will be temporary changes in the rate of production of hair. This will be evident as local alterations in the hair thickness, i.e. hairs will be produced with thin regions and consequently with kinks and twists. This is the basis of the end point referred to as *hair dysplasia* [Williams, 1966; Van Scott and Reinertson, 1957; Malkinson and Griem, 1965; Griem and Malkinson, 1967]. This response is dose dependent with an initial threshold of about 300 rad. It is clear that the upper limit for the end point is the dose at which cell production is reduced to the point where the hair is so thin that it breaks and falls out (temporary growing follicle epilation) (about 900–1000 rad).

As discussed above there are difficulties in defining epilation particularly if regenerating epithelium is capable of forming new hair follicles (i.e. mimicking embryological sequences). The time interval between irradiation of the tissue and its observation becomes important as does the precise definition of epilation. A resting hair may persist for some time in a sterilised hair follicle. Alternatively a hair may be lost from a resting or growing follicle that subsequently reorganises and reforms a functionally competent follicle. The reformation of a functionally competent follicle may take longer at high doses than at lower doses. With these limitations in mind some recent studies have suggested that the hair follicles survive according to a dose response curve with a slope (D_0 value) [Alper, Gillies and Elkind, 1960; Alper et al., 1962] of about 200 rad [Dubravsky et al., 1976]. This D_0 value must also express the sensitivity of the individual hair follicle stem cells. The assay was conducted on skin 3 days after hair follicle stimulation by plucking, and surviving follicles were defined as hairs present 17–19 days after irradiation. There may be large variations in the D_0 of follicle survivors if different stages of hair growth are studied (492 rad for resting and 174 rad for growing follicles) [Griem et al., 1973].

There is an apparent correlation between radiation-induced damage to the follicles and the incidence of skin tumours [Albert, Burns and Heimbach,

1967a; Albert, Burns and Bennett, 1972]. In mice a dose of 2,500 rad induces the largest number of atropic follicles and a high yield of sarcomas and carcinomas. The response in rat differs slightly in its fine details but follows a similar general relationship. This may indicate that the follicles contain a carcinogen-responsive cell type (stem cells?): a fact further suggested by the dependence of tumour yield on radiation penetration to depths encompassing the entire follicles [Albert, Burns and Heimbach, 1967b, 1968; Burns et al., 1976]. In fact it appears that the target cells for follicle tumour induction are located about halfway down the follicles while follicle non-survival depends on irradiation of most of the follicle [Burns et al., 1976]. Many, but not all, chemical carcinogen-induced tumours may also have a hair follicle origin [Howell, 1962; Bennington, Holmes and Combs, 1976].

2.4. Effects on sebaceous glands

The sebaceous glands are also affected by irradiation [Strauss and Kligman, 1959; Kurban and Farah, 1969; Hambrick and Blank, 1956]. They are in fact quite sensitive and also possess a considerable capacity for regeneration [Strauss and Kligman, 1959]. There appear to be some differences in radiosensitivity between epidermis, hair follicle and sebaceous glands even though all three regions are developmentally derived from the same cell source. Sebaceous gland cells are also capable of regenerating epidermis [Eisen et al., 1956]. The sebaceous glands are sensitive to the cytotoxic effect of methyl-cholanthrene [Chase and Montagna, 1951; Montagna and Chase, 1950; Cramer and Stowell, 1943] and are probably regenerated after either X-irradiation or methyl-cholanthrene destruction by upper hair follicle canal cells [Chase and Montagna, 1951; Montagna and Chase, 1950].

2.5. Effects on epidermal vasculature

It was obvious from studies on erythema and skin reaction that radiation could cause significant changes to the circulatory system though it is by no means certain what mechanisms are involved in this type of damage [Reinhold, 1972]. This is also true for ultraviolet-induced vascular damage [Ramsay and Challoner, 1976]. UV can cause significant rapid vascular changes (dilation) [Forbes and Urbach, 1969].

Damage to the endothelial system is probably a significant factor in the late effects of radiation but there are equally clear early changes, e.g. changes in permeability [Jolles and Harrison, 1963, 1965, 1966; Maisin, Oledzka-Slotwinska and Lambiet-Collier, 1973]. The effects on vasculature (dye leakage) can be detected as early as 3 h post-irradiation and after doses as low as 450 rad [Eassa and Casarett, 1973]. It is not clear to what extent endothelial cell killing plays a role in determining the vascular reaction [Reinhold, 1972, 1974]. Endothelial cells have radiation survival curves with D_0 values of about 170–240 rad [Reinhold 1973, 1974; Reinhold and Buisman, 1973; Van den Brenk, 1972]; values consistent with the post-irradiation capillary growth rate studies of Yamaura, Yamada and Matzuzawa [1976]. Effects on the vasculature are variable, but significant in the overall response. They were effectively reviewed by Dunjic, Reinhold and others in a 1974 publication [Dunjic, 1974].

The early work of Jolles indicated that diffusable factors were responsible for at least some of the early inflammatory-type reaction though the precise nature of the factor and its effect on capillaries is not clear [Jolles, 1949, 1950]. In chick embryos damage to the circulatory system may be a major factor in determining the embryo's post-irradiation wellbeing [Stearner and Christian, 1968, 1969]. Devik concluded that vascular damage is of secondary importance in acute skin reaction; however, the survival and behaviour of epithelial cells are strongly influenced by their environment [Devik, 1955].

2.6. Effects on skin pigment cells (melanocytes)

Changes in skin pigmentation was another of the very early radiobiological effects recorded [Ellinger, 1951, 1957]. These early reports were concerned mainly with pigmentation changes resulting in increased pigment (hyperpigmentation). This could be observed in the skin after doses that caused erythema and it was variable in its duration. The process is similar to the sun tanning observed after ultraviolet exposure. There were also several early reports on the production of pigmented hairs in areas that were grey or white before irradiation [reported in Ellinger, 1951, 1957]. In 1926 Hance and Murphy reported the production of depigmented (white) hairs on pigmented mice. This end point was subsequently studied in some detail by Chase [1949, 1951; Chase and Rauch, 1950; Chase, Straile and Arsenault, 1963] and later by Potten [1968, 1970, 1972a; Potten and Chase, 1970].

It is clear that two quite different responses of melanocytes can be detected. The first involves an increase in melanin synthesis and its distribution to other cells and possibly an increase in melanocyte numbers by increased division of precursor melanocytes (melanoblasts). This process can occur in the epidermis, in dermal pigment cells and also in suitably "pale" hair follicles (this type of hyperpigmentation could not be detected in pigmented follicles where it would normally be masked by the abundant melanin). Hyperpigmentation occurs after low or chronic X-ray treatment as well as after UV [Quevedo and Smith, 1963; Quevedo and Grahn, 1958; Quevedo et al., 1965].

The second response which tends to occur at higher doses is an inactivation of melanocytes or melanoblasts resulting in depigmentation (hypopigmentation) of the epidermis or hair follicle. With suitable "hot" point radiation sources it may be possible to see both hypopigmentation and hyperpigmentation as concentric rings about a central radiation-induced ulcer. The depigmentation reaction of the hair follicles has provided a "clone" cell survival assay for melanoblasts (resting hair follicles) which has a D_0 value of about 180–200 rad [Potten, 1968]. The sensitivity of the cells is dependent on the hair growth stage [Potten, 1970; Chase, 1949; Chase and Rauch, 1950] and also on the level of oxygenation at the critical site within the follicle [Chase and Hunt, 1959; Potten and Howard, 1969]. The oxygen level within the follicle seems to be controlled by both vascular oxygen and also oxygen diffusion from the atmosphere down the follicle canal [Potten and Howard, 1969].

There is one final type of hyperpigmentation that may contribute in part to the phenomenon already discussed but is detectable at low doses within growing hair follicles. Skin with growing follicles irradiated with doses within the range 25–250 rad shows excessive pigment in the matrix of the follicles within 72 h. The precise mechanism involved is unknown but it would appear that melanin is synthesised, or at least deposited in large amounts, within non-melanocytes [Potten, 1972b; Potten and Forbes, 1972; Potten, Merkow and Pardo, 1971].

2.7. Effects on the dermis

There are very few data available on the radiation effects on dermis or dermal cell populations [see also von Essen, 1972]. However, dermal capillaries and melanocytes do exhibit changes after irradiation. There is some indication from various wounding experiments that the dermis can influence the course of re-epithelialisation. Biochemical effects on collagen metabolism can be detected after doses of about 1500 rad [De Loecker et al., 1976] and are detectable after UV irradiation [Shuster and Bottoms, 1963]. These changes indicate an effect on fibroblasts. The radiation chemistry and biochemistry of collagen has been reviewed by Bailey [1968] and Hall [1973]. After kilorad to megarad doses chemical changes are detectable in vitro. However the response in vivo could be quite different. Post-wounding fibroblast proliferation also appears to be reduced by irradiation [Grillo, 1963].

Fibrosis and contraction of irradiated areas is a common observation and this is probably the consequence of connective tissue and vascular damage in the dermis. The contraction is dose dependent [1,600–4,000 rad] and follows a time course similar to skin reaction [Hayashi and Suit, 1972].

Collagen and dermal fibroblasts appear to be important in the long term maintenance of epidermal cells in vitro. The fibroblasts appear to produce a factor conducive to epidermal cell proliferation [Melbye and Karasek, 1973; Rheinwald and Green, 1975; see also Green, Rheinwald and Sun, 1976] and it seems reasonable to expect that in vivo fibroblast damage may well have an effect on the subsequent behaviour of damages and surviving epithelial cells.

3. Radiation survival curve data for epidermal cells

3.1. Review

It is common to regard most of the radiation effects on skin as attributable to effects on epidermal basal cells. For this reason, and also because suitable quantitative end point assays have been developed, epidermal cell response has been relatively well studied. The development of these methods was due to the ingenuity of Withers [1965] who devised a technique whereby epidermal cell survival could be assayed by applying an approach similar to that used in determining radiation survival curves in vitro; namely an assay of intact reproductive integrity. In irradiated skin, cells were defined as survivors if they were capable of sufficient cell divisions to produce a nodule of cells (a clone or macrocolony) visible to the naked eye in an otherwise sterilised area 10–20 days after irradiation.

A typical nodule might measure 2 mm across and could contain 3–4 × 10^4 cells (assuming that the regenerating epithelial cells are cuboid with a surface area (in the plane of the skin surface) of 10 × 10 μm: in fact the cells are smaller than this and a cube is not the most efficient method of density packing cells, so the cell numbers might be even higher). This represents a minimum of 15–16 doublings in 10–12 days, which, considering the points outlined above, the fact that the epithelium is not a single layer of cells, and the fact that some cells are lost to keratinisation [Withers, 1967a] would suggest cell cycle times of about 12 h even though the doubling time estimates may be 20–24 h [Withers, 1967c].

In order to minimise the occurrence of confluent clones, and confusion due to migration from the edge of the irradiated field, only high test doses could be studied inside a sterilised (3,000 rad) "moat" of skin. Thus only a portion of the cell survival curve could be obtained but nevertheless enough to enable an accurate measurement of the D_0 value. The technique involved the use of test areas of varying sizes inside the moat, the size depending on the dose (being adjusted to give only 0, or 1–4 survivors within the area). The skin was invariably plucked and thus hair growth initiated 0–24 h before irradiation. The epidermal and hair follicle cells were thus in the first postplucking burst of DNA synthetic activity when irradiated. The technique has been used by several groups since Withers first described his results [1965,

Table 3
Epidermal cell survival experiments

Reference	Strain	Mice sex	Age (wk)	Skin	Time post-plucking (h)	Day of scoring	Anaesthetised during irradiation
Withers [1967a] (1)	WH/Ht	♀ & ♂	6–56	Albino haired	1–12	Macrocolony 10–20	Yes
Withers [1967b] (2) (3)	WH/Ht	♀ & ♂	5–19	Albino haired	1–12	Macrocolony 10–20 Macrocolony 10–20	Yes Yes
Emery et al. [1970] (4)	SAS/TO	♂	8	Albino haired	17–21	Macrocolony 10–28	Yes
Denekamp et al. [1971] (5)	SAS/TO	♂	8	Albino haired	16	Macrocolony 14–21	Yes
Leith et al. [1971] (6)	CDI	♂	10–14	Hairless	–	Macrocolony 10–20	Yes
Al-Barwari and Potten [1976] (7)	DBA2	♂	7–8	Pigmented haired	20	Microcolony 3	No

Table 3 (continued)

Reference	Oxygenation	Quality	Dose rate (rad/min)	Comments	Results D_0(rad)	N(cells/cm^2)
Withers [1967a] (1)	Air	29 kV X-rays	770	Most γ 9–14 wk old 845 R/min	135	1.7×10^6
Withers [1967b] (2)	Air	150 kV X-rays	195	Most 29 kV	140	1.4×10^6
(3)	Oxygen (3 atm)	150 kV X-rays	195	Most 29 kV	107	1.0×10^7
				2 and 3 Common N(2)	134	2.7×10^6
				N(3)	112	2.7×10^6
Emery et al. [1970] (4)	Air	250 kV X-rays	300	Most 14 MeV deutrons, 4 h after X-rays	135	1.2×10^7
Denekamp et al. [1971] (5)	Air	8 MeV neutrons	–	A mean energy	109	2.3×10^5
Leith et al. [1971] (6)	Air	28 MeV helium ions	3000		95	3.1×10^7
Al-Barwari and Potten [1976] (7)	Oxygen	290 kV X-rays	147		233	1.2×10^4

Unirradiated basal cell density $\sim 1.5 \times 10^6$ cells/cm^2.
Overall $D_0 \sim 135$ rad. Overall N adjusted for RBE $\sim 1 \times 10^6$ cells/cm^2.
D_0 [Alper et al., 1960] N, extrapolation number [Alper et al., 1962].

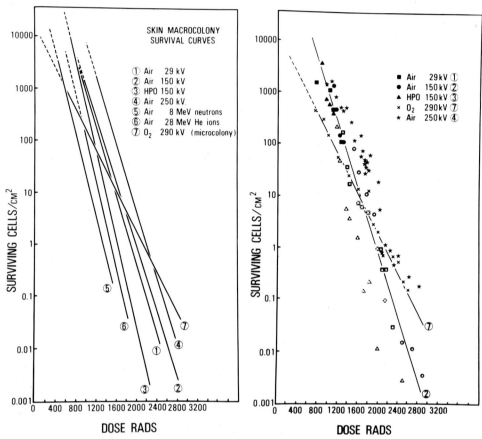

Fig. 4. Cell survival curves for mouse epidermal cells. Surviving cells/cm² plotted on a logarithmic scale against radiation dose in rad. For details of the references see table 3.

Fig. 5. Actual data points for five of the X-ray experiments reported in the literature for epidermal cell survival. For experiments 1 (squares), 2(circles) and 3 (triangles) the different test area sizes are shown by the differences within the symbols: in some cases the slopes of the lines drawn through points with a common test area size would differ from that of the overall best fit line. For experiment 4 (stars) the slope suggested by the low dose and high dose values is different from the overall best fit line. The best fit lines for experiments 2 and 7 are shown.

1967a, b, c] but all subsequent workers have continued to use plucked skin. The assay technique has been used for a variety of radiobiological experiments [Withers, 1967a, b, c; Emery, Denekamp and Ball, 1970; Denekamp, Emery and Field, 1971; Leith, Schilling and Welch, 1971]. The results of these experiments are summarised in table 3 and figs. 4 and 5. All the macrocolony curves provide D_0 values between 100 and 140 rad with zero dose extrapolates of up to 3×10^7 cells per cm^2.

3.2. Microcolonies or macrocolonies?

If the macrocolonies grow from single cells and divide with cycle times in the order of 12 h then these developing foci should be histologically detectable at times earlier than 10 days, particularly if autoradiography is used (fig. 6). The earlier the samples are taken the less is the problem of confluence and lower doses can then be studied. In section material the chances of detecting early foci of regeneration (microcolonies) are less than in sheets of epidermis separated at the basement membrane [Hamilton and Potten, 1972]. A micro-

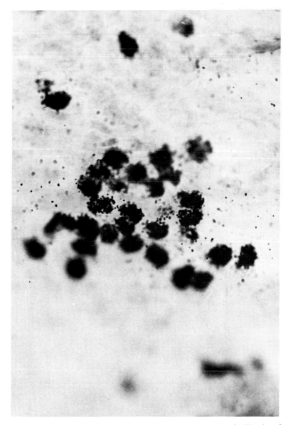

Fig. 6. Mouse epidermal sheet autoradiograph showing an interfollicular focus of regeneration (microcolony) consisting of about 30 closely associated labelled cells. A few isolated non-clone labelled cells can also be seen. 1,000 rad day 3. ×900.

colony approach has been used [Al-Barwari and Potten, 1976] and the results are summarised in table 3 and figs. 4 and 5. The slope of the survival curve differs somewhat from the earlier macrocolony data but as can be seen from figs. 5 and 7 both micro- and macrocolony points could all scatter about a common line with a D_0 of 135 rad. However, the microcolony data alone are fitted by a line with a D_0 of 233 rad and it is impossible to fit a line with a D_0 of 135 rad. There is one other preliminary report which suggests a high D_0 of 345 R [Archambeau, 1971]. The reasons for the differences in D_0 are not altogether clear but may be found in one or more of the following:

1. The macrocolony assay is likely to detect only the most rapidly growing foci and a considerable range in individual clone growth rates seems likely [Al-Barwari and Potten, 1976].

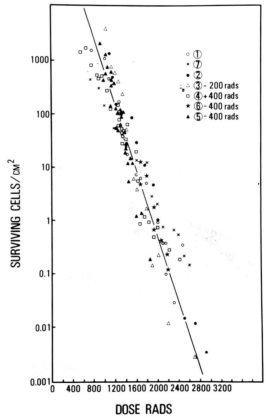

Fig. 7. Actual data points for all 7 skin survival experiments adjusted to a common line with a D_0 of 135 rad. The actual values can be obtained by subtracting 200 rad for experiment 3, 400 rad for experiments 5 and 6 and by adding 400 rad for experiment 4.

2. Clones derived from surviving clonogenic cells far down the follicle will take longer to reach the surface and form a nodule than those on or near the surface. The macrocolony assay will detect more of these follicle derived clones (see fig. 2).

3. Migration from the edge of the irradiated field, either from clones themselves as they spread, or from hair follicles and sebaceous glands, might be influenced directly or indirectly by the presence of heavily irradiated dermis. This might be more significant at high test doses than low ones in the macrocolony assay (see section 3.4).

4. Differences in the biological and physical aspects of the experiments, strain, age, sex, stage of hair growth (time after plucking), levels of oxygenation, dose rate, and radiation quality (see table 3).

5. The possibility that clonogenic cells do not begin regeneration rapidly after irradiation but do so only after the basal layer has been depleted of cells. In this case the microcolony assay may be measuring the radiosensitivity of a more differentiated proliferative population of cells. However, as discussed elsewhere, the reverse may be more likely, i.e. microcolonies representing stem cell clones.

6. There is some evidence of grouped labelled cells 3 days after plucking alone (grouping is also seen in other cell kinetic studies using epidermal sheets). It is possible that the microcolony assay is to some extent contaminated by this phenomenon. However, labelling studies on both plucked and unplucked, irradiated and unirradiated skin all suggest that this effect is small It is not clear to what extent plucking is essential in both the macro- and microcolony techniques and as already discussed it is not clear why all workers with haired animals find it necessary to use plucked skin.

3.3. Regenerating cell doubling time

The published data all seem to indicate that the clones or clonogenic cell numbers have a doubling time of 20–36 h [Withers, 1967c; Emery et al., 1970; Dutreix et al., 1971; Al-Barwari and Potten, 1976] though this value may increase after the 5th post-irradiation day [Al-Barwari and Potten, 1976] or after longer times [Denekamp, Ball and Fowler, 1969]. As discussed elsewhere these estimates are likely to be greater than the cell cycle times, possibly by as much as a factor of 2–3.

3.4. Cell migration during epithelialisation

The importance of migration of epithelial cells in the restoration of a denuded epidermis is evident from many studies on wound healing [Rovee et al., 1972; Winter, 1972; Krawczyk, 1971; Odland and Ross, 1968; Winter, 1964]. It is also very evident, to those with experience of irradiating test areas on experimental animals, that the field shrinks fairly rapidly as new epidermis is formed from the edges and the dermis contracts.

Migration proceeds most effectively when the surface is moist [Rovee et al., 1972; Winter and Scales, 1963; Winter, 1972]. If the denuded area dries, epithelial migration tends to be slower and may only proceed beneath a thick scab including some dermal tissue [Winter, 1972; Winter and Scales, 1963]. After some types of epithelial damage the dead or dying cells may actually release materials [Jolles, 1949, 1950] that cause damage to otherwise healthy cells and this may also influence migration. The fact that irradiation can cause wounds that never heal (ulcerated) indicates that migration from the edges of these wounds is prevented by either dehydration, release of toxic byproducts from damaged cells or the production of a dermis that is not conducive to cell migration for some other reason. The importance of the dermis is further suggested by early experiments using 440 times the threshold erythema dose (i.e. a very severe reaction) of soft radiation which causes less damage to the dermis. These severe reactions heal completely (Raper et al., 1951).

It is not clear at present how dependent the regenerating epidermis is on an intact healthy dermis or any one of its specific cell populations (fibroblast, nerve fibres etc.).

3.5. Epithelial cell numbers

It is clear that the response of skin to radiation is the consequence of the combined effects on all the various cellular constituents in epidermis, hair follicles and dermis (table 1). However it is probably reasonable to assume that the most important cell population determining overall skin response is the epithelial cells (epidermis, hair follicles and sebaceous glands). The important cells of the epidermis are those involved in cell proliferation which in most, but not all, skin are the single layer of cells attached to the basement membrane (basal cells) (fig. 1). The number of basal cells per cm^2 in mouse is between $1.2-2.0 \times 10^6$ [Potten, 1974, 1975a; Potten and Hendry, 1973; Hamilton and Potten 1972]. There are about 4,000 hair follicles and sebaceous glands/cm^2 [Hamilton and Potten, 1974; Potten, Jessup and Croxson, 1971; Dubravsky et al., 1976]. Each sebaceous gland may contain 40 basal cells [Hamilton, 1974] and a resting hair follicle may contain a few hundred cells. Thus there are about $3-3.5 \times 10^6$ epithelial basal cells/cm^2 in a typical region of haired mouse skin. When the hair follicles are actively producing new hair the number may rise to $1-3 \times 10^7/cm^2$ since growing follicles contain $1-3 \times 10^3$ cells [Withers, 1965; Potten, 1967].

3.6. Clonogenic cell numbers

There are about 1.5×10^6 basal cells per cm^2 in normal mouse interfollicular epidermis. This value may vary slightly from mouse strain to strain, during the 24 h day, or even with season; it certainly varies after wounding or plucking [Hamilton and Potten, 1972]. However under all conditions it is relatively accurately and easily determined from epidermal sheet preparations. All the survival curves using the macrocolony assay technique extrapolate to less than 1.0×10^7 cells/cm^2 with the exception of the experiments using hairless mice where the value is 3×10^7 (the basal cell density for hairless mice is slightly higher than the usual value for haired mice) (table 3). For several experiments the zero dose extrapolate is considerably lower than this. Unless the survival curves are biphasic with a steep initial portion (for which there is no evidence or suggestion at present) then this extrapolation value must represent the highest possible origin of the survival curve and for several cases this is clearly less than or equal to the observed number of basal cells. Mammalian radiation cell survival curves tend to have a shoulder region [Alper et al., 1960, 1962] which is commonly attributed to the necessity to accumulate sublethal damage within individual cells and the fact that given time, cells may be able to repair or recover from this sublethal damage. Experiments where this has been studied in skin [Withers, 1967c] indicate that skin cells are capable of considerable sublethal damage repair. If the recovery factor can be determined it can be used to estimate the origin of the survival curve by equating it with the extrapolation number of the individual cell survival curves [Alper et al., 1960; Potten and Hendry, 1973]. This approach has been used to estimate the true origin of the survival curves [Potten and Hendry, 1973] (fig. 8). There are three experiments where an estimate can be made of the recovery potential of the epidermal basal cells and where this can be related to a comparable cell survival curve [Withers, 1967c; Emery et al., 1970; Denekamp et al., 1971] and in all three cases (using very different quality of radiation and different recovery potentials) the estimate for the cell survival curve origin ranges between 3×10^4 and 1.1×10^5 cells per cm^2, i.e. less than 10% of the basal cells. For the other experiments there are no suitable estimates for the recovery factor but if comparable recovery factors apply to the microcolony assay then the estimates for the survival curve origin, and thus the number of clonogenic cells/cm^2 of normal mouse epidermis could be extremely low, particularly since the total epithelial cell population/cm^2 may be 200 times greater than the number of epidermal basal cells (see section 3.5).

Thus this type of interpretation of the skin macroclone data would suggest that less than 10% (in fact 2–7%) of the basal cells are clonogenic, i.e. possess the ability to regenerate the epidermis if called upon to do so. The microcolony data suggest that this fraction might be even lower. Since the growth fraction is believed to be not lower than 60% many of these proliferative cells do not possess the ability to regenerate, i.e. divide rapidly and extensively and thus re-epithelialise the skin. Clonogenicity can be taken as one attribute of

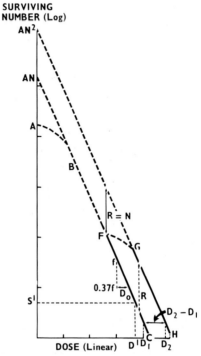

Fig. 8. A typical mammalian cell survival curve (ABC) is shown with a split-dose curve (FGH). If N, the extrapolation number, can be determined then the initial number of cells (A) can be deduced from the single dose extrapolation to zero dose (AN). N can be equal to or slightly greater than the recovery factor (R) deduced from the two dose curves and calculated from the equation: $\log_{10} N = 0.4343 \, (D_2 - D_1)/D_0$, where $(D_2 - D_1)$ is the additional dose required to achieve the same survival level when a dose is given in two fractions separated in time (see figure). The point of extrapolation of the single dose curve on the ordinate axis was calculated from the equation $\log_{10}(AN = \log S^1 + 0.4343 \, D^1/D_0)$, and the survival number, S^1, at a given dose, D^1, was read off published curves. $A = AN/N$. (Taken from Int. J. Radiat. Biol. 24; 538 [Potten and Hendry, 1973] by courtesy of Taylor and Francies, Ltd.)

stem cells, and thus few basal cells are stem cells and the epidermis must be composed of at least two types of proliferative cells, namely stem (clonogenic) cells and more differentiated stem cell-derived proliferative cells. A comparison of absolute survival levels, and histologically identifiable dead cells also suggest the existence of two or more subpopulations of basal cells with different radiation responses (section 4).

The difficulty with this type of approach is that the zero dose extrapolates on the experimental data are poorly defined and extremely dependent on small changes on the D_0. However, as can be seen from table 3, all the data, with the possible exception of the helium ion data, extrapolate to values equal to, or less than, the observed basal cell density without any consideration of the recovery factor. The helium ion value is based on only 8 experimental points and thus the zero dose extrapolate is poorly defined.

The calculations might be expected to give low estimates for clonogenic cells if the survival curve data were selecting a more resistant subpopulation of cells. It is not clear what effect the pre-irradiation plucking would have on these estimates nor is it completely clear whether it is valid to equate the recovery potential determined on pre-irradiated cells (split-dose technique) to the recovery potential (extrapolation number) of unirradiated cells. Within these limitations it is clear that the data are not consistent with a model where all basal cells are clonogenic but rather suggest that less than 10% possess this ability. The values are better correlated with the number of EPU [Potten, 1974] than basal cells, i.e. if each EPU is to some extent autonomous then each EPU may contain one cell with regenerative capacity.

Since the original discussion on the numbers and distribution of clonogenic cells and EPUs [Potten and Hendry, 1973, Potten, 1974] one further possible explanation has arisen which is outlined below. There are about 4×10^3 hair follicles per cm^2. The estimates of clonogenic cells/cm^2 are between 3×10^4 and 1.1×10^5 with the microcolony data (based on the assumption of a similar recovery factor) giving a value possibly as low as 3×10^2 cells/cm^2. The hair follicles vary in size and cellularity but are known to contain cells capable of re-epithelialisation (see section 1.2). The follicle survival curve D_0 values might be similar [Dubravsky et al., 1976] to the microcolony value of 230 rad. There is an apparent correlation between hair follicle damage and skin tumour induction (see section 2.3) and the hair follicles appear to be important in determining the response and healing of irradiated tail skin [Hendry, unpublished data; personal communication]. It is possible that the hair follicles may contain *the* vital epithelial stem cells, or at least *some* epithelial stem cells (perhaps some protected by hypoxia or non-cycling status). In this case, if the EPU are partially autonomous then they may contain one early differentiated proliferative cell that is also capable of some measure of regeneration. If this is so then the numbers suggest that the macrocolony assay may measure differentiated cell survival with the possibility of some stem cell contamination. The differentiation-committed clones may begin later, only after severe basal cell depletion (day 3 post-irradiation) and cells may have a greater tendency to "pile up" and thus form macroscopic nodules consisting of many maturing cells. The microcolony assay may thus measure upper follicle stem cell survival. About half the microcolonies seen on the 3rd day (44%) surround or are immediately adjacent to the hair follicle (see fig. 2). It is impossible to say how many of these are indeed derived from cells of the follicle but it seems likely that many are. It is also possible that some or many of the remaining 56% are also follicle derived and have moved further from their origin. The carcinogen target cells are also believed to exist in the upper regions of the follicle [Burns et al., 1976].

The stem cell re-epithelialisation would not be expected to show any piling up of cells, and thus not to be readily observed as macroscopic nodules. The cells may be more concerned with rapid spreading than with differentiation and maturation. This might also explain why the re-epithelialised skin has many deep follicle-like rete pegs or ridges (fig. 9) as the stem cells relocate themselves at the base of these pegs. However this hypothesis is contrary to the results obtained in the haematopoietic system where the stem cells (CFU-S) are more sensitive (have a smaller D_0: 92–94 rad) than the early differentiated cells (CFU-C) (D_0 160–290 rad) [Testa, Hendry and Lajtha, 1973]. There are clearly some points still to be resolved concerning the micro- and macrocolony assays.

It should be noted that this hypothesis (outlined above) is contrary to a previous explanation of the data, where it was suggested that perhaps the microcolony data were measuring differentiated cell activity [Al-Barwari and

Fig. 9. Mouse epidermal sheet (Feulgen stained) obtained 35 days after a dose of 2,000 rad. The complex three dimensional arrangement of the basal cells can clearly be seen with many deep follicle like rete pegs or ridges. ×140.

Potten, 1976], but the hypothesis outlined above seems to fit more of the observations than the earlier hypothesis. That there is no conclusive explanation for the clone assay technique and its results illustrates one of the drawbacks in cutaneous radiobiology. Until we know exactly what is being measured it is difficult to define the processes and mechanisms involved in skin response.

3.7. Variability in survival curves

As can be seen from fig. 4 there is some variability in the position of the curves (variability in D_q [Alper et al., 1962]). The first two curves of Withers [1967a, b] are quite similar even though the quality of the X-rays was changed from 29 kV to 150 kV. If irradiated under high pressure oxygen (HPO) the results are shifted by about 200 rad to the left (figs. 4 and 7 and table 3). Irradiation by 250 kV X-rays in air produces results to the right of the 150 kV data by about 250 rad [Emery et al., 1970] and about 400 rad to the right of the 29 kV data (figs. 4 and 7). It is probable that much of this could be due to the RBE of the different LET radiation [Leith et al., 1971; Emery et al., 1970]. This is certainly the case with the neutron and helium ion data which lie about 400 rad to the left of the 29 kV data curve (figs. 4 and 7) [Denekamp et al., 1971]. It is somewhat surprising, from the RBE point of view, that the high LET results do not have curves with different D_0 values (table 3 and figs. 4 and 7) but merely exhibit a reduction in shoulder width or extrapolation number. The microcolony data without any dose adjustments cross the macrocolony data curves at about 1,600 rad (figs. 4 and 7) [Al-Barwari and Potten 1976]. Thus at lower doses more macro- than microcolonies are scored (more colonies derived from differentiated than from stem cells?) while at high doses the reverse is true (fewer macro-differentiated than micro-stem colonies?).

3.8. Epithelial cell D_0 values

As can be seen from table 3 the epidermal cell survival curves consistently provide values for the D_0 of 100–140 rad (the only exception being the microcolony data). This range is remarkably similar to the range of D_0 values found in the small intestine (109–161 rad, see table 4) [Potten and Hendry, 1975; Withers and Elkind 1970; Withers, Brennan and Elkind, 1970; Broerse et al., 1971; Hornsey, 1973; Hendry and Potten, 1974; Hendry, Major and Greene, 1975]. The proliferating cells in these two tissues have cell cycle times differing by a factor of 7–10 [Cleaver, 1967; Cameron, 1971; Potten, 1975a]. The length of G_1 accounts for most of the difference (crypt cell G_1 is about 0.25 of T_c while in skin cells it might be as much as 0.8).

Table 4
Epithelial cell survival D_0 values (rad) for mice

Skin	macrocolony		135	Table 3 or fig. 7
Skin	macrocolony	1–7 days after 900 rad	182–218	Emery et al. [1970]
	microcolony		233	Al-Barwari and Potten [1976]
Gut	macrocolony		100	Withers and Elkind [1969]
	microcolony		100–160	Withers and Elkind [1970] Potten and Hendry [1975]
	microcolony	1–4 days after 900 rad	175–245	Potten and Hendry [1975]
Hair follicle survival			174–500	Dubravsky et al. [1976]; Griem et al. [1973]
Melanoblasts [a]			180–200	Potten [1968]
Endothelial cells [a]			170–240	Reinhold [1974]

[a] For comparison.

Since cell cycle phases have different radiosensitivities [Sinclair and Morton, 1966; Terasima and Tolmach, 1963] it is a little surprising that these two epithelial tissues have such similar radiosensitivities. This could be explained if in both tissues the radiosensitivity being measured was that of a functionally similar subpopulation of cells with more similar cycle times or cycle phase distributions, i.e. stem cells or an early differentiated cell population. After one burst of crypt regeneration subsequent survival curves tend to have a higher D_0 (closer to 200 rad) (table 4) 175–245 rad: [table 1 in Potten and Hendry, 1975], i.e. the D_0 is 100 rad if the cryptogenic cells are forced to regenerate once, but the D_0 is higher if the test is applied again. This is also apparent with the skin macrocolony assay where second survival curves, determined 1–7 days after a first, provide curves with D_0 values ranging between 180–220 rad [Emery et al., 1970] (table 4). These data could be explained if the initial cryptogenic or clonogenic data are largely the consequence of early differentiated cell regenerative activity which is lost (used up) if a second test is applied. The second dose may thus measure stem cell response more accurately (see also section 3.6). Alternatively there could have been different cycle phase distributions at the time of the second test. This could explain the data if G_1 were sensitive and occupied similar proportions of the cell cycle in both epidermis and intestine.

One final point relevant to this discussion is that the skin survival curves tend to show a slightly different slope if doses above 1,600 rad are considered (see fig. 4 in Emery et al. [1970]). It is probable that the higher the dose the more likely it is that the stem cells, rather than early differentiated cells, are being assayed.

3.9. Summary of epidermal cell survival data

It is clear that further work is needed to define precisely what is being measured by the various clone assays not only in epidermis but also in intestine. A number of interpretations can be applied to the data and various models suggested which clearly need further experimental testing. It is not clear to what extent the various epidermal cell populations and subpopulations contribute to the experimental results (melanocytes, Langerhans cells etc.) neither is it clear what influence the dermis and its cell populations, and the variations in field size of the macrocolony test area, and also the difference between partial body and whole body irradiation, may have on the experimental results.

4. Radiation induced cell death in epidermis

4.1. Pycnotic or apoptotic cells

If one considers the survival curve data it is clear that a moderate single dose of 1,200 rad results in either 70–75 [Withers, 1967b; Al-Barwari and Potten, 1976], 200–250 [Withers, 1967a, b] or about 2,000 [Emery et al., 1970] surviving cells/cm^2. Taking a value of 1.5×10^6 basal cells/cm^2 this could indicate surviving fractions of 0.0013–0.000047 (or killed fractions of at least 0.999). It would seem likely therefore that a large number of histologically dead cells should be observed and that the number of dead cells should be clearly related to the dose delivered. Even though all reproductively sterilised cells may not become histologically identifiably dead, many might be expected to do so and the numbers would be expected to increase with dose. Dead cells can be recognised and scored as pycnotic nuclei [Fowler and Denekamp, 1976; Hegazy and Fowler, 1973a,b] or as apoptotic bodies [Kerr, Wyllie and Currie, 1972; Searle et al., 1974; Olson and Everett 1975] (fig. 10). It is clear from the available data in the literature and also the data in table 5 that the apoptotic/pycnotic yield rarely rises above a few per cent and shows no clear relationship to dose. Precisely similar results have been obtained for the apoptotic yield in the small intestinal crypt after various doses of X-rays which reproductively sterilise many cells [Potten, Al-Barwari and Searle, unpublished data].

There are a number of problems concerned with pycnotic–apoptotic cell yields, many of which remain unresolved. With X-ray doses that sterilise 90–99.99% of the epidermal cells it would be expected that dead cells appear (in frequencies related to the dose) and that the number of non-sterilised viable basal cells decrease. The rate of decrease will depend on whether the

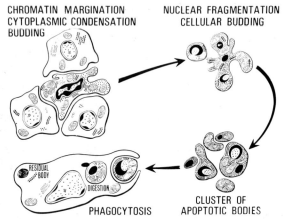

Fig. 10. Schematic representation of the sequence of events during apoptosis (redrawn after Kerr et al. [1972; Searle et al [1975] after discussions with Dr. J. Searle). The sequence shows the fate of a single apoptotic cell surrounded by three normal cells, one of which may ultimately engulf one or more of the apoptotic fragments.

cells become pycnotic when they attempt mitosis or whether they become pycnotic from any stage of the cell cycle. The majority of epidermal cells are in G_1/G_0. The pycnotic/apoptotic incidence does not show an early peak when there is a burst in labelling [Al-Barwari and Potten, 1976] or when the labelling index *starts* to rise after its initial inhibition [Hegazy and Fowler

Table 5
Apoptotic index (% of basal cells) after irradiation

Time after irradiation (days)	X-ray dose (rad)			
	500	1200	2000	2700
1	0.32 ± 0.03	0.53 ± 0.07	0.88 ± 0.09	–
2	0.42 ± 0.02	0.72 ± 0.09	1.17 ± 0.17	–
2.5	–	1.18 ± 0.14	–	–
3	0.57 ± 0.08	2.96 ± 0.73	1.65 ± 0.34	0.86 ± 0.09
3.5	–	1.46 ± 0.19	–	–
4	0.60 ± 0.04	1.63 ± 0.24	2.78 ± 0.40	–
5	–	1.16 ± 0.12	–	–
6	–	1.77 ± 0.21	–	–
9	–	0.54 ± 0.02	–	–

Control values range between 0.31 ± 0.02 to 0.42 ± 0.02 (immediately or 20 h after plucking). Counts on epidermal sheets. 290 kV X-rays delivered (with mice in oxygen) 20 h after plucking (147 rad/min.). Mean and standard error of individual mouse values.

1973b; Fowler and Denekamp, 1976] nor is it correlated with the mitotic activity in plucked or unplucked irradiated skin [Hegazy and Fowler, 1973b; Fowler and Denekamp, 1976] when the basal cell population as a whole is considered. Thus the pycnotic cells do not appear to be derived from aborted mitoses but are formed from some other point in the cell cycle (interphase death). However, the pycnotic cells may be derived from some special small subpopulation of cells that abort in mitosis; this effect being masked by the proliferative activity of the majority of cells.

The pycnotic/apoptotic yield could be explained on the basis of a special subpopulation of cells which responds to irradiation by this clear manifestation of death. The subpopulation could either be the stem cell/early differentiated cell population or cells at a specific phase of the cell cycle (e.g. G_2 which is known to be a sensitive phase; [Sincleair and Morton, 1966]). In either case the numbers are consistent with the number of pycnotic/apoptotic cells (i.e. less than 10%) [Gelfant, 1966; Gelfant and Candelas, 1972; Potten and Hendry, 1973]. Since the cell cycle behaviour after irradiation of either of these subpopulations is unknown it is impossible to say whether or not the cells die when attempting mitosis.

If either of these two subpopulations (stem cells or G_2 cells) was sensitive to irradiation, then the lack of overall dose-dependence could be due to the fact that only large changes in the fraction of killed cells within a subpopulation making up less than 10% of the basal cells could be detected.

In both sheet preparations of epidermis and section material there are severe limitations to the accuracy with which one can determine the fraction of basal cells that are either mitotic or pycnotic/apoptotic. Apart from the difficulties inherent in defining mitosis and pycnosis (where they begin and end) there are difficulties because both these processes involve chromatin condensation. There are difficulties in distinguishing these cell types from (1) cells in G_2 or any other cell with a possible DNA content greater than that of G_1 cells; (2) infiltrating lymphocytes or macrophages which can have small dense nuclei or multi-lobed nuclei (like a fragmenting apoptotic body) and finally, (3) clear cells of the epidermis which often have rather small nuclei with a tendency for some chromatin condensation (marginal) (section 5.1). This is particularly true in regenerating or damaged epidermis where all these factors may be applicable.

A second problem is that it is difficult in those samples that are from partial-body irradiated animals (most high-dose long-time interval groups) to distinguish reliably between clone regeneration and field-margin cell migration (re-epithelialisation) and regions lacking obvious signs of either of these but still possessing a spreading basal cell population.

4.2. Changes in basal cell density

Further evidence that radiation is predominantly acting only on a subpopulation of the basal cells comes from a consideration of the overall number of basal cells (basal cell density per unit area) and the behaviour and appearance of these cells.

An estimate of the density of basal cells can be obtained from sectioned material [Etoh et al., 1975; de Rey and Klein-Szanto 1972] but can be more easily obtained from sheets of epidermis separated at the basement membrane [Hamilton and Potten, 1972; Al-Barwari and Potten, 1976] (table 5 and fig. 11). Plucking results in a transient fall in basal cell density with normal values restored by the 3–4th day after plucking. This is then followed by a transient hyperplasia and overshoot in basal cell density [Hamilton and Potten, 1972; Potten and Allen, 1975a, b]. Irradiation with 1,200 rad (99.9% kill) results in a basal cell depletion which reaches a minimum between days 3–5 (table 6, figs. 11 and 12; [de Rey and Klein-Szanto, 1972]). The maximum fraction of cells lost after 1,200 rad is 32–40% and even after 16 krad to rat skin 50% of the cells are still present on the 5th post-irradiation day [de Rey and Klein-Szanto, 1972].

The basal cell density is obviously the result of summation of (1) cell loss by death, (2) continued cell loss by differentiation and migration, (3) inhibition of cell division, (4) limited division amongst surviving differentiated cells, (5) a few residual divisions in otherwise sterilised stem cells, (6) regeneration by stem cells, (7) migration of viable cells from unirradiated regions and their subsequent division, (8) infiltrating non-dividing cells. The final balance sheet of all these processes is the basal cell density. The relative contribution of these eight processes will depend on dose and time after irradiation. After even higher doses the basal cell density changes do not change over the first few days post-irradiation in proportion to the relative cell sterilisation suggested by the survival curves with their D_0 values of about 135 rad [Al-Barwari and Potten, unpublished data]. This is also evident from the data of de Rey and Klein-Szanto [1972] where there is little difference in relative

---

Fig. 11. Mouse epidermal sheet preparations (Feulgen stained) showing changes in basal cell density and nuclear size at various times after irradiation. All photomicrographs in fig. 11 are at the same magnification of ×650. A. Unirradiated immediately after plucking showing typical unirradiated basal cell density and nuclear size with the outlines of the EPU just visible. B. Unirradiated 20 h after plucking. The nuclei are slightly enlarged and the density is usually slightly decreased. C. Epidermal sheet taken 6 days after a dose of 1,200 rad to hairless mouse epidermis showing extensive repopulation with a region of residual cells (out of focus) in a line diagonally across the field (arrows). D. Three closely associated mitoses within an irradiated (3,500 rad) area (day 1). Possibly representing an early regeneration focus. From unpublished data of Al-Barwari.

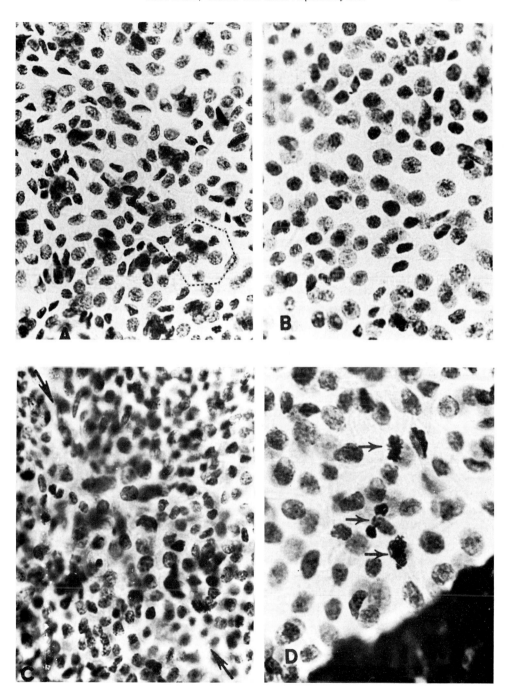

Table 6
Basal cell density (cells/cm^2) in epidermal sheets after a dose of 1200 rad X-rays

	×10^6
Control 1	1.27 ± 0.03
Control 2	1.02 ± 0.03
Time after irradiation (days)	
1	1.09 ± 0.05
2	0.98 ± 0.02
2.5	1.03 ± 0.01
3	0.88 ± 0.02
3.5	0.95 ± 0.02
4	0.77 ± 0.03
5	0.74 ± 0.07
6	0.79 ± 0.02
9	1.38 ± 0.04

Control 1 unplucked unirradiated dorsal skin.
Control 2 unirradiated skin 20 h after plucking. Animals irradiated in oxygen 20 h after plucking (290 kV X-rays; 147 rad/min). Mean and standard error of individual mouse values. Cell densities derived from 10 fields (each 0.019 mm^2) from each of 4 mice.

cellularity on days 1—5 post-irradiation after doses from 4—16 krad [see also Etoh et al., 1975, using 3 krad to guinea pigs]. Fowler and Denekamp [1976], however, believe that some dose dependence in basal cell depletion is to be seen though the evidence they use is not unequivocal.

Thus relatively few of the basal cells are lost due to direct cell killing. Given time, points 4—8 listed above can dominate the picture and the basal cell density can be restored and even overshoot (table 6 and fig. 2 in de Rey and Klein-Szanto [1972]). With sufficient time the surface can be completely re-epithelialised, providing the doses were not too high, and a hyperplastic epithelium is established (figs. 9 and 11C). This epithelium can be very thick with many deep folds, rete pegs or follicle-like projections (compare figs. 9 and 11). There is some indication that epidermal columnar organisation can

Fig. 12. Mouse epidermal sheets taken day 3 or 4 after a dose of 2,000 rad (same magnification as fig. 11, ×650). A. Day 3 showing enlarged nuclei and many apoptotic bodies (arrowed). B. Day 4 showing some very large nuclei and a greatly reduced basal cell density. C. Day 4 showing enlarged nuclei with a few small nuclei of a size typical of epidermal basal cells. It is not possible to say whether these small nuclei are migrating epidermal cells or some other infiltrating cells. D. Day 4 showing another area where there are only a few enlarged, in some cases multi-lobed, nuclei. From unpublished data of Al-Barwari. B—D all from one irradiated sample illustrating an extreme case of regional variability.

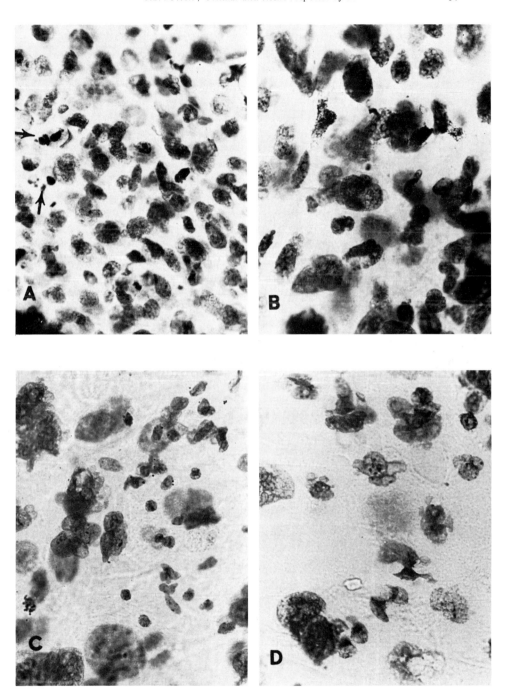

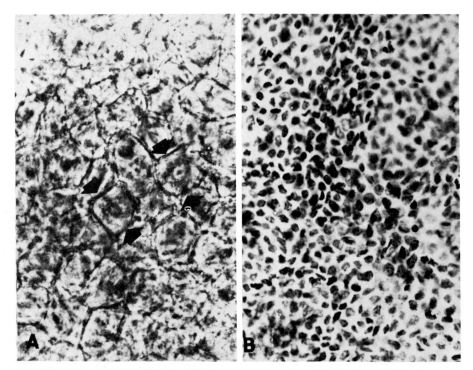

Fig. 13. Mouse epidermal sheet (Feulgen stained) 30 days after 2,000 rad. Even though the basal cell density is higher than normal (compare with fig. 11) there is evidence of columns of cornified cells in some regions. The presence of individual polygonal cornified cell outlines with minimal overlap with neighbouring cells indicates the presence of columns. Right hand picture focussed on basal cells. Left hand picture (phase contrast) on cornified columns. ×650.

be re-established eventually (fig. 13) and that suprabasal migration of epidermal cells over the first week or so is still controlled so that many cells align themselves correctly (fig. 14).

4.3. Behaviour of non-pycnotic/apoptotic cells

After doses in the range 500–2,000 rad few cells appear as pycnotic/apoptotic bodies, the basal cell densities do not show rapid decreases and neither end point shows clear dose dependence. Within this dose range (or even up to 3,000 rad [Etoh et al., 1975]) cell migration off the basal layer seems to continue at close to the normal rate [Etoh et al., 1975; Fowler and Denekamp, 1976]. This is evident from the data of Etoh et al. [1975] and also from the fact that the basal cell density decreases slowly (in fact at a rate

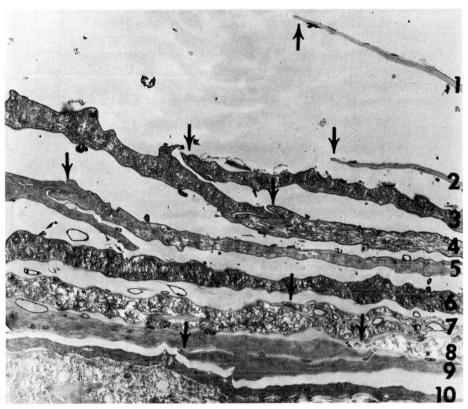

Fig. 14. Electron micrograph section of mouse dorsum cornified cells 7 days after 3,800 rad. Ten cells (numbered on the right) can be seen. The top 2 are typical of the thickness of normal dorsal epidermis and probably represent cells present at the time of irradiation. The remaining 8 are thicker, probably as a consequence of more rapid, or inadequate keratinisation. The edges of many of these cells are roughly coincident with the original top 2 cells. Only cells numbered 6 and 10 lack this alignment. The region shown represents only a small fraction of the total profile length of these cells (compare with figs. 13 and 15). × 12,000.

consistent with about 10% loss per day, i.e. the normal cell migration rate [Potten, 1976a], and also because the thickness does not decrease for some time in spite of presumed continued cell desquamation. At high doses the suprabasal migration may be stimulated and may account for much of the basal cell depletion [de Rey and Klein-Szanto, 1972]. Even after high doses to mouse dorsum (3,500 rad) there is some indication that migration not only continues but many (but not all) of the migrating cells manage to position themselves correctly relative to the epidermal columns (fig. 14).

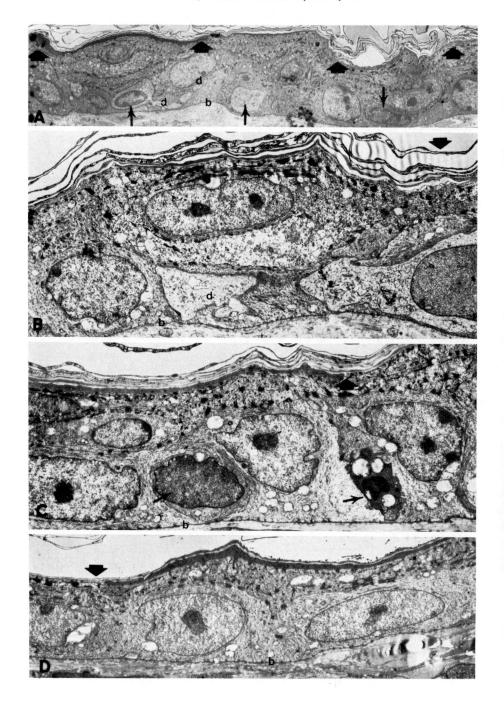

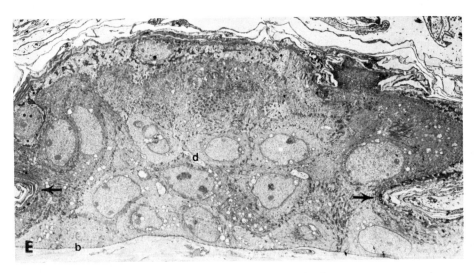

Fig. 15. Low power electron micrographs of mouse skin after irradiation. A. Unplucked, unirradiated skin showing overall thickness, width of epidermal columns (large arrows), and regular spacing of basal clear cells (small arrows). The left-hand clear cell is dendritic (d) while the right hand one though positioned similarly to the other two is a less likely Langerhans cell candidate. Though it is impossible to state the cell type accurately this cell is either a Langerhans cell, an amelanotic melanocyte, or an indeterminate cell. (b) basement membrane. ×2500. B. Two days after 500 rad. There is evidence of some basal cell depletion with fewer basal cells per EPU profile and a very extended presumed Langerhans cell on the right with a dendritic extension (d) seen in other profiles to be connected with the main body of the cell. ×6000. C. Three days after 900 rad. Two cells lacking desmosomes can be seen (arrows), the one on the right suggestive of early apoptosis. ×7500. D. Five days after 500 rad. The epidermis is significantly thinner in places with the basal cells clearly having increased in size laterally to compensate for basal cell depletion. ×6000. E. Four days after 900 rad. Showing a possible small region of regeneration with a local hyperplasia and an overall mushroom shape. Though no clear cells are evident in this profile, dendrites can be seen (d). ×3000.

At the level of both the light and electron microscope many of the basal cells look normal after doses in the range 900–1,200 rad (figs. 12 and 15). Many of the suprabasal keratinising cells also look normal. However in both basal and suprabasal layers some ultrastructural abnormalities can be seen. These become more apparent at higher doses. However it should be noted that according to the survival curve data more than 90% of the cells have been reproductively sterilised by these doses and yet many or most of the cells look normal, migrate and differentiate normally.

At higher doses (2,000–3,500 rad) there are considerably more light microscope and ultrastructural abnormalities (figs. 12, 15 and 16) but even after these doses there are many normal-looking cells at the earlier times. At

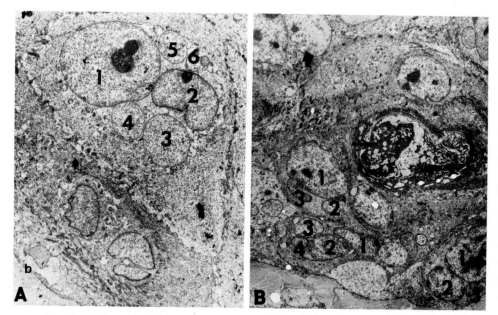

Fig. 16. Epidermal cells seven days after a dose of 3,500 rad. A. Region showing one large epidermal cell with 6 apparent nuclear bodies (numbered). These may be all part of one multi-lobed nucleus or separate pieces. ×3700. B. Another region showing damage and many apparent multi-nucleate cells with fragments numbered. ×3000.

these doses there are clear changes in nuclear morphology. The nuclei are enlarged and can become multi-lobed (figs. 15 and 16). Some cells even appear multi-nucleate [see also Gonsalves and Chase, 1975]. There is considerable variability in the basal cell density from area to area (fig. 12) making section studies difficult. This might be the reason for the subsequent thickness variability [de Rey and Klein-Szanto, 1972].

Much of the labelling and mitotic data suggest that many cells after doses in the range 900–1,200 rad continue to replicate DNA and divide [Hegazy and Fowler, 1973b; Fowler and Denekamp, 1976; Etoh et al. 1975; Potten, Al-Barwari and Searle, unpublished data]. Although the labelling index is reduced after irradiation it rises steadily with time [Fowler and Denekamp, 1976]. The mitotic index is more variable but is not reduced to zero. The rate of loss of pre-labelled basal cells with time after irradiation follows much the same slope as for unirradiated skin, as does the mean grain density of the labelled cells [Etoh et al., 1975]. After 3,000 rad the mitotic activity is greatly affected for the first few days. The labelling index of non-regenerating (non-clone) areas in epidermal sheets is fairly constant (days 3–6) and not dissimilar to that for unirradiated skin (table 7). The absolute number of

Table 7
Labelling index (% of basal cells) after irradiation (1,200 rad)

	Overall LI	
Control 1	3.9 ± 0.3	
Control 2	26.7 ± 1.4	
Time after irradiation (days)		Non-clone area LI
1	18.1 ± 1.4	
2		8.6 ± 1.0
3		3.5 ± 0.3
4		2.9 ± 0.1
5		2.8 ± 0.1
6		2.2 ± 0.1

Control 1 Immediately after plucking (no irradiation).
Control 2 20 h after plucking (no irradiation). (290 kV X-rays 147 rad/min) 1,000 basal cells scored in epidermal sheets per mouse; mean and standard error of individual mouse values.

labelled cells per unit area is constant (tables 6 and 7) between days 3 and 6 at a level that is 0.5–0.7 of the immediately post-pluck value and 0.65–0.9 of the value in unirradiated unplucked skin. Continuous labelling studied (repeated tritiated thymidine injection) on repeatedly irradiated skin (4 or 9 daily doses of 300 rad) do not show very different rates of accumulation of labelled cells when compared with unirradiated skin [Denekamp, Stewart and Douglas, 1976]. Preliminary studies on the non-regenerating regions of epidermis after a single dose of 1,200 rad indicates a rate of accumulation of labelled cells during continuous labelling that is identical with the rate for unirradiated skin [Al-Barwari and Potten, unpublished].

Many of the data on post-irradiation mitotic and labelling behaviour of basal cells are based on sectioned material. The results obtained would tend to be the consequence of proliferative behaviour in many regeneratively sterilised cells with the occasional inclusion of regenerative foci. These foci will be small at early times but will increase in size with time and their frequency is clearly dose dependent.

There is one further, as yet unexplained, complication in the post-irradiation response of mouse epidermal cells, namely that the variations in basal cell density, nuclear appearance and regenerative foci are not uniformly distributed throughout the irradiated field.

It is hard to explain this non-random distribution but the following are possibilities: 1) non-uniform stimulation of basal cells by plucking: this is difficult to reconcile with the long term regeneration which is still non-random; 2) local variation in radiosensitivity because of vascularity and oxygen tension; 3) a spatial patterning of clonogenic cells, i.e. areas of more or less differentiated cells possibly related to a clonal patterning laid down in development [Mintz, 1970].

It is clear that as the basal cell density decreases, the labelling index of the non-regenerative areas remains fairly constant, and the basal cell nuclei increase in size, often becoming multi-nucleate (see figs. 12 and 16). It is probable that some of these sterile differentiated cells continue to replicate their DNA after irradiation but have a lower probability of subsequent mitosis. This results in changes in DNA content and nuclear size and the number of nuclear bodies. This hypothesis has yet to be experimentally investigated. The process may be more likely at high doses (above 2,000 rad) and may account for some of the labelled nuclei.

However, bearing in mind these points the labelling data all seem to indicate that many basal cells after doses such as 1,200 rad (which results in 99.9% reproductive sterilisation) enter DNA synthesis, divide normally and subsequently migrate normally. This is easily explained on the basis of a two compartment proliferative model (see section 1.4 and Potten [1976a, 1975a]) where the differentiated proliferative cells (particularly the late differentiated proliferative cells which greatly outnumber all other stages) are relatively unaffected by even large doses of radiation. When affected by even higher doses they still tend to migrate and mature normally and do not become pycnotic/apoptotic. It is possible that only the stem cells or very early differentiated cells manifest radiation damage by pycnotic/apoptotic changes.

It seems quite reasonable to assume that gross skin reaction begins when regions of the basal layer have lost most of their nuclei and many suprabasal cells are lost and that this may take 5—7 days to occur. It is not altogether clear why the time when skin reaction begins does not vary with dose. This again might be better explained by the two-compartment model where the skin reaction is dependent on stem cell destruction and not general basal cell killing. Where many stem cells are killed a patchy cell depletion is inevitable once the differentiated cells have completed their divisions and migration, and the skin reaction begins when some areas are depleted in this way.

4.4. Summary

It is clear that neither the number of pycnotic/apoptotic cells nor the overall number of basal cells alters with dose in a manner to be expected on the basis of the survival curves. It appears possible that only a fraction of the basal cells responds to irradiation by dying in an apoptotic sense. The survival curves indicate that most observed basal cells are reproductively sterilised after moderate doses (1,200 rad) but a more detailed consideration of the origin of these curves leads to the conclusion that it is a large fraction of a small subpopulation of basal cells that are sterilised by the radiation, most of the basal cells being reproductively ineffective anyway and therefore relatively unaffected by the radiation. Labelling, mitotic, migration, and light and electron microscopic studies, all indicate that most cells continue, at least for a short time, fairly normally.

These observations all strongly suggest the existence of two epidermal subpopulations; one that is present in small numbers, capable of extensive regeneration, which if killed by radiation, dies rapidly by proceeding through apoptosis; and the second predominant population (that may be incapable of extensive regeneration) which is relatively unaffected by irradiation.

It is clear that much further work is needed to establish the validity of this concept and to determine what role, if any, G_0, G_2, Langerhans cells, etc. play in the response.

5. Langerhans cells

5.1. Description

Twenty-seven years before Roentgen described the production of X-rays Langerhans [1868] advanced the scope of skin biology by describing a new cell type in epidermis. Since then there have been many suggestions as to the function of these cells but up to the present their origin, function and fate remain a mystery [see review by Wolff, 1972]. Langerhans cells have been observed in many locations (epidermis, oesophagus, stomach, thymus, dermal lymphatic vessels and epidermoid metaplastic trachea and bladder; cells with many Langerhans features are also seen in histiocytosis X [see Wolff, 1972; Riley, 1974; Potten and Allen, 1976]. Langerhans cells replicate their DNA, albeit rarely [Giacometti, 1969; Giacometti and Montagna, 1967; MacKenzie, 1975c]. Mitoses are extremely rare, possibly because the cells lose some of their characteristic features when they divide [Potten and Allen, 1975a, b]. The occasional divisions seen have been in conjunction with abnormal cellular proliferation (Histiocytosis X) [Gianotti and Caputo, 1969].

The number of Langerhans cells in the epidermis remains remarkably constant [Wolff, 1972; Riley, 1974] within a given species. After wounding, their frequency drops but is ultimately restored to its normal level [Lessard, Wolff and Winkelmann, 1966, 1968; Wolff, 1967]. This may be in part due to changes in the histological appearance of the cells [Potten and Allen, 1975b, 1976]. In mouse dorsal skin the fraction of Langerhans cells in the epidermal basal layer, where they are specifically located, is approximately 0.1. In mouse dorsum they are regularly spaced with a single Langerhans cell cen-

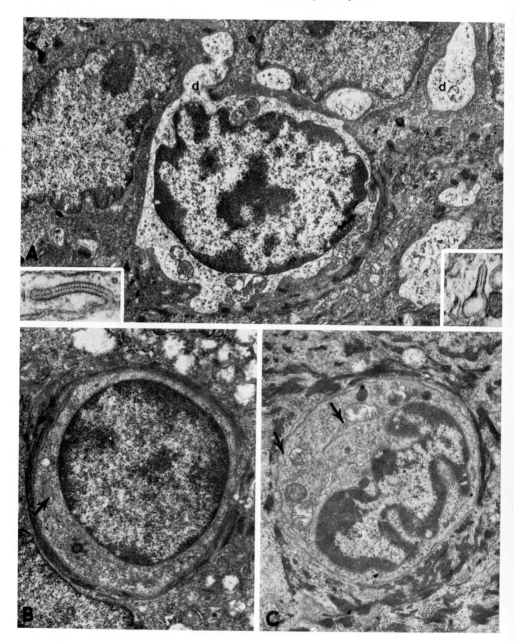

Fig. 17. Langerhans cells in unirradiated mouse epidermis as seen in the electron microscope. A. A typical epidermal dendritic Langerhans cell with a clear cytoplasm (sectioned parallel to skin surface). ×12,000. This particular profile lacks granules but typ-

trally positioned beneath each cornified cell column (EPU) [MacKenzie, 1972; Allen and Potten, 1974]. The ratio of Langerhans cells to basal epidermal cells, their basal position and their relationship to cornified cell columns appear to vary with species [MacKenzie, 1975a, b] and also with different sites even within one species [Potten, 1975a].

Langerhans cells are characterised in the light microscope by their dendritic form and chromophobic nature, often appearing as "clear" cells; by being rich in ATPase and having some specificity for gold, silver or osmium staining [Wolff, 1972; Riley, 1974; Allen and Potten, 1974]. In the electron microscope the cells have a relatively electron-lucent undifferentiated cytoplasm (again appearing as "clear" cells); they often have an irregularly shaped nucleus with some marginal chromatin condensation; they lack desmosomes or hemidesmosomes and any extensive contact with the basement membrane; they frequently possess a characteristic racket-shaped or rod-shaped cytoplasmic organelle [Birbeck, Breathnach and Everall, 1961; Sagebiel and Reed, 1968] which is believed to be the most specific identifying feature (fig. 17) [Wolff, 1972; Riley, 1974]. However, in mouse, the Langerhans cells have few granules and thus many profiles lack this "specific" marker. The granules have been observed in other cell types including keratinocytes and melanocytes where they may have been engulfed [Wolff, 1972; Riley, 1974; Potten and Allen, 1976]. Thus none of the identifying features is entirely specific, making studies of these cells difficult, particularly if any of the features outlined above change with the behaviour of the cell (e.g. during the cell cycle, after wounding or after irradiation). In normal mouse epidermis the cells are most likely to be confused with amelanotic melanocytes, indeterminate cells and Merkel cells, all of which are "clear" cells with a dendritic tendency and similar staining characteristics. However, with the possibility of a change in Langerhans cell morphology and the possible presence of Langerhans granules in keratinocytes [Bell, 1969] it is possible that Langerhans cells might also under some circumstances be difficult to distinguish from other basal cells (i.e. keratinocytes).

There is some suggestion that Langerhans cells are migratory and may even cross the basement membrane but this is contested by others and at present is unresolved. In this laboratory we have yet to see any evidence of migration

ical granules are shown in the inserts. Left-hand insert a typical rod granule in contact with the cell membrane. ×90,000. Right-hand insert a typical racket granule. ×80,000 (d) dendrites. B. Langerhans cell seen three days after plucking. A rounded cell enclosed by many filaments in the keratinocyte neighbours. The cell possesses a granule (arrow), a prominent centriole and lacks a "clear" cytoplasm. ×15,000. C. Langerhans cell from mouse plantar skin with a rounded shape, a not particularly "clear" cytoplasm, several granules (arrows) and a very indented nuclear profile with marginal chromatin condensation.

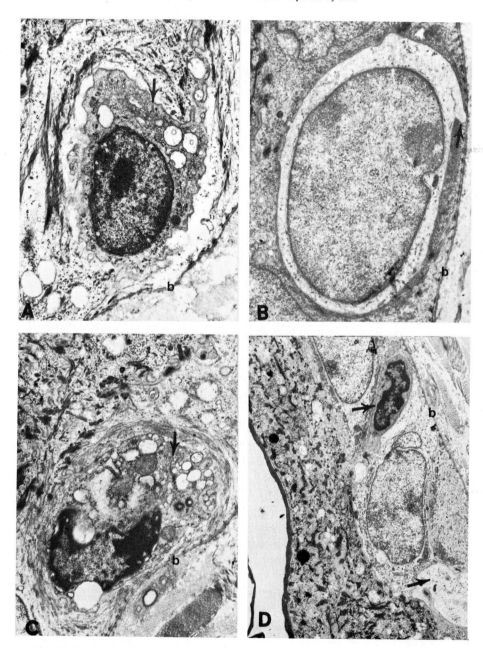

Fig. 18. Presumed Langerhans cells (many with granules) seen at various times after irradiation some of which may show signs of damage. A. 500 rad day 1 a cell with a Langerhans granule (arrow), a dark cytoplasm and a cell membrane drawn away slightly from

across the basement membrane though there have been many instances when movement within the epidermis seems to have been likely.

The dendritic morphology, the cytoplasmic appearance (density), nuclear morphology and granular frequency can all vary considerably (figs. 17–19) [Potten and Allen, 1975b]. These changes appear to be correlated with changes in the proliferative activity of the surrrounding keratinocytes ([Potten and Allen, 1975] and fig. 17).

Although the Langerhans cell is centrally positioned in relation to the EPU basal cells and divides infrequently, it is not clear whether the slowly cycling (and yet stimulus-responsive) more central EPU cells observed in light microscope autoradiographs are in fact the Langerhans cells. There are usually 2–4 cells that are in a non-peripheral position which makes it difficult to define these relationships.

5.2. Effects of irradiation

There are very few published data on the effects of ionizing radiation on the Langerhans cells which is perhaps a little surprising since they constitute up to 10% of the basal cells and usually outnumber the melanocytes, the response of which has been studied. Ultraviolet light irradiation of guinea pig skin causes significant changes in the melanocyte population and the quantity of melanin but no changes could be detected in the Langerhans cells [Wolff, 1972, Wolff and Winkelmann, 1967]. Thorium-X exposure of skin (predominantly alpha particle irradiation) appeared to alter the number of gold-staining cells [Fan, Schoenfeld and Hunter, 1959] and electron-microscopically detectable Langerhans cells [Breathnach, Birbeck and Everall, 1963]. The authors point out that in both cases the results might be explained by changes in the metabolic state of the cells.

Zelickson and Mottaz [1970] found the Langerhans cell numbers decreased after UV irradiation of human skin. Minor wounding of the skin (such as that caused by plucking or adhesive tape stripping: both of which remove cornified cells) causes changes in the location, numbers and appear-

---

neighbouring cells. ×13,000. (b) basement membrane. B. 900 rad day 5, a cell lacking granules but believed to be a Langerhans cell, with a very "clear" cytoplasm and a typical "foot" to the basement membrane (arrowed). ×14,000. C. 500 rad day 7. Basal cell with a granule (arrow), lacking a clear cytoplasm but with considerable nuclear chromatin condensation. ×13,000. D. 500 rad day 10. Basal cell region showing one dark cell with microvillar projections and much condensed chromatin, a second cell with clear cytoplasm nearby. Neither of these cells possessed granules but both lacked desmosomes and were clearly not keratinocytes. It was unclear how these cells should be classified and indeed whether they were both healthy undamaged cells of the same or differing cell types. Pairs of dark and light cells close together were seen on several occasions.

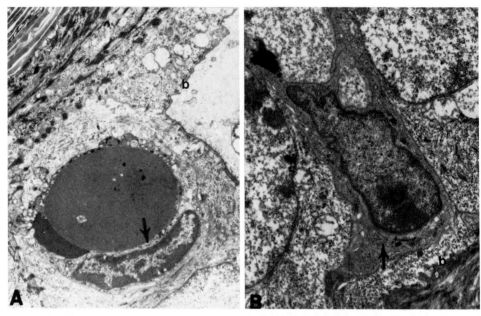

Fig. 19. Langerhans cells at various times after irradiation (continued from fig. 17). A. 900 rad day 5. A Langerhans cell with a granule (arrowed), observed at the neck of a hair follicle (see right angle in basement membrane (b)) with a large cytoplasmic inclusion possibly representing the remains of a dead cell (apottotic body). ×10,000. B. 3,500 rad day 7. A Langerhans cell with several granules in the basal layer of an otherwise hyperplastic region of re-epithelialisation. This cell was one of many similar cells (many lacking granules) regularly spaced in this re-epithelialising region. This particular cell had a fairly dark cytoplasm and long dendrites not all of which are seen. ×10,000.

ance of Langerhans cells in both guinea pig [Lessard et al., 1966, 1968] and mouse epidermis [Potten and Allen, 1975b]. In mouse the cells may change their position (become suprabasal): appearance (become more rounded and less "clear"); and apparent numbers. Part of the change in numbers may be due to a change in the appearance of the cells and thus their detectability. As can be seen from figs. 17–19 similar, or even more dramatic, changes can be observed after X-irradiation. Cells lacking desmosomes with Langerhans granules can be observed with either extremely "clear" cytoplasm or quite "dark" cytoplasm. These cells can be rounded, dendritic or even possess microvillar-like projections. It is interesting that Langerhans cells are present in significant numbers in regenerated epidermis after X-ray doses that resulted in enormous epidermal cell killing (fig. 19). The Langerhans granule frequency in these cells is low and in some cases it is impossible to identify positively the cells. However, regularly-spaced identical-looking cells lacking desmosomes and with occasional Langerhans granules can be seen in both basal and suprabasal positions in the multi-layered hyperplastic regenerating epithelium. These cells may in some cases have extremely long dendritic processes and are clearly present at a time when columnar organisation is re-establishing. These apparent Langerhans cells are common in regenerating epidermis while melanocytes are rare in these samples.

5.3. Possible functions

The function of Langerhans cells is unclear and has been extensively discussed elsewhere [Wolff, 1972; Riley, 1974; Potten and Allen, 1976]. For some considerable time they were thought to be related to the melanocyte population, as either immature melanocytes or, more probably, old effete melanocytes. This hypothesis has now been generally rejected on many grounds. It appears that at present there are two possible roles that would be proposed for the Langerhans cell: that of (1) a controller or nurse cell for keratinocyte behaviour, (2) a stem cell (or very early differentiated cell) for the keratinocyte cells.

From its numerical and spatial distribution in mouse skin, its universal occurrence in external surface epidermis, and its reappearance or persistence in regenerated epidermis, it seems likely that the Langerhans cell plays an important role in skin biology. For this reason its radiobiological response should be more extensively investigated.

In mouse dorsum Langerhans cells tend to be centrally positioned within an EPU with radiating dendrites (when they happen to have a dendritic morphology) which appear to reach to the periphery of the EPU and probably have contact with all 9–11 other basal cells, as well as with the suprabasal cells. In thicker epithelia the cells tend to occur in suprabasal positions and can possess long dendritic processes. It is a little hard to see how these cells, which lack desmosomal contacts and have an extensive surface area, are not carried into the cornified layer with the general flow of cells particularly in thicker more rapidly dividing epithelia. Their lack of apparent replacement at levels comparable to keratinocytes and their constancy in numbers indicates that they are not lost with corneocytes unless, however, they are derived directly from keratinocytes or have a precursor pool that is indistinguishable from keratinocytes. Their contact with all or most of the proliferative keratinocytes would make them suited to receiving or transmitting proliferation or differentiation messages. Their apparent behavioural changes running parallel to keratinocyte proliferative changes would further support this idea.

The function of the Langerhans granules is equally unclear. They are often observed in some numbers, with many presumed immature forms, in the Golgi region, and can also be seen to have contact with the intercellular space by opening through the cell membrane (see fig. 17). Unfortunately it is impossible at present to say whether the granule might be bringing a message or some factor into the cell or taking one out. It is obviously tempting to speculate that the granule may be a keratinocyte-controlling message and this hypothesis has been previously suggested [MacKenzie, 1972; Sagebiel, Clarke and Hutchens, 1971; Breathnach and Wyllie, 1967; Lisi, 1973; Prunieras, 1969; Potten and Allen, 1976]. If the granules do contain a message which controls proliferation via inhibition, i.e. acts somewhat like a chalone [Bullough, 1965, 1973a, b] then it could partially explain why Langerhans cells (i.e. cells with granules) are not seen in division — cell division and granule presence may be mutually exclusive.

The second suggested role for the Langerhans cell is more heretical: namely, that the Langehans cells may act as a source of new young keratinocytes, i.e. are keratinocyte stem cells. This hypothesis could explain (1) the rough correlation between the estimates for clonogenic cells and Langerhans cells [Potten and Hendry 1973; Potten, 1974]; (2) the concomitant changes in Langerhans cell behaviour and keratinocyte proliferation in wounding [Potten and Allen, 1975b] and (3) the persistence or reappearance of Langerhans cells in regenerated epithelium. If these cells are keratinocyte stem cells then they may not exhibit many Langerhans cell features during phases of rapid epithelial repopulation and thus be difficult to identify and count. If they play a stem cell role then the number of cells relative to the proliferating keratinocytes may vary from site to site and species to species and there would be no need for a close correlation with cornified cell columns. The lack of correlation between apparent Langerhans cell numbers and cornified columns in some species (e.g. hamsters) [MacKenzie, 1975b] may indicate that they do not act as differentiation-controlling cells but this need not prevent them from having a stem cell function.

One final point that needs to be resolved in connection with the stem cell role for Langerhans cells is the discrepancy between the micro- and macrocolony data (see section 3.6) where one possible explanation was that the macrocolony assay might be the consequence of early differentiated cell growth while the microcolonies might be the result of stem cell derived "spreading" repopulation. If this is the case then the estimates for clonogenic cell numbers would suggest that Langerhans cells are not stem cells but early differentiated precursors. It might be argued that it is unlikely that the early differentiated cells would be morphologically distinct from the stem cells. However, if there is a morphological distinction between stem cells and differentiated cells it seems possible that the young differentiated cells might retain some stem cell morphology. It has also been suggested that perhaps full stem cell activity may be restricted to cells in the follicles (section 3.6) and it is interesting that Langerhans cells are frequently to be found in and around the follicle neck in mouse dorsum (fig. 19) and appear in tail skin to be more specifically located in the hair follicles [Riley, 1966, 1974].

In the tail, the post-irradiation fate of the skin appears to depend on a sub-epidermal cell population (?follicular), as shown by oxygenation and clamping experiments on wound healing [Hendry, unpublished data, personal communication].

6. General conclusions and future objectives

In spite of 80 years of radiobiological studies on skin it is clear that many questions remain unanswered and with the advances in skin biology new questions arise. Before interpreting the results and predicting the consequences of multiple dose schedules it is essential that the tissue and cellular response of skin to single doses of varying size be completely understood particularly the low doses commonly used in radiotherapy. The problems are not only those associated with the radiation response of skin but there are many unanswered questions connected with basic cutaneous biology. One of the major problems involves the fact that skin is a complex tissue made up of many cell types, most of which are replacing cells lost by death or differentiation. Some of the replacement rates may be very slow and are difficult to study. Each cell type has a role to play within the tissue and each cell is intimately inter-related with other cells. Damage, its repair and expression is almost certainly influenced by the microenvironment within which the damaged cell finds itself. This type of interdependence is almost completely neglected in radiobiological studies in vivo and needs to be experimentally investigated. There are very few data available on the effects of skin irradiation on fibroblasts, nerve fibres, capillary cells, and Langerhans cells to mention only a few of the more obvious participants in cellular communication processes and the control of the microenvironment.

At a higher (more macroscopic) level it is by no means clear what role the hair follicle and its associated structures and regions play in skin biology and the response of the skin to damage including that induced by irradiation. There are a number of indications that it might be a very significant region in connection with re-epithelialisation. At a similar tissue level the full role played by the dermis or any of its constituent cell populations in regeneration and healing is not clear.

In studies of this sort one is obviously interested in the response of human skin to the doses commonly used in radiotherapy. Although radiotherapy is ultimately concerned with cell killing processes, there is a wide discrepancy between the size of single doses used to measure cell survival experimentally, and the individual dose size commonly used in fractionation regimes. There are other considerations such as repair of sublethal damage, the modification of potentially lethal damage, repopulation and cell kinetic changes and changes in the oxygenation of crucial cells (reoxygenation), that have to be taken into account in the understanding of the overall response. All these factors have not been within the scope of this review but clearly cannot be neglected in considering skin radiobiology. It is clear that the response of the tissue and its individual cell populations to doses less than 500 rad should be more extensively studied. In terms of basic skin biology it is clear that the implications of the columnar organisation and the homogeneity of the basal cell population has to be further investigated since both affect the interpretation of the cell kinetic and radiobiological data. The determination of the function of the Langerhans cell population is also of paramount importance.

The overall behaviour of the epidermal basal cells seems, for a number of reasons, to suggest that there are two or more subpopulations of cells. This clearly needs to be further resolved before the micro- and macrocolony and skin reaction data can be fully understood.

Acknowledgments

Some of the work reported here has been conducted at the Paterson Laboratories and supported by grants from the Medical Research Council and the Cancer Research Campaign. I am grateful for help and advice on the electron microscopy from Dr. T. Allen and Mr. Gordon Bennion and to Mr. S.E. Al-Barwari for allowing me to quote from his unpublished work. My thanks also go to Hilary Goodwin for her help in preparing this report.

References

Al-Barwari, S.E. and Potten, C.S. 1976. Int. J. Radiat. Biol. 30, 201.
Albert, R.E., Burns, F.J. and Bennett, P. 1972. J. Natl. Cancer Inst. 49, 1131.
Albert, R.E., Burns, F.J. and Heimbach, R.D. 1967a. Radiat. Res. 30, 590.
Albert, R.E., Burns, F.J. and Heimbach, R.D. 1967b. Radiat. Res. 30, 515.
Albert, R.E., Burns, F.J. and Heimbach, R.D. 1968. Radiat. Res. 36, 225.
Allen, T.D. and Potten, C.S. 1974. J. Cell. Sci. 15, 291.
Allen, T.D. and Potten, C.S. 1976. J. Cell. Sci. 21, 341.
Alper, T., Fowler, J.F., Morgan, R.L., Vonberg, D.D., Ellis, F. and Oliver R. 1962. Br. J. Radiol. 35, 722.
Alper, T., Gillies, N.E. and Elkind, M. 1960. Nature 186, 1062.
Archambeau, J.O. 1971. Radiat. Res. 47, 320 abs.
Argyris, T.S. 1976. Am. J. Pathol. 83, 329.
Bailey, A.J. 1968. Int. Rev. Connect. Tissue Res. 4, 233.
Bell, M. 1969. J. Cell. Biol. 41, 914.

Bennington, J.L., Holmes, E.J. and Combs, J.W. 1976. J. Invest. Dermatol. 66, 183.
Birbeck, M.S.C., Breathnach, A.S. and Everall, J.D. 1961. J. Invest. Dermatol. 37, 51.
Bishop, G.H. 1945. Am. J. Anat. 76, 153.
Breathnach, A.S., Birbeck, M.S.C. and Everall, J.D. 1963. Ann. N.Y. Acad. Sci. 100, 223.
Breathnach, A.S. and Wyllie, L.M.A. 1967. In: Advances in Biology of the Skin: The Pigmentary System, Vol. 8. Eds. Montagna, W. and Hu, F. (Pergamon Press, Oxford) p. 97.
Breit, R. and Kligman, A.M. 1969. In: The Biologic Effects of Ultraviolet Radiation. Ed. Urbach, F. (Pergamon Press, Oxford) p. 267.
Broerse, J.J., Barendsen, G.W., Freriks, G. and Van Putten, L.M. 1971. Eur. J. Cancer 7, 171.
Bruce, W.R. and McCulloch, E.A. 1964. Blood 23, 216.
Bullough, W.S. 1965. Cancer Res. 25, 1683.
Bullough, W.S. 1973a. J. Natl. Cancer Inst. Monogr. 38, 5.
Bullough, W.S. 1973b. J. Natl. Cancer Inst. Monogr. 38, 99.
Burns, F.J., Sinclair, I.P., Albert, R.E. and Vanderlaan, M. 1976. Radiat. Res. 67, 474.
Cameron, I.L. 1971. In: Cellular and Molecular Renewal in the Mammalian Body. Eds. Cameron, I.K. and Thrasher, J.D. (Academic Press, New York) p. 45.
Chase, H.B. 1949. J. Morphol. 84, 57.
Chase, H.B. 1951. Science 113, 714.
Chase, H.B. and Hunt, J.W. 1959. In: Pigment Cell Biology. Ed. Gordon, M. (Academic Press, New York) p. 537.
Chase, H.B. and Montagna, W. 1951. Proc. Soc. Exp. Biol. Med. 76, 35.
Chase, H.B., Montagna, W. and Malone, J.D. 1953. Anat. Rec. 116, 75.
Chase, H.B. and Rauch, H. 1950. J. Morphol. 87, 381.
Chase, H.B., Rauch, H. and Smith, V.W. 1951. Physiol. Zool. 24, 1.
Chase, H.B., Straile, W.E. and Arsenault, C. 1963. Ann. N.Y. Acad. Sci. 100, 390.
Cheng, H. and Leblond, C.P. 1974. Am. J. Anat. 141, 537.
Christophers, E. 1971a. Z. Zellforsch. Mikrosk. Anat. 114, 441.
Christophers, E. 1971b. J. Invest. Dermatol. 57, 241.
Christophers, E. 1971c. J. Invest. Dermatol. 56, 165.
Christophers, E., Wolff, H.H. and Laurance, E.B. 1974. J. Invest. Dermatol. 62, 555.
Claxton, J.H. 1966. Anat. Rec. 154, 195.
Cleaver, J.E. 1967. Thymidine metabolism and cell kinetics (North-Holland, Amsterdam) 259 pp.
Cohen, J. 1969. In: Biology of the Skin, Vol. 9. Eds. Montagna, W. and Hu, F. (Pergamon Press, Oxford) p. 1.
Cramer, W. and Stowell, R.E. 1943. Cancer Res. 3, 36.
De Loecker, W., Van der Schueren, E., Stass, M.L. and Doms, D. 1976. Int. J. Radiat. Biol. 29, 351.
Denekamp, J., Ball, M.M. and Fowler, J.F. 1969. Radiat. Res. 37, 361.
Denekamp, J., Emery, E.W. and Field, S.B. 1971. Radiat. Res. 45, 80.
Denekamp, J. and Fowler, J.F. 1966. Int. J. Radiat. Biol. 10, 135.
Denekamp, J., Fowler, J.F., Kragt, K., Parnell, C.J. and Field, S.B. 1966. Radiat. Res. 29, 71.
Denekamp, J., Stewart, F.A. and Douglas, B.G. 1976. Cell Tissue Kinet. 9, 19.
de Rey, B.L.M. and Klein-Szanto, A.J.P. 1972. Strahlentherapie 143, 699.
Devik, F. 1955. Acta Radiol. (Stockholm) Suppl. 119, 1.
Devik, F. 1962. Int. J. Radiat. Biol. 5, 59.

Dry, F.W. 1926. J. Genet. 16, 287.
Dubravsky, N., Hunter, N. and Withers, H.R. 1976. Radiat. Res. 65, 481.
Dunjic, A. 1974. Curr. Top. Radiat. Res. 10, 151.
Dutreix, J., Tubiana, M., Wambersie, A. and Malaise, E. 1971. Eur. J. Cancer 7, 206.
Eassa, E.M. and Casarett, G.W. 1973. Radiology 106, 679.
Eisen, A.Z., Holyoke, J.B. and Lobitz, W.C. 1956. J. Invest. Dermatol. 25, 145.
Elgjo, K. 1973. J. Natl. Cancer Inst. Monogr. 38, 71.
Elgjo, K., Laerum, O.D. and Edgehill, W. 1971. Virchows Arch. B. Zellpathol. 8, 277.
Elgjo, K., Laerum, O.D. and Edgehill, W. 1972. Virschows Arch. B. Zellpathol. 10, 229.
Ellinger, F. 1941. The Biologic Fundamentals of Radiation Therapy. (Elsevier Publ. Co., Amsterdam).
Ellinger, F. 1951. Ann. N.Y. Acad. Sci. 53, 682.
Ellinger, F. 1957. Medical Radiation Biology (Thomas, Springfield, Ill.).
Emery, E.W., Denekamp, J. and Ball, M.M. 1970. Radiat. Res. 41, 450.
Etoh, H., Taguchi, Y.H. and Tabachnick, J. 1975. J. Invest. Dermatol. 64, 431.
Fan, J., Schoenfeld, R.J. and Hunter, R. 1959. J. Invest. Dermatol. 32, 445.
Field, S.B., Morris, C., Denekamp, J. and Fowler, J.F. 1975. Eur. J. Cancer 11, 291.
Forbes, P.D. 1966. Radiat. Res. 27, 521 abs.
Forbes, P.D. and Urbach, F. 1969. In: Biological Effects of Ultraviolet Irradiation. Ed. Urbach, F. (Pergamon Press, Oxford) p. 279.
Fowler, J.F. and Denekamp, J. 1976. In: Stem Cells of Renewing Cell Populations. Eds. Cairnie, A.B., Lala, P.K. and Osmond, D.G. (Academic Press, New York) p. 117.
Fowler, J.F., Denekamp, J., Delapeyre, C., Harris, S.R. and Sheldon, P.W. 1974. Int. J. Radiat. Biol. 25, 213.
Gelfant, S. 1966. In: Methods in Cell Physiology. Ed. Prescott, D.M. (Academic Press New York) p. 359.
Gelfant, S. and Candelas, G.C. 1972. J. Invest. Dermatol. 59, 7.
Giacometti, L. 1969. J. Invest. Dermatol. 53, 151.
Giacometti, L. and Montagna, W. 1967. Science 157, 439.
Gianotti, F. and Caputo, R. 1969. Arch. Dermatol. 100, 342.
Gonsalves, N.I. and Chase, H.B. 1975. TIT J. Life Sci. 5, 77.
Green, H., Rheinwald, J.G. and Sun, T.T. 1976. J. Supramol. Struct. In press.
Griem, M.L. and Malkinson, F.D. 1967. Radiat. Res. 30, 431.
Griem, M.L., Malkinson, F.D., Marianovic, R. and Kessler, D. 1973. Advances in Radiat. Res. Eds. Duplan, J.F. and Chapiro, A. (Gordon and Breach, New York) p. 845.
Grillo, H.C. 1963. Ann. Surg. 157, 453.
Grube, D.D., Auerbach, H. and Brues, A.M. 1970. Cell Tissue Kinet. 3, 363.
Hall, D.A. 1973. Advances in Radiat. Res. Eds. Duplan, J.F. and Chapiro, A. (Gordon and Breach, New York) p. 1331.
Hambrick, G.W. and Blank, H. 1956. J. Invest. Dermatol. 326, 185.
Hamilton, E. 1974. Cell Tissue Kinet. 7, 389.
Hamilton, E., Howard, A. and Potten, C.S. 1974. Cell Tissue Kinet. 7, 399.
Hamilton, E. and Potten, C.S. 1972. Cell Tissue Kinet. 5, 505.
Hamilton, E. and Potten, C.S. 1974. J. Invest. Dermatol. 62, 560.
Hance, R.I. and Murphy, J.B. 1926. J. Exp. Med. 44, 339.
Hayashi, S. and Suit, H.D. 1972. Radiology 103, 431.
Hegazy, M.A.H. and Fowler, J.F. 1973a. Cell Tissue Kinet. 6, 17.
Hegazy, M.A.H. and Fowler, J.F. 1973b. Cell Tissue Kinet. 6, 587.
Hendry, J.H., Major, D. and Greene, D. 1975. Radiat. Res. 63. 149.
Hendry, J.H. and Potten, C.S. 1974. Int. J. Radiat. Biol. 25, 583.

Holti, G. 1955. Clin. Sci. 14, 143.
Hornsey, S. 1973. Radiat. Res. 55, 58.
Houck, J. ed. 1976. Chalones. (North-Holland, Amsterdam) 510 pp.
Howard, A. and Pelc, S.R. 1953. Heredity 6, suppl., 261.
Howell, J.S. 1962. Br. J. Cancer 16, 101.
Hume, W.J. and Potten, C.S. 1976. J. Cell. Sci. 22, 149.
Iversen, O.H., Bjerkness, R. and Devik, F. 1968. Cell Tissue Kinet. 1, 351.
Jarrett, A. ed. 1974, The Physiology and Pathophysiology of the Skin, Vol. 3. (Academic Press, New York).
Johnson, E. and Ebling, F.J. 1964. J. Embryol. Exp. Morphol. 12, 465.
Jolles, B. 1949. Nature 164, 63.
Jolles, B. 1950. Br. J. Radiol. 23, 18.
Jolles, B. and Harrison, R.G. 1963. Nature 198, 1216.
Jolles, B. and Harrison, R.G. 1965. Nature 205, 920.
Jolles, B. and Harrison, R.G. 1966. Br. J. Radiol. 39, 12.
Kerr, J.F.R., Wyllies, A.H. and Currie, A.R. 1972. Br. J. Cancer 26, 587.
Kligman, A.M. 1959. Ann. N.Y. Acad. Sci. 83, 507.
Krawczyk, W. 1971. J. Cell. Biol. 49, 247.
Krawczyk, W. 1972. In: Epidermal Wound Healing. Eds. Maibach, H.I. and Rovee, D.T. (Year Book Med. Publ., Chicago) p. 123.
Kurban, A.K. and Farah, F.S. 1969. Acta Derm. Venereol. 49, 64.
Lajtha, L.G. 1963. J. Cell. Comp. Physiol. 60, suppl., 143.
Lajtha, L.G. 1964. Medicine (Baltimore) 43, 625.
Langerhans, P. 1868. Virshows Arch. A. Pathol. Anat. 44, 325.
Leblond, C.P. and Cheng, H. 1976. In: Stem Cells of Renewing Cell Population. Eds. Cairnie, A.B., Lalz, P.K. and Osmond D.G. (Academic Press, New York) p. 7.
Leblond, C.P., Clermont, Y. and Nadler, N.J. 1967. In: Canadian Cancer Conf. (Pergamon Press, Oxford) p. 3.
Leith, J.T., Schilling, W.A., Lyman, J.T. and Howard, J. 1975. Radiat. Res. 62, 195.
Leith, J.T., Schilling, W.A. and Welch, G.P. 1971. Int. J. Radiat. Biol. 19, 603.
Lessard, R.J., Wolff, K. and Winkelmann, R.K. 1966. Nature 212, 628.
Lessard, R.J., Wolff, K. and Winkelmann, R.K. 1968. J. Invest. Dermatol 50, 171.
Lisi, P. 1973. Acta Derm. Venereol. 53, 425.
Mackenzie, I.C. 1969. Nature 222, 881.
Mackenzie, I.C. 1970. Nature 226, 653.
Mackenzie, I.C. 1972. In: Epidermal Wound Healing. Eds. Maibach, H.I. and Rovee, D.T. (Year Book Med. Publ., Chicago).
Mackenzie, I.C. 1975a. J. Invest. Dermatol. 65, 45.
Mackenzie, I.C. 1975b. Anat. Rec. 181, 705.
Mackenzie, I.C. 1975c. Am. J. Anat. 144, 127.
Maisin, J.R., Oledzka-Slotwinska, H. and Lambiet-Collier, M. 1973. Advances in Radiat. Res. Eds. Duplan, J.F. and Chapiro, A. (Gordon and Breach, New York) p. 1347.
Malkinson, F.D. and Griem, M.L. 1965. In: Biology of the Skin and Hair Growth Eds. Lyne, A.G. and Short, B.F. (Angus and Robertson, Sydney) p. 755.
Melbye, S.W. and Karasek, M.A. 1973. Exp. Cell Res. 79, 279.
Mendelsohn, M.L. 1962. J. Natl. Cancer Inst. 28, 1015.
Mintz, B. 1970. Symp. Int. Soc. Cell Biol. Vol. 9. (Academic Press, New York) p. 15.
Moffat, G.H. 1968. J. Anat. 102, 527.
Montagna, W. 1962. The Structure and Function of Skin. (Academic Press, New York) 454 pp.

Montagna, W. and Chase, H.G. 1950. Anat. Rec. 107, 83.
Moretti, G., Baccaredda-Boy, A. and Rebora, A. 1969. In: Advances in the Biology of the Skin, Vol. 9. Hair Growth. Eds. Montagna, W. and Dobson, R.K. (Pergamon Press, Oxford) p. 535.
Odland, G. and Ross, R. 1968. J. Cell. Biol. 39, 135.
Oduye, O.O. 1975. Res. Vet. Sci. 19, 245.
Oliver, R.F. 1969. In: Biology of the Skin, Vol. 9. Eds. Montagna, W. and Hu, F. (Pergamon Press, Oxford) p. 19.
Olson, R.L. and Everett, M.A. 1975. J. Cutaneous Pathol. 2, 53.
Osgood, E.S. 1957. J. Natl. Cancer Inst. 18, 155.
Potten, C.S. 1967. Ph.D. Thesis University of London.
Potten, C.S. 1968. Cell Tissue Kinet. 1, 239.
Potten, C.S. 1970. J. Invest. Dermatol. 55, 410.
Potten, C.S. 1971. J. Invest. Dermatol. 56, 311.
Potten, C.S. 1972a. In: Pigmentation – its Genesis and Biologic Control. Ed. Riley, V. (Appleton Century-Crofts, New York) p. 433.
Potten, C.S. 1972b. Radiat. Res. 51, 167.
Potten, C.S. 1974. Cell Tissue Kinet. 7, 77.
Potten, C.S. 1975a. J. Invest. Dermatol. 65, 488.
Potten, C.S. 1975b. Br. J. Dermatol. 93, 649.
Potten, C.S. 1976a. In: Stem Cells of Renewing Cell Populations. Eds. Cairnie, A.B., Lala, P.K. and Osmond, D.G. (Academic Press, New York) p. 91.
Potten, C.S. 1976b. Bull. Cancer 62, 419.
Potten, C.S. 1976c. In: Stem Cells of Renewing Cell Populations. Eds. Cairnie, A.B., Lala, P.K. and Osmond, D.G. (Academic Press, New York) p. 79.
Potten, C.S. and Allen, T.D. 1975a. Differentiation 3, 161.
Potten, C.S. and Allen, T.D. 1975b. J. Cell Sci. 17, 413.
Potten, C.S. and Allen, T.D. 1976. Differentiation 5, 43.
Potten, C.S. and Chase, H.B. 1970. Radiat. Res. 42, 305.
Potten, C.S. and Forbes, P.D. 1972. Int. J. Radiat. Biol. 22, 337.
Potten, C.S. and Hendry, J.H. 1973. Int. J. Radiat. Biol. 24, 537.
Potten, C.S. and Hendry, J.H. 1975. Int. J. Radiat. Biol. 27, 413.
Potten, C.S. and Howard, A. 1969. Radiat. Res. 38, 65.
Potten, C.S., Jessup, B.A. and Croxson, B.M. 1971. Cell Tissue Kinet. 4, 241.
Potten, C.S., Merkow, L.P. and Pardo, M. 1971. Lab. Invest. 25, 607.
Prunieras, M. 1969. J. Invest. Dermatol. 52, 1.
Quevedo, W.C. and Grahn, D. 1958. Radiat. Res. 8, 254.
Quevedo, W.C. and Smith, J.A. 1963. Ann. N.Y. Acad. Sci. 100, 364.
Quevedo, W.C., Szabo, G., Virks, J. and Sinesi, S.J. 1965. J. Invest. Dermatol. 45, 295.
Ramsay, C.A. and Challoner, A.V.J. 1976. Br. J. Dermatol. 94, 487.
Raper, J.R., Wirth, J.E. and Barnes, K.K. 1951. Effects of external beta radiation. Ed. Zirkle. R.E. p. 42.
Reinhold, H.S. 1972. Front. Radiat. Ther. Oncol. 6, 44.
Reinhold, H.S. 1973. In: Advances in Radiat. Res. Eds. Duplan, J.F. and Chapiro, A. (Gordon and Breach, New York, 1315.
Reinhold, H.S. 1974. Curr. Top. Radiat. Res. 10, 9.
Reinhold, H.S. and Buisman, G.H. 1973. Br. J. Radiol. 46, 54.
Rheinwald, J.G. and Green, H. 1975. Cell 6, 331.
Riley, P.A. 1966. Br. J. Dermatol. 78, 388.
Riley, P.A. 1974. In: Physiology and Pathophysiology of the Skin. Ed. Jarrett, A. (Academic Press, London) p. 1199.

Roberts, D.S. 1965. J. Pathol. Bacteriol. 90, 213.
Roentgen, W.K. 1895. Sitzungsber. Würzburger Physik. med. Gesellschaft.
Rovee, D.T., Karowsky, C.A., Labun, J. and Downes, A.M. 1972. In: Epidermal Wound Healing. Eds. Maibach, H.I. and Rovee, D.T. (Year Book Med. Publ., Chicago) p. 159.
Sagebiel, R.W. Clarke, M.A. and Hutchens, L.H. 1971. The Histology of Oral Mucosa. Eds. Squier, C.A. and Meyer, J. (Thomas, Springfield, Ill.) p. 143.
Sagebiel, R.W. and Reed, T.H. 1968. J. Cell Biol. 36, 595.
Searle, J., Lawson, T.A., Abbott, P.J., Harmon, B. and Kerr, J.F.R. 1975. J. Pathol. 116, 129.
Shuster, S. and Bottoms, E. 1963. Clin. Sci. 25, 487.
Sinclair, W.K. and Morton, R.A. 1966. Radiat. Res. 29, 450.
Spear, F.G. 1953. Radiations and Living Cells. (Chapman and Hall, London) 222 pp.
Stearner, S.B. and Christian, E.J.B. 1968. Radiat. Res. 34, 138.
Stearner, S.B. and Christian, E.J.B. 1969. Radiat. Res. 38, 153.
Strauss, J.S. and Kligman, A.M. 1959. J. Invest. Dermatol. 33, 347.
Terasima, T. and Tolmach, L.J. 1963. Biophys. J. 3, 11.
Testa, N.G., Hendry, J.H. and Lajtha, L.G. 1973. Biomedicine 19, 183.
Tvermyr, E.M.F. 1972. Virchows Arch. B. Zellpathol. 11, 43.
Van den Brenk H.A.S. 1972. Int. J. Radiat. Biol. 21, 607.
Van Scott, E.J. and Reinertson, R.P. 1957. J. Invest. Dermatol. 29, 205.
Von Essen, C.F. 1972. Front. Radiat. Ther. Oncol. 6, 148.
Weinstein, G.D. and Frost, P. 1969. Natl. Cancer Inst. Monogr. 30, 225.
Williams, A.W. 1906. Br. J. Dermatol. 18, 63.
Winter, G.D. and Scales, J.T. 1963. Nature 197, 91.
Winter, G.D. 1964. In: Advances in the Biology of the Skin, Vol. 5. Wound Healing, Eds Montagna, W. and Billingham, R.E. (Pergamon Press, Oxford) p. 113.
Winter, G.D. 1972. In: Epidermal Wound Healing. Eds. Maibach, H.I. and Rovee, D.T. (Year Book Med. Publ., Chicago) p. 71.
Withers, H.R. 1965. Ph.D. Thesis University of London.
Withers, H.R. 1967a. Br. J. Radiol. 40, 187.
Withers, H.R. 1967b. Br. J. Radiol. 40, 335.
Withers, H.R. 1967c. Radiat. Res. 32, 227.
Withers, H.R., Brennan, J.T. and Elkind, M.M. 1970. Br. J. Radiol. 43, 796.
Withers, H.R. and Elkind, M.M. 1969. Radiat. Res. 38, 598.
Withers, H.R. and Elkind, M.M. 1970. Int. J. Radiat. Biol. 17, 261.
Wolff, K. 1967. Arch. Klin. Exp. Dermatol. 229, 54.
Wolff, K. 1972. Curr. Probl. Dermatol. 4, 79.
Wolff, K. and Winkelmann, R.K. 1967. J. Invest. Dermatol 48, 531.
Yamaura, H., Yamada, K. and Matzuzawa T. 1976. Int. J. Radiat. Biol. 30, 179.
Zelickson, A.S. and Mottaz, J. 1970. Arch. Dermatol. 101, 312.

TRANSMUTATION EFFECTS OF ^{32}P AND ^{33}P INCORPORATED IN DNA

S. APELGOT and J.P. ADLOFF

Laboratoire Curie de la Fondation Curie-Institut du Radium, and Laboratoire de Chimie Nucléaire du Centre de Recherches Nucléaires de Strasbourg, France

Mindful of the data in Hot Atom Chemistry, we have tried to postulate various hypotheses allowing an explanation of the processes involved in the lethal effect due to the decay of radiophosphorus atoms incorporated in cell or micro-organism DNA. When, in such DNA, a stable sulphur atom takes the place of a radioactive phosphorus atom, many primary events follow this sudden substitution. Our study suggests that the primary events of this transmutation end in many different secondary events, due to various possible pathways. The DNA molecule will undergo diverse localized changes; each change is characterized by a lethal probability. The final lethal event will depend on the experimental conditions. There is no unequivocal relationship between a given localized modification and a precise secondary event.

CONTENTS

PART ONE: CHEMICAL EFFECTS, J.P. ADLOFF and S. APELGOT

1. INTRODUCTION 63

2. PRIMARY EFFECTS OF ^{32}P AND ^{33}P DISINTEGRATION 64
   2.1. Transmutation 64
   2.2. Electronic excitation 64
   2.3. Recoil energy 65
   2.4 Autoionization 67

3. POSSIBLE TRANSMUTATION EFFECTS OF ^{32}P AND ^{33}P IN DNA 67

4. CONCLUSIONS 70

PART TWO: BIOLOGICAL CONSEQUENCES (LETHAL EFFECT),
S. APELGOT

1. INTRODUCTION 71

2. CHANGES IN DNA AFTER TRANSMUTATION OF INCORPORATED
   ^{32}P AND ^{33}P 73
   2.1. Experiments in vitro 73
   2.2. Experiments in vivo 77

3. TRANSMUTATION LETHAL EFFECT OF ^{32}P AND ^{33}P INCORPORATED
   IN BACTERIOPHAGES 79

4. TRANSMUTATION LETHAL EFFECT OF ^{32}P AND ^{33}P INCORPORATED
   IN BACTERIA 82

5. TRANSMUTATION LETHAL EFFECT OF ^{32}P INCORPORATED IN YEAST
   OR MAMMALIAN CELLS 85

6. DISCUSSION 89

7. CONCLUSIONS 93

ACKNOWLEDGEMENTS 94

REFERENCES 94

PART ONE: CHEMICAL EFFECTS, J.P. ADLOFF and S. APELGOT

1. Introduction

Phosphorus has two suitable radioactive tracers, ^{32}P and ^{33}P, which decay by β^--emission into stable sulphur nuclides with respective half-lives of 14.3 and 25.3 days.

The chemical effects of β decay have been observed since 1934, but they are still lacking a complete rationale [Nefedov et al., 1963; Wexler, 1965; IAEA, 1975]. The difficulties in assessing the atomic and molecular consequences of the transmutation are twofold. Firstly, the nuclear process itself is complex, since two particles (an electron and an antineutrino) are emitted simultaneously with a broad energy spectrum. Secondly, it is extremely difficult to measure the primary effects of the decay before their alteration, due to secondary interactions. Only with the particular working conditions of the mass spectrometer, i.e. in the gas phase at low pressure, has it been possible to measure the distribution of the charged species formed by β decay of noble gases, or by disintegration of bonded radionuclides in very simple molecules.

In more complex molecular structures and/or in the condensed phase, intra- and intermolecular interactions completely whip off the primary aftereffects of the decay. The current experimental makeshift, relevant only to radioactive daughter nuclides, is the radiochemical determination of the chemical state(i.e. valency, bondings, labelled molecules) of the atoms formed in the disintegration. It is thus restricted to the investigation of radioactive pairs which do not compulsorily represent the most valuable models. An interesting, but thus far undervalued variant, relies on molecules doubly labelled with similar or distinct radionuclides [e.g. Wolfgang et al., 1956; Manning and Monk, 1962; Skorobogatov and Nefedov, 1966; Jiang et al., 1975; Den Hollander et al., 1975; Van der Jagt et al., 1975]. The changes induced in the system after β decay of one of the tracers are revealed in the spectrum of the chemical forms of the second radionuclide. However, the labelling of a same molecule with two radioactive atoms is not as straightforward and requires unusually high specific activities. In a few instances, physical methods, and in particular Mössbauer emission spectroscopy, provide some information on the chemical configuration of nucleogenic atoms ten to one hundred ns after the decay process.

The conclusions drawn from these experiments, which invariably report on relatively simple molecules, must be used cautiously in interpreting the consequences of ^{32}P and ^{33}P decays in a structure as complicated as DNA. However, they are the only available bases, since ^{32}S and ^{33}S formed in the decays are stable nuclides produced in amounts which are presently undetectable (the complete decay of one mCi ^{32}P produces 3.5 ng ^{32}S).

Nonetheless, the resources of radiation chemistry must not be ignored. The effects of ionizing radiations on biomolecules, and especially on DNA, have been widely investigated [Scholes, 1963; Latarjet, 1974]. Here again, the data must be used with some caution. In fact, it is well proved that the biological effects of ^{32}P and ^{33}P disintegrations result from the transmutation itself and not from the cell's self-radiolysis due to the emitted β rays. On the other hand, the primary excitations and ionizations produced by an internal or external radiation source are distributed statistically along the radiation paths over the entire irradiated volume, whereas the primary events resulting from the transmutation are initially located entirely on the nucleogenic atom.

2. Primary effects of ^{32}P and ^{33}P disintegration

Despite the above mentioned caution, a comparative study of the decay effects of the two isotopes ^{32}P and ^{33}P is of real interest. The characteristics of both nuclides are given in table 1. The decay schemes are quite similar, and differ only in the energy of the β rays. All decays lead to the ground state of the daughter nuclei, thus avoiding any complications which might be induced by the emission of photons and by internal conversion processes. All primary effects which do not depend on the decay energy are the same for the two phosphorus isotopes *.

2.1. Transmutation

The very first consequence is obviously the transformation of a phosphorus atom belonging to the V A group of the periodic chart into a sulphur atom with chemical properties of a VI A group element.

2.2. Electronic excitation

The sudden change of the nuclear charge, which increases by one unit at the time of the transmutation, must be followed by a rearrangement of the atomic shell, i.e. by an excitation of the electron cloud. In the case of the phosphorus isotopes, the mean value of the excitation energy, usually termed "shake off", amounts to 60 eV. This energy is more than sufficient to excite

---

* The spin change in ^{32}P and ^{33}P decays is $\Delta I = 1$, without change of parity. The log ft values are respectively 7.9 and 5.0. It can be assumed that both transitions are equally allowed.

Table 1
Useful characteristics of ^{32}P and ^{33}P decays

Decays characteristics	$^{32}_{15}P \rightarrow ^{32}_{16}S$	$^{33}_{15}P \rightarrow ^{33}_{16}S$
Half-life, days	14.3	25.2
E_β (maximum), MeV	1.710	0.248
E_β (mean), MeV	0.695	0.0903
E_{ex} shake-off energy (mean), eV	≈60	≈60
E_R (maximum), eV	77.3	5.1
E_R (mean), eV	20	1.7

Data taken from Lederer et al. [1967]. The shake-off energy is calculated from $E_{ex} = 24.47\, Z^{1/3}$ [Serber and Snyder, 1952], the recoil energies from E_R (eV) = $536(E_\beta/M)(E_\beta + 1.02)$ with E_β in MeV and M in atomic mass units. The calculations are relevant to free atoms.

the electrons of the resulting sulphur atom into higher states or even into the continuum.

Several authors have calculated the probability of electronic excitation of the molecular ion formed in the β decay of a bonded radionuclide into a diatomic molecule such as ^{3}HH or ^{3}H$_2$. The computation technique always considers a sudden-perturbation in the Born-Oppenheimer approximation. Seemingly theoretical attempts have not been made concerning the decay of phosphorus isotopes. However it might be of interest to mention recent calculations of the effects of ^{14}C β decay (for which recoil effects are inoperative) in C$_1$ to C$_{10}$ hydrocarbons, including aromatics [Raadschelders-Buijze et al., 1976]. The electronic retention expressing the yield of molecular ions in the ground electronic state (and consequently the survival probability of these ions) is largely independent of the specific compound and averages 63%. Thus *the electronic transmutation effects are localized in the immediate surroundings of the decay site.* A similar shake-off behavior can be accepted, with some certainty, for the decay of radiophosphorus, and one can guess that excitation energy is taken initially by the sulphur atom. The rotational and vibrational excitation probabilities of the molecular ions resulting from shake-off are several orders of magnitude lower than those due to the recoil effect.

2.3. Recoil energy

The specific effects of ^{32}P and ^{33}P decays must be due to the amount of the recoil energies of the daughter nuclei. Calculations of the recoil energy spectra should take the β spectrum shape and the β-antineutrino angular cor-

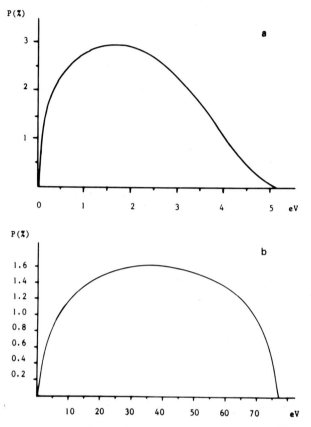

Fig. 1. Recoil spectrum of ^{32}S and ^{33}S formed by the β decay of ^{32}P and ^{33}P. The probability of the energy, P (%), is given in ordinates and the energy in eV in the abssicca. The cumulative probability of all events is 100%. Calculation for free atoms performed by P. Watson, in (a) case of ^{33}S, and (b) case of ^{32}S.

relation into account. The most recent calculations performed by P. Watson * give the recoil spectra shown on fig. 1. The ^{32}S spectrum is symmetrical at about the average energy of 38.5 eV. About 90% of the ^{32}S atoms acquire a recoil energy higher than 10 eV. The spectrum of ^{33}S recoils is less symmetrical and tails towards the higher energies; the average energy is 2.07 eV and about 90% of the recoil have an energy of less than 3.7 eV.

The maximum and mean values of the recoil energies calculated from well-

---

* We thank Dr. Watson from Carleton University in Ottawa (Canada) for providing the recoil spectra. We also thank Professor D.R. Wiles, from the same University, who made us aware of the calculations performed by Dr. Watson.

known simplified expressions are given in table 1. The recoil provided by the antineutrino is neglected.

All these calculations refer to free atoms. Actually, the parent atoms are bound in molecules, and one should estimate how and by what amount the recoil energy is transferred into the rotation and vibration levels of the molecular ion. This is hardly feasible for the complex structure of DNA. However, theoretical calculations, in accordance with the reported experimental data, indicate that the consequences of β decays *are confined to a small space*, so that the available energy will mainly be shared among the bonds of the newly formed sulphur atom. Quite reasonably, breakages must be much more frequent in ^{32}P decays.

2.4 Autoionization

In β^- decays, the charge of the daughter atom invariably increases by one unit as against the charge of the parent. However, in a few instances, secondary processes pursue further ionization. As mentioned before, the shake-off energy may be high enough to eject one or more electrons. Or, the β particle emitted by the nucleus may suffer a direct ionizing collision with the electronic cloud. Theoretical estimates of the probability of these autoionization processes vary considerably among different authors, but must not exceed 10^{-4}. In particular for ^{32}P, the shake-off ionization probability is 0.30% and the collision ionization probability is 0.10%. The experimental values lie between 0.47 and 0.99% [Weiner, 1966; Carlson et al., 1968]. Similar values should fit for ^{33}P.

Despite their low probabilities, the secondary ionization processes may well have a significant role in the consequence of β^- decays. They mainly affect internal electronic shells, K or L. As the fluorescence yield of the sulphur K shell is 10%, the vacancy triggers an Auger charging cascade which ends in a highly charged sulphur atom. In that case (i.e. about 0.1% of the events) the chemical effects of the β^- decay are analogous to those following an electron capture decay as for instance ^{125}I.

3. Possible transmutation effects of ^{32}P and ^{33}P in DNA

In DNA, phosphorus is found as phosphate in the pentavalent state (fig. 2a). All the primary events following ^{32}P and ^{33}P disintegrations are achieved in 10^{-13} s or less. In the vast majority of the decays, the phosphorus atom of the phosphodiester bond is replaced by a sulphur atom with a positive charge, excited to few tens of eV. Various rearrangements must be considered: replacement of the phosphorus by hexavalent sulphur atom, reorganization of

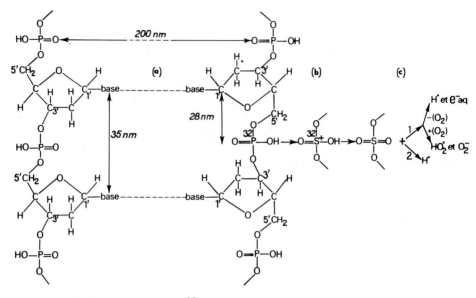

Fig. 2. Transmutation of ^{32}P bound in DNA. Pathway 2 is less probable.

the initial phosphate into sulphate (fig. 3), neutralization of the sulphur charge, elimination of the excedent H atom (fig. 2b), dissipation of the excitation and recoil energies. Complex processes must be operative, for which a few plausible mechanisms can be suggested.

(a) The sulphur atom in the sulphate, like the phosphorus atom in the phosphate, sits at the center of a tetrahedron. Interatomic distances in both configurations are not very different, so that the substitution of both structures should occur rather freely, inasmuch as the DNA molecule is not partic-

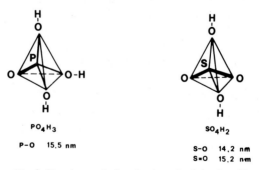

Fig. 3. Structures of phosphoric and sulphuric acid.

ularly rigid. However sulphur-diester bond ruptures cannot be formally excluded.

(b) The neutralization of the sulphur atom's positive charge requires an electron from a sugar molecule, or more likely, from a base, since the π bonds facilitate the electron transfer. This process is of course equivalent to a positive charge migration. According to Forster, a resonant charge transfer occurs over a distance of 600 nm. In the DNA molecule it may possibly reach the strand opposite that affected by the transmutation. A bond- or even a strand-rupture can result from the ultimate neutralization. Free bases and even free nucleosides have been observed in DNA irradiated with an external radiation source [Ullrich and Hagen, 1971; Teoule et al., 1974]. An alternative neutralization mechanism by a proton transfer from the sulphate to a base is less likely for kinetic reasons.

The change of phosphate (or of PO_4H_3) into sulphate (SO_4H_2) leaves an excedent hydrogen atom. When the sulphur atom is neutralized by an electron transfer (fig. 2c), this freed hydrogen atom acts as a free radical whose behavior in an aqueous medium is well known from radiation chemistry. In the neutral medium, characteristic of biological systems, the H atom coexists with the aqueous electron e_{aq}^- since the pK of the equilibrium:

$$H \rightleftharpoons H^+ + e_{aq}^-$$

is 9.7.

However, we are dealing essentially with aerated systems, so that both forms rapidly react with oxygen:

$$e_{aq}^- + O_2 \rightarrow O_2^- \qquad k = 2.9 \times 10^{10} \, M^{-1} s^{-1}$$
$$H + O_2 \rightarrow HO_2 \qquad k = 2 \times 10^{10} \, M^{-1} s^{-1},$$

where the HO_2 hydroperoxide radical is the acidic form of O_2^-. The predominate species at pH 7, i.e. e_{aq}^- and O_2^- are known for their high reactivity; the rate constant of e_{aq}^- with DNA is higher than $10^{12} \, M^{-1} s^{-1}$ [Hart and Anbar, 1970].

(c) The shake-off energy excites rotational and vibrational levels of the sulphur bonds. From the geometry of the sulphate configuration, it is obvious that the internal energy will be equally distributed among the four S-O bonds, probably without ruptures.

However the recoil energy of ^{32}S may be expected to break chemical bonds at positions 3' and 5' of the deoxyribose, the residual energy still remaining available for excitation. On the other hand, the much weaker recoil energy of ^{33}S is entirely dissipated as internal energy, with a very low probability of bond breakage.

(d) When the sulphur atom is multionized, the loss of several valence elec-

Table 2
Synopsis of after-effects of ^{32}P and ^{33}P decays in DNA

Primary events	Consequences
Substitution $^{32-33}PO_4^{\equiv} \rightarrow {}^{32-33}SO_4^{=}$	Possible breakage of sulphate-deoxyribose bond
Shake-off excitation	Excitation of rotation-vibration in S-O bonds
Kinetic recoil energy of ^{32}S and ^{33}S	Breakage of sulphate deoxyribose 3' or 5' bonds: high probability for ^{32}S low probability for ^{33}S
Positive charge of $^{32-33}S$ after β^- emission	Neutralization by electron transfer (or proton transfer) from base (or sugar) with succeeding bond or strand ruptures *and* radiolytic processes from H, e^-_{aq}, HO_2, O_2^- species with possible bond and strand ruptures
Multi-onization of $^{32-33}S$ after shake-off or direct interaction	Complete rupture of sulphate (probability $\approx 10^{-3}$)

Items in brackets refer to less likely processes.

trons, due to the Auger effect, sharply reduces the survival probability of the sulphate. The complete collapse of a molecule, when one of its constituent atom suffers a deep-shell ionization, has been observed in many cases and accounted for by the coulombic explosion model [Wexler, 1965].

A synopsis of the various possible events following the decay of radiophosphorus in DNA is represented in table 2.

4. Conclusion

The previous general discussion points out that the DNA molecule will experience some localized changes after the decay of incorporated ^{32}P and ^{33}P. Specific effects for each radionuclide, as well as for different structures, are expected. Detailed analysis of the experimental data on the biological consequences of the decays (Part II) will assess the role of the various physico-chemical after-effects of the transmutation. It is hoped that the observed biological effects, in turn, will support some of the proposed mechanisms.

PART TWO: BIOLOGICAL CONSEQUENCES (LETHAL EFFECT), S. APELGOT

1. Introduction

The lethal effect which occurs after ^{32}P transmutation was first observed in 1951 [Hershey et al.] in bacteriophage T4. This effect was also seen in different bacteriophages [Stent and Fuerst, 1955; Tessman, 1959], bacteria [Fuerst and Stent, 1956; Rörsch and Van der Kamp, 1961; Apelgot and Latarjet, 1962], yeasts [Moustacchi, 1959, 1968; Drobnick, 1964] and mammalian cells [Ragni and Szybalski, 1962]. In all these experiments it was noted that:

(1) The lethal effect depends on the transmutation itself, and not on the irradiation by the emitted β particle;

(2) The transmutation efficiency depends on the experimental conditions, and, in particular, on storage temperature ($-196°$ to $4°C$), the biological material [Stent and Fuerst, 1955; Latarjet et al., 1961], the presence or absence of oxygen [Van Dyke, 1960], and the radical scavengers such as 2-aminoethylisothiouronium bromide (AET) [Matheson and Thomas, 1960].

Since this efficiency depends on the experimental conditions, it is impossible to suppose that the lethal effect is a direct effect of the transmutation itself, as in the case of the lethal effect of ^{125}I incorporated in micro-organisms or cells. For this radionuclide, in all cases studied, the lethal effect is independent of storage temperature ($-196°$ to $37°C$) of micro-organisms [Schmidt and Hotz, 1973] or cells [Burki et al., 1973; Feindendegen et al., 1977], as well as of the presence or absence of radical scavengers such as cysteamine [Schmidt and Hotz, 1973]. In addition, the inactivation probability α per transmutation for ^{125}I, is clearly greater than that for ^{32}P. If, for each of these radioelements, α diminishes when biological complexity increases, the efficiency of ^{125}I decreases less than that of ^{32}P (table 3).

In order to better define the role of recoil energy in the transmutation of ^{32}P atoms, experiments were performed with ^{33}P. The disintegrations of these two isotopes differ only in the energy of the emitted β particle, therefore in the associated recoil energies (table 1 of Part I). However, these experiments did not permit a clear conclusion, since the killing efficiency of these two isotopes is the same for bacteria [Apelgot and Latarjet, 1966; Krisch, 1970], while in the case of bacteriophages [Steinberg and von Borstel, 1966; Krisch, 1970], that of ^{32}P is greater at $-196°C$, but similar to that of ^{33}P at $4°C$ (table 3).

Finally, in experiments with bacteria and yeasts, it has been found that certain effects of ^{32}P transmutation parallel those of ultraviolet radiation, e.g.

Table 3
α values

	^{32}P −196° (A)	+4°	^{33}P −196° (B)	+4°	^{125}I −196° (C)	+4°	B/A	C/A
Phages T1	1/17	1/7	1/30	1/8	1/2	1/2	1/8	8.5 (4)
Bacteria B_{s-1}	1/20		1/20		1/2		1	10 (5)
Bacteria B/r	1/50		1/50		1/5		1	10 (5)
Mammalian cells Human D98/AG	1/4000 1/8000 (*)							
Hamster V79					1/39 (*)	1/39 (*)		200 (100)

α = inactivation probability per transmutation. The α values were determined by Krisch [1970] for ^{32}P or ^{33}P labelled T1 and bacteria; Krisch [1972] for the same cells labelled with ^{125}I; Ragni and Szybalski [1962] for human cells, and Burki et al. [1973] for hamster cells.

Hamster cells have a different chromosome number (48) from that the human cells (42), but the difference between these numbers themselves is too small to allow an explanation of the difference between α values of ^{32}P and ^{125}I.

We know that cellular radiosensitivity is, generally, increased by a factor of about 2, when its DNA contains halogenated analogues, as is the case in experiments with ^{125}I. The ratio C/A therefore compares cells which have a higher radiosensitivity due to this labelling with ^{125}I. The values of this ratio, in brackets, indicate the values expected if this labelling had not modified the radiosensitivity.

The α values marked with (*) correspond to experiments where the radioactive atoms were exclusively incorporated in one of the two DNA strands.

the sensitivity of strains to these agents (^{32}P and UV) [Hill, 1960; Hatzfeld and Moustacchi, 1969], and the influence of caffeine [Apelgot, 1972].

These considerations taken together led us to re-examine this lethal effect in order to better understand the way in which ^{32}P transmutation might produce the observed effects.

The phenomena which accompany the transmutation of radioactive phosphorus isotopes are discussed in the first part of this paper.

2. Changes in DNA after transmutation of incorporated ^{32}P and ^{33}P

2.1. Experiments in vitro

When these transmutations occur in pneumococcus transforming DNA, Fox has shown that they bring about a decrease in its transforming activity [Fox, 1963]. In 1970, he demonstrated breaks in DNA, either of one strand (single-strand break) or of the two strands (double-strand breaks). When transforming DNA is stored at $-30°C$, the transmutation efficiency is 1 for single-strand breaks, and 1/20 for double-strand breaks [Rosenthal and Fox, 1970a, b]. It can thus be seen that if each transmutation brings about a break of one DNA strand, only one in 20 brings about a break in both strands. These double-strand breaks were again observed for DNA stored at different temperatures, having first been extracted from various ^{32}P labelled micro-organisms such as T2 [Davison et al., 1961], λ [Tomizawa and Ogawa, 1967], and T4 [Ley and Krisch, 1974]. It has been shown that each double-strand break is brought about by a single event, that is, the transmutation of only one ^{32}P atom [Miller, 1970]. Such a double-strand break, due to a single event, can be induced only when the sulphur recoil energy is sufficient to induce a break in the sulfate-deoxyribose bond, with a simultaneous break in the opposite phospho-diester strand. Such a break can be produced by two different events

Table 4
ϵ'_{db} values

DNA of	M.W. of DNA (daltons)	P	ϵ'_{db} 4–0°C −AET (A)	+AET (a)	A/a	ϵ'_{db} −20 to −196°C −AET (B)	A/B
Transforming principle	$4.8 \cdot 10^6$	32				1/20	
Phage λ	$3.3 \cdot 10^7$	32	1/18	1/29	1.6	1/23	1.3
Phage T4	$9 \cdot 10^7$	32	1/15			1/20	1.3
		33	1/40 (2.7)			1/77 (3.8)	1.9

ϵ'_{db} = efficiency of ^{32}P or ^{33}P transmutations for double-strand breaks in DNA stored after extraction, therefore in vitro. These values and the DNA molecular weight (M.W.) were established by Rosenthal and Fox (1970a,b) for the transforming principle; Tomizawa and Ogawa [1967] for phage λ, and by Ley and Krisch [1974] for T4. The figures in brackets represent the $^{32}P/^{33}P$ efficiency ratios. A/B represents the variation in temperature of this efficiency ϵ'_{db}; it is the same for the two phages studied, and greater for ^{33}P than for ^{32}P. The results show that, as compared with ^{32}P, ^{33}P efficiency is 2.7 times lower at 4°C, but 3.8 times lower at −196°C.

(see Part I): either as a consequence of S^+ neutralization by the charge transfer to a base or elimination of a proton, or by an attack by one of the radicals formed by the elimination of the excedent hydrogen atom; the breaks brought about in the strand by this procedure have every chance of being analogous to those produced during irradiation by ionizing radiations. Since, a priori, these mechanisms can depend on the nature and the properties of the medium, it must be expected that the transmutation's efficiency, ϵ'_{db}, to break the double-strand depends on the experimental conditions. In addition, the discussions in Part I indicate that this efficiency, ϵ'_{db}, must be lower in the case of ^{33}P. These two hypotheses proved to be true in the experiments, which indeed show that, for the DNA studied, the efficiency, ϵ'_{db} (table 4):

(1) depends neither on the origin of the DNA, nor its molecular weight; the *phenomena are thus localized in the immediate environment of the disintegration site*, as has frequently been observed in the chemistry of hot atoms produced by β disintegration;

(2) diminishes in the presence of AET. This compound seems to intercept the radicals capable of inducing the break in the opposite strand. In the presence of the scavenger, the observed double-strand breaks would occur essentially by transfer of an electron or a proton (see Part I).

(3) is lower in the case of ^{33}P; the maximum recoil energy of this isotope (lower than 5.1 eV, table 1 in Part I) is too low to induce the breakage of a sulphate-deoxyribose bond. Since these double-strand breaks do occur even with this isotope, it must be admitted that it is the reorganization of the sulphate structure which induces the break of a sulphate-deoxyribose bond, and that the latter is associated with a phosphate-deoxyribose break in the opposite strand, with a probability of 1/40 at 4°C (liquid state) and of 1/77 at −196°C (solid state).

(4) diminishes by about 30% for ^{32}P when the storage temperature changes from 4° (liquid state) to −196°C (solid state); this temperature effect is the same for the two bacteriophages studied — λ and T4; it is weaker than the action of AET, and weaker also than the action of temperature observed in the case of ^{33}P. Concerning ^{32}P, there is a break in the sulphate-deoxyribose bond by reorganization of the sulphur molecule, as in the case of ^{33}P, but also by use of the recoil energy; with respect to ^{33}P, the probability of double-strand breakage is multiplied by about 2.7 at 4°, and by 3.8 at −196°C. For ^{32}P, the recoil energy is, therefore, as expected (see Part I), more efficient in creating these breaks in the sulphate-deoxyribose bonds.

Hypotheses on the mechanism of breakage

Beginning with the biological results just recalled, and mindful of the data from hot atom chemistry (discussed in Part I), we propose to postulate some hypotheses on the procedures which might combine to induce double-strand breaks. These events require the simultaneity of breaks in a sulphate-deoxyribose bond and a phosphate-deoxyribose bond, in the opposite strand. The latter break appears as a result of different events — either the action of one of the radicals formed during the elimination of the excedent hydrogen atom, or S^+ neutralization when this process is brought about by the transfer of either a proton or an electron from a neighbouring base (table 2, Part I). *The break in this opposite strand therefore has a probability which is not 1.* In effect, the radicals can participate in other reactions, and the S neutralization can occur without a break in the opposite strand. For ^{32}P, we have observed that, at $-30°C$, the DNA single-strand breaks occur with a probability of 1 [Rosenthal and Fox, 1970a, b]. Such a probability can correspond only to the brakage of the sulphate-deoxyribose bond. It is thus noted that, *for ^{32}P, the recoil energy allows the breakage of sulphate-deoxyribose bonds with a probability of 1 at $-30°C$.* This result agrees with Watson's calculation (see Part I), which shows that at least 90% of the ^{32}S atoms acquire a recoil energy which is higher than, or equal to, 10 eV. At 4°C, the breakage probability must not be very different. Since, at $-30°C$, the DNA double-strand breaks occur with a probability of 1/20 [Rosenthal and Fox, 1970a, b] and of 1/15 at 4°C (T4 bacteriophage [Ley and Krisch, 1974], it can be seen that, in fact, these probabilities correspond to those of the phosphate-deoxyribose breaks in the opposite strand — the probability of the double-strand breaks being the product of the probability of the breaks of each of the two strands.

For ^{33}P, the breaks of the phosphate-deoxyribose bonds are brought about by the same mechanisms as for ^{32}P, and therefore their probabilities are the same. Knowing the probabilities of DNA double-strand breaks for T4 bacteriophages (table 4), it is possible to calculate the probabilities of breaks in the sulphate-deoxyribose bonds for this isotope. This computation gives a probability of 1/2.7 at 4° and 1/3.8 at $-196°C$, which corresponds to the breaks resulting from a reorganization of the sulphate structure when recoil energy plays a negligible role.

Role of temperature

It is noted that this role is similar for single-strand breaks in sulphate- or in phosphate-deoxyribose bonds (table 5). For double-strand breaks, its role is more important, since it introduces phenomena which apply to both strands (multiplication of the probabilities).

Role of AET

This compound seems to act on the radicals which, during irradiation with ionizing radiations, for example, bring about single-strand breaks; it thus acts on the number of breaks in phosphate-deoxyribose bonds.

Table 5
Probability of single- and double-strand breaks in DNA (in vitro experiments)

Phages	Breaks	4–0°C			−20 to −196°C (−AET)		Role of temperature	
		^{32}P −AET (A)	^{33}P +AET (a)	^{33}P −AET (A')	^{32}P (B)	^{33}P (B')	A/B	A'/B'
T4	ϵ'_{sb}	1/1		1/2.7	1/1	1/3.8	1	1.4
	ϵ'_{pb}	1/15		1/15	1/20	1/20	1.3	1.3
	$\epsilon'_{db} = \epsilon'_{sb} \cdot \epsilon'_{pb}$	1/15		1/40	1/20	1/77	1.3	1.8
λ	ϵ'_{sb}	1/1	1/1	1	1/1		1	
	ϵ'_{pb}	1/18	1/29	1.6	1/23		1.3	
	$\epsilon'_{db} = \epsilon'_{sb} \cdot \epsilon'_{pb}$	1/18	1/29	1.6	1/23		1.3	

ϵ'_{sb} and ϵ'_{pb} represent respectively the probabilities of single-strand breaks of the sulphate-deoxyribose bond, and the opposite phosphate-deoxyribose bond; ϵ'_{db} represents the probability of the double-strand break following both previous breaks. Except where AET is concerned, these values have to be of the same order for DNA in vivo.

2.2. Experiments in vivo

In some cases, various authors have determined, during the course of the same experiment, bacteriophage survival and the number of the double-strand breaks in extracted DNA. They have thus been able to calculate the inactivation probability, α, of the bacteriophage and the efficiency, ϵ_{db}, of the transmutations, to produce such breaks in vivo. For T4 bacteriophage, Ley and Krisch [1974] simultaneously determined the values of α, ϵ_{db}, and ϵ'_{db} (tables 4 and 6). A comparison of ϵ'_{db} and ϵ_{db} shows that when these efficiencies are

Table 6
Phages: ϵ_{db} and α values

Phage	P	Efficiency	4–0°C −AET (A)	+AET (a)	A/a	−20 to −196°C −AET (B)	+AET (b)	B/b	Role of temperature A/B	a/b
φX174 S13	32	α	1/1 (M_{sb})			1/1 (M_{sb})			1	
λ	32	α	1/7 (M_{r1})	1/25 (M_2)	3.6	1/19 (M_{db})			2.7	
T1	32	α	1/7 (M_{r1})			1/17 (M_{db})			2.4	
	33		1/8 (M_{r1})			1/30 (M_1)			3.7	
T7	32	α	1/7 (M_{r1})			1/15 (M_{db})			2.1	
	33		1/10 (M_{r1})			1/30 (M_1)			3	
SP82G	32	α	1/13 (M_{db})			1/19 (M_{db})			1.5	
		ϵ_{db}	1/12			1/18			1.5	
	33	α	1/18 (M_{r2})			1/37 (M_2)			2.1	
T4	32	α	1/11 (M_{db})	1/17 (M_{db})	1.5	1/17 (M_{db})	1/19 (M_{db})	1.1	1.5	1.1
		ϵ_{db}	1/14	1/15	1.1	1/18	1/18	1	1.3	1.2
	33	α	1/17 (M_{r2})	1/26 (M_2)	1.5	1/45 (M_2)	1/50 (M_2)	1.1	2.6	1.9
		ϵ_{db}	1/37			1/71			1.9	

ϵ_{db} = efficiency of ^{32}P or ^{33}P transmutations for double-strand breaks in DNA stored in vivo as phage; α = inactivation probability per transmutation. The letters in brackets represent the symbols of the likely modifications as explained in the text. ϵ_{db} and α values were established by Tessman [1959] for φX174 and S13, by Tomizawa and Ogawa [1967] for λ, by Krisch [1970] for T1, Krisch [1974] for T7 and SP82G, and Ley and Krisch [1974] for T4.

of the same order, ϵ_{db} is always greater by about 7–10% (table 7). In situ, the DNA bacteriophage (in vivo) produces double-strand breaks more easily than in vitro. This sensitivity could be related to the particular structure of DNA inside the bacteriophage [Tikchonenko, 1969; Gabrilovich and Romanovkaja, 1970]. This structure depends on its in vivo or in vitro state; it is independent of temperature, and must play the same role in either ^{32}P- or ^{33}P-induced breaks. It should be noted that $\epsilon_{db}/\epsilon'_{db}$ depends neither on the temperature nor on the phosphorus isotope employed (table 7).

Role of AET

The experiments show that, for T4 phage, the ϵ_{db} efficiencies are the same in the presence or the absence of this compound (relationship A/a relative to ϵ_{db}, table 6). *Thus, AET does not act, in vivo, in the double-strand breaks, while it does act in DNA double-strand breaks in vitro* (tables 4 and 5). Since these breaks are the results of indirect effects of radicals, the DNA environment must play an important role. The DNA environment is different in vivo or in vitro. This might be the reason for the different roles of AET under these two experimental conditions.

However, except where the role of AET is concerned, the great similarity between the results obtained with DNA in vivo and in vitro leads us to believe that the results computed from the data shown by DNA in vitro, and presented in table 5, remain valid for DNA in vivo. However, in this latter case, AET has no effect.

Table 7
Comparison of ϵ'_{db}, ϵ_{db} and α

| Phages | P | $\epsilon_{db}/\epsilon'_{db}$ | | ϵ_{db}/α | | | |
| | | 4° | –196°C | 4° | | –196°C | |
				–AET	+AET	–AET	+AET
SP82G	32			1.08		1.05	
T4	32	1.07	1.1	0.79	1.01	0.94	1.06
	33	1.08	1.08	0.46		0.63	
λ	32			0.39	1.39	0.82	

ϵ'_{db}, ϵ_{db} and α values are those of the preceding tables 4 and 6. For the λ phage, the values of ϵ_{db} were not determined. The experiments performed with the T4 phage having shown that, in the absence of AET, the ϵ_{db} and ϵ'_{db} values are of the same order, we took $\epsilon_{db} = \epsilon'_{db} = 1/18$ (table 4). In the presence of AET, we saw that the ϵ_{db} value was higher than the ϵ'_{db} value, but of the same order as in the absence of this compound. Therefore, in the presence of AET, we again used $\epsilon_{db} = 1/18$ for λ at 4°C.

3. Transmutation lethal effect of ^{32}P and ^{33}P incorporated in bacteriophages

When these transmutations occur in the DNA of a bacteriophage, experiments have shown that their infectious activity is destroyed with a probability, α, which depends on the bacteriophage and on the experimental conditions. Bacteriophages are classified in two groups, those for which α is equal to 1, and those for which α is less than 1.

$\alpha = 1$: This group comprises essentially the two bacteriophages ϕX174 and S13, whose DNA is single-stranded [Sinsheimer, 1959; Tessman, 1959]. The value of α is about 1 at 4° and at $-196°$C [Tessman, 1959]. We have previously seen that it is the break in the sulphate-deoxyribose bond which, at -30^3, takes place for ^{32}P, with a probability of 1. Our assumption that this probability would be of the same order at 4° was confirmed by the experiments with these two bacteriophages, since the value of α is independent of temperature between 4° and $-196°$C. In addition, experiments show that for single-stranded DNA, the integrity of the strand is necessary for maintaining its infectious activity; this infectivity disappears when the first break occurs.

$\alpha < 1$: The DNA of this group's bacteriophages is double-stranded. In order to understand the mechanisms of the lethal effect, we have selected only those experiments in which a same bacteriophage was studied, under various experimental conditions. This is the case for bacteriophages λ, T4, T7 and SP82G.

We noted that, for phages whose DNA is single-stranded, the infectious activity disappears with the first strand-break. For the phages whose DNA is double-stranded, this infectious activity might have disappeared with the first double-strand break. In order for this hypothesis to be correct, the efficiency, ϵ_{db}, needed to form this double-strand break must be of the same order as the inactivation probability, α. Thus, we should have $\epsilon_{db}/\alpha = 1$. The experimental results demonstrate that this is true, with ^{32}P, for certain bacteriophages and under certain experimental conditions, but false with ^{33}P, in all cases studied (table 7).

(a) $\epsilon_{db}/\alpha \simeq 1$

The lethal modification is the double-strand break in DNA; in this case, the role of AET and of temperature must be weak (values of ϵ_{db}, table 6).

(b) $\epsilon_{db}/\alpha < 1$

This indicates that fewer disintegrations are required to inactivate a bacteriophage than to induce a double-strand break. Here, the lethal modification cannot be such a break. *It is thus seen that the transmutations of ^{32}P and ^{33}P can lead to different localized modifications, one being double-strand* breakage in DNA; each of these modifications is characterized by a lethal probability. Analysis of the consequences of ^{32}P transmutation in DNA (see Part I) permits a forecast on a certain number of possible local modifications. By M_i, we shall designate the possible modifications, and by ϵ_i, their probability of being lethal; that is, M_{db} will represent the double-strand break, and ϵ'_{db} or ϵ_{db} its lethal efficiency for DNA stored respectively in vitro or in vivo.

(c) Hypotheses concerning lethal events
^{32}P *at* $-196°C$: The values of $\alpha(1/15$ to $1/19)$ and of ϵ_{db} $(1/18)$ are of the same order, and it can reasonably be admitted that, under these conditions, the lethal modification is the double-strand break, M_{db} (table 6).

^{32}P *at* $4°C$: For SP 82G and T4 phages, the values of $\alpha(1/12)$ and of ϵ_{db} $(1/13)$ are of the same order (table 6). The lethal effect can be attributed to the M_{db} modification, that is, the double-strand break.

For the other bacteriophages studied (λ, T1, and T7), the value of α is too great $(1/7)$ to attribute its lethal effect to the M_{db} modification ($\epsilon_{db} = 1/13$, table 6). M_{r1} will be the corresponding modification, and $\epsilon_{r1} = 1/7$ its average lethal efficiency.

For λ phage, the experiments show that $\alpha = 1/25$ in the presence of AET; this value of α is lower than those for ϵ_{r1} $(1/7)$ and ϵ'_{db} $(1/18)$ (table 7). In the presence of AET, the lethal modification seems to be neither the M_{r1} modification nor the double-strand break, M_{db}. Thus, M_2 will be the modification which, in the presence of AET, has the lethal probability of $\epsilon_2 = 1/25$.

The M_{r1} modification, whose effect diminishes considerably in the presence of AET, seems to result from the attack of a radical sensitive, in vivo, to AET. This radical must be one of those formed around the sulfate, when the excedent hydrogen atom is eliminated (see Part I). At $-196°C$, this M_{r1} modification is of little consequence, since the lethal modification is M_{db} (efficiency = $1/19$). Thus, at $-196°C$, it should be that $\epsilon_{r1} < 1/19$.

^{33}P *at* $4°C$: Two values for α are again observed, one corresponding to SP82G and T4 phages, and the other corresponding to T1 and T7 phages. If, for these latter phages, the M_{r1} modification introduced at $4°C$ with ^{32}P is due to a radical which appears when the excedent hydrogen atom disappears, it should again be observed in the case of ^{33}P. And in fact this is demonstrated by the experiments, since for these last two phages, the values of α are the same for both of these isotopes (table 6).

For SP82G and T4 phages, the lethal modification is, with ^{32}P, the M_{db} modification whose efficiencies ϵ_{db} and ϵ'_{db} are very different for each phosphorus isotope. However, for ^{32}P, the M_{db} modification, whose efficiency ϵ_{db} is $1/37$, cannot account for the lethality, whose probability α is about $1/8$. These two phages do not appear to be sensitive to the radical which gives the M_{r1} modification (efficiency $\epsilon_{r1} = 1/7$), and by M_{r2} we will designate this

Table 8
Likely modifications and their lethal efficiencies

Modifications		Efficiency (±15%)				Cells	Remarks
Symbols	nature of agent	Symbols	0–4° (A)	–20 to –196°C (B)	A/B		
M_{sb}	single-strand break	ε_{sb} ^{32}P	1/1	1/1	1	φ_s	Calculated efficiency
		ε_{sb} ^{33}P	1/3	1/4	1/4		does not exist with N_2 or AET
M_{r1}	radical	ε_{r1}	1/7			$\varphi_d + B$	low (10%) sensitivity to AET
M_{db}	double-strand break	ε_{db} ^{32}P	1/13	1/19	1.3	φ_d	does not exist with AET
		ε_{db} ^{33}P	1/37	1/70	1.8		
M_{r2}	radical	ε_{d2}	1/17			$\varphi_d + B$	
M_1		ε_1		1/25			exists with AET
M_2		ε_2	1/25	1/45	1.8		
M_3		ε_3		1/100		$B + y$	exist with caffeine
M_4		ε_4	1/40				does not exist with caffeine
M_{b1}	base modified	ε_{b1}	1/48				exists with and without caffeine
M_5		ε_5		1/270			exist with caffeine
M_6		ε_6	1/80	1/380	4.8		does not exist with caffeine
M_{b2}	base modified	ε_{b2}	1/100	1/560	5.6		
M_7		ε_7		1/1500		C_M	with halogenated analogues in DNA
M_n		ε_n		1/4000			non-reparable

φ_s represents the phages with single-stranded DNA and φ_d those with double-stranded DNA; B represents bacteria, y yeasts, and C_M mammalian cells.

new modification whose lethal efficiency ϵ_{r2} is about 1/18 (table 6). In the presence of AET, α is equal to 1/26 for phage T4. This probability, which is greater than that of M_{db} (ϵ_{db} = 1/37), cannot be related to a double-strand break; it is of the same order as that observed for λ phage, under identical experimental conditions, but with ^{33}P. We attributed such a probability to the M_2 modification. This modification is not connected to a strand-break and thus is found with the same probability in both of the phosphorus isotopes — as is also the case in the M_{r1} event. Since AET modifies the efficiency ϵ_{r2}, the M_{r2} modification must be caused by radicals. With ^{32}P, this modification gives no important effect, since it is less probable than the M_{db} modification, (ϵ_{r2} = 1/18, while ϵ_{db} = 1/13).

For ^{33}P at $-196°C$: Values of α = 1/30 for T1 and T7, 1/37 for SP82G and 1/45 for T4 were obtained. For T4, the role of AET is weak, and this leads us to assume that the lethal modification which corresponds to this probability is the M_2 modification observed at 4°C, in the presence of AET. At $-196°C$, the same modification seems to have the probability ϵ_2 = 1/45. For the other phages, the results are not numerous enough to allow a hypothesis concerning lethal modification.

Role of temperature: When, at 4°C, and $-196°C$, the lethal modification is M_{db}, its lethal probability is little modified (10–15%) by temperature (Values A/B and a/b, table 6). This result is analogous to that observed where breaks are produced in DNA (table 5). In short, our analysis shows that the lethal modification is not always the same at 4° or $-196°C$.

These findings on the possible lethal modifications are reported in table 8.

4. Transmutation lethal effect of ^{32}P and ^{33}P incorporated in bacteria

When these transmutations occur in bacterial DNA, the bacteria lose their ability to divide indefinitely (lethal effect). However, for the bacteriophages, if the probability varies from 1–1/50 (table 6), the probability varies from 1/6–1/560 for bacteria (table 9). As in the bacteriophage, this variation depends not only on the strain employed, but also on the experimental conditions. The amplitude of these variations is clearly greater for bacteria.

For no bacterial strain, no matter what the experimental conditions, is efficiency α equal to 1, as is the case for the bacteriophages whose DNA is single-stranded. For bacteria, the highest efficiency is 1/6 (table 9). This efficiency is of the same order as that found for various bacteriophages whose DNA is double-stranded (table 6). For bacteria, as for the bacteriophages with a double-stranded DNA, the sulfate-deoxyribose break produced after each disintegration of ^{32}P is not lethal. We now know that, for bacteria, repair phenomena exist which determine their radiosensitivity. This ability to repair

Table 9
Bacteria: α values

Strains	P	Growth conditions	4–0°C	−20 to −196°C	References
Pneumococcus	32			1/25 (M_1)	Fox [1963]
B_{s-1}	32	I^+		1/22 (M_1)	Krisch [1970]
	33			1/22 (M_1)	
15 THU (B/r type)	32	I^+		1/40 (M_2)	
	33	I^-		1/40 (M_2)	
W 3623 rec$^+$	32			1/100 (M_3)	Tomizawa et Ogawa [1968]
N 23–53 rec A41				1/20 (M_1)	
B/r	32	I^+	$+O_2 = 1/8$ (Mr_1)		Van Dyke [1960]
			$+N_2 = 1/14$		
B_3 thy$^-$ S/r (B/r type)	32	I^+C^-f	1/6 (Mr_1)	1/25 (M_1)	Apelgot [1972]
		I^+C^-h	1/16 (Mr_2)	1/48 (M_2)	
		I^-C^+f	1/40 (M_4)	1/270 (M_5)	
		I^-C^+h		1/270 (M_5)	
		I^-C^-f	1/48 (M_{b1})	1/380 (M_6)	
		I^-C^-h	1/80 (M_6)	1/560 (M_{b2})	
		I^+C^-m	1/100 (M_{b2})	1/34	Apelgot and Latarjet [1966]
	33	I^+C^-m		1/36	

In the Van Dyke experiments, the storage medium for bacteria was saturated either with oxygen ($+O_2$) or nitrogen ($+N_2$). The growth medium described by Stent and Fuerst [1955] contained either glycerol and sodium lactate (I^+) or glycerol alone (I^-) for carbon source; the plating medium was supplemented (C^+) or not (C^-) with caffeine, the radiophosphorus specific activity was either high (h) or low (f) or medium (m); in this last case, the α value was between the observed with high or low specific activities. The letters in brackets represent the likely modification symbols as defined in the text.

the lesions created by different types of radiations (UV and ionizing) depends on two parameters, one of which is genetic and the other physiological — therefore, on growth conditions [Town et al., 1974]. Experiments prove that the value of α varies by a factor of about 5 for genetic characteristics (1/20–1/100 [Tomizawa and Ogawa, 1968]), but by a factor of about 22 for growth conditions (1/25–1/560 [Apelgot, 1972]). This variation of α in genetic and physiological characteristics suggests that the repair phenomena observed in the case of irradiations also play an important role in the lethal effect of ^{32}P transmutations.

Microscopic examination of bacterial development, in a solid nutritive medium, and having accumulated ^{32}P transmutations, permitted us to demonstrate that this lethal effect is not an "all or nothing" effect. The lesions brought about by the transmutations leave the bacteria with their ability to synthesize and even to divide two or three times, before their sterilization becomes effective [Apelgot and Thiery, 1974].

In a comparison of the values of α found in bacteria and in the bacteriophages, it is seen that many of the values are the same (tables 6 and 9). Thus, these values are connected to definite modifications, which must be the same for certain types of cells. This assumption led us to re-introduce, in bacteria, the M_{r1}, M_{r2}, M_1, M_2 and M_{db} modifications defined in the preceding section on bacteriophages, and to add new ones (table 9). The latter are less probable than those evidenced in the bacteriophages, and can only be observed in those bacteria whose genetic equipment and growth conditions allow repair of the more probable modifications.

The M_{b1} and M_{b2} modifications are lethal when bacteria which have accumulated ^{32}P transmutations are allowed to grow in the absence, but not the presence, of caffeine (table 9). Caffeine plays multiple roles, the principal one being the excision inhibition of pyrimidine dimers induced by UV radiation [Witkin and Farquharson, 1969]. Experiments with ^3H labelled caffeine have shown that, in vitro, this compound is specifically fixed on those sections of DNA which were denatured by UV radiation [Domon et al., 1970]. It can also be supposed that fixation of caffeine on DNA sites modified by UV radiation induces an inhibition of certain repair procedures. Such a hypothesis was suggested by Grigg [1968] to explain the lethal effect of high concentrations of caffeine. It is also known that caffeine does not affect the survival of bacteria irradiated with ionizing radiations [Mouton and Treneau, 1969; Vizdalov et al., 1971; Apelgot, 1972]. The role of caffeine in the lethal effect of ^{32}P can be explained if Mb and Mc are *local modifications of DNA bases* located near the ^{32}P atom which disintegrates, caffeine seeming to be able to attach itself to the thus modified zones.

In their experiments, Tomizawa and Ogawa [1968] studied simultaneously the lethal effect and the modifications in bacterial DNA from W3623 and

N23-53 strains. They showed that, given their experimental conditions, the values of ϵ_{db} (efficiency in vivo, to bring about a double-strand break in DNA) are the same for these two cultures: $\epsilon_{db} = 1/20$, while the values of α are different (table 9). For the W3623 rec$^+$-strain, which can bring about repair of such breaks, the value of α is 5 times lower than that of ϵ_{db}, that is, $\alpha = 1/100$. For the N23-53 recA-strain which is incapable of such repairs, one sees that $\alpha = \epsilon_{db} = 1/20$, and the lethal modification could be the double-strand break M_{db}. Further, the experiments of Krisch [1970] with B_{s-1} bacteria, labelled with either one of the two phosphorus isotopes, show that the values of α are 1/20 and are independent of the isotope employed, as in the preceding case of recA-strain (table 9). This value $\alpha = 1/20$ is of the same order as that of ϵ_{db} for ^{32}P, but is greater than that of ^{33}P, which, at the same temperature, is 1/70 (table 6). These considerations lead us to postulate that *for bacteria, and contrary to the case for bacteriophages, the lethal modification is never a double-strand break in DNA, but a modification which is produced with the same probability for ^{32}P or ^{33}P.* Moreover, the probability of this modification is close to ϵ_{db} in the case of ^{32}P; M_1 will be this modification. At $-196°C$, its probability, ϵ_1, is 1/25. This probability is the same as that observed in the case of ^{32}P-labelled T2 and T7 bacteriophage stored at $-196°C$ (table 6). For B/r bacteria strain, Van Dyke [1960] demonstrated an oxygen effect, comparable to that which exists in irradiations by ionizing radiations. In the presence of oxygen, the value of α is 1/8 (table 9), and by M_{11} we have designated the modification which corresponds to this probability (table 8). In the presence of nitrogen, the value of α becomes 1/14 (table 9); the modification corresponding to this probability is difficult to characterize; it could be a double-strand break which brings about a particular termination. It has been seen that the M_{r1} modification is sensitive to AET, a radical scavenger; we can now state that it no longer exists in the presence of nitrogen.

When the specific activity of ^{32}P increases, the value of α decreases (table 9); the bacteria are thus more resistant, but this particular role of the specific activity may not yet have been interpreted [Apelgot, 1972].

5. Transmutation lethal effect of ^{32}P incorporated in yeasts or mammalian cells

In the case of yeasts, it is essentially the role of genetic factors which has been studied [Hatzfeld and Moustacchi, 1969]; ^{33}P was not employed. When, beginning with wild saccaromyces cervisiae strains, they isolated mutants showing modified sensitivities to ionizing or ultraviolet radiations, their sensitivity to ^{32}P was also modified (table 10). For yeasts, one observes α values

Table 10
Yeast radiosensitivity

Strains	Sensitivity			$\alpha(^{32}P)$ (0°C)
	X	UV	^{32}P	
WT (Wild)	1	1	1	1/62
UVS$_6$ (Rad. 6)	2	10	2	1/31
N123 (Wild)	0.75	0.65	0.5	1/123
UVS$_z$ (Rad. 1–3)	0.75	18.5	1.15	1/54

These results are those of Hatzfeld and Moustacchi [1969]. The values corresponding to the X, UV and ^{32}P sensitivity are given in arbitrary units.

which are lower than those observed in the bacteriophages, but of the same order as those found in bacteria, when the growth medium described by Stent and Fuerst [1955] contains only glycerol as carbon source (tables 9 and 10). It therefore seems that, for yeasts, as for bacteria, the double-strand break is not a lethal modification. In fact, it is known that these breaks are reparable in yeast [Ho, 1975]. Yeasts and bacteria are able to repair the most probable modifications for which α is greater than, or equal to, 1/17. The low probability modifications can also be repaired, but this ability depends on the strain. Thus, the wild WT strain, which is more sensitive to X and UV radiations, is also more sensitive to ^{32}P; $\alpha = 1/62$, while $\alpha = 1/123$ for the N 123 wild culture. The two mutants studied are deficient in at least one of their repair systems [Haynes, 1975; Waters and Moustacchi, 1974], and are also more sensitive to ^{32}P than the strains from which they are derived.

In 1962, Ragni and Szybalski studied the lethal effect of ^{32}P incorporated in human cells (strain D 98/AG), cells which are even more complex than yeasts. These cells showed high resistance, since, stored at $-70°C$, $\alpha = 1/4000$. The reasons for their resistance were not explained. One of their hypotheses might suggest that only a small portion of the cells' SNA was related to the growth and replication phenomena. This hypothesis is disproved by the facts one of which is related to the results obtained with ^{125}I. In hamster cells (whose complexity is of the same order as human cells), the lethal effect of ^{125}I is characterized by a high value of α: 1/39 [Burki et al., 1973 and table 3]. In DNA, the presence of an analogue such as iododeoxyuridine (which carries ^{125}I) in the place of thymidine molecules, increases the radiosensitivity of the cells by about only 2 [Burki et al., 1973]. The difference between the values α ^{125}I (1/39) and α ^{32}P (1/4000) can only be explained by a difference in the seriousness, for the cells, of the modifications which result from the transmutation of these two isotopes, whose disintegration schemes are very different (electronic capture for ^{125}I and β^- for ^{32}P). The high efficiency of ^{125}I demonstrates that cellular DNA always participates almost entirely in the lethal effect, in the generally accepted hypothesis according to which iododeoxyuridine can be substituted with equal probability for any thymidine molecule, whatever its location on the cells' chromosome may be. In addition, if only part of these cells' DNA participated in this effect, the α value could not be modified, as it is, by the different characteristics in the labelling of DNA (table 11). If, finally, we compare the macimum and minimum values of α obtained at low temperature (the only ones studied for human cells) for human cells and bacteria, we note that the human cells may be 160 times more resistant than bacteria ($\alpha = 1/4000$ as opposed to $\alpha = 1/25$), or only twice ($\alpha = 1/1210$, versus $\alpha = 1/560$) (tables 9 and 11). This comparison confirms that, *even in a cell as complex as a mammalian cell, almost all the DNA is involved in the lethal effect.* The low value of α corresponds to the fact that, in these cells, the lethal effect of ^{32}P corresponds to a rare modification, the more probable one having been repaired.

The experiments undertaken by Ragni and Szybalski [1962] also demonstrate that the greatest resistance in the human cells studied is observed when

Table 11
^{32}P lethal effect in human cells ($-196°C$)

no.	DNA	$\alpha(-70°)$
1	—*—*—*—*—	1/8000
2	—*—○—*—○—*—*	1/8000
3	—*—*— —*—*—	1/4000
4	—*—*—*—* —○—○—	1/4000
5	—*—○—*—○— —○—*—○—*—	1/1905
6	—▲—*—▲—*— ▲—*—▲—*—	1/1210

These results are those of Ragni and Szybalski [1962]. The ^{32}P atoms are symbolized by *, the bromodeoxyuridine or chlorodeoxyuridine molecules by ○ and those of iododeoxyuridine by ▲. The position of these symbols on the DNA strands are arbitrary and show only that the ^{32}P atoms or the analogous molecules are located on one or the other strand, according to a random distribution. The experiments with chloroanalogue were performed according to the No. 5 diagram only. In experiments No. 1 and 3, the ^{32}P specific activities used were of the same order, that is, the same total number of ^{32}P atoms was distributed on either one or both of the DNA strands. In these experiments, the thymidine molecules were replaced by the analogues in a proportion of about 60%.

^{32}P is exclusively incorporated in only one of the two DNA strands, for which case, $\alpha = 1/8000$ the greatest sensitivity is observed when ^{32}P and one halogenated analogue of thymidine are randomly incorporated in the two DNA strands, in which case, α varies from 1/1905 to 1/1210, depending on the halogen of the analogue. The iodized derivative provides the greatest sensitivity (table 11). It is known that the substitution of such analogues in DNA increases the cells' radiosensitivity [Szybalski, 1962]. With the ionizing radiations, the iodized derivative gives the highest sensitivity; those exhibited by bromide and chloride derivatives are of the same order. With ultraviolet radiation, the bromide derivative shows the highest sensitivity. With ^{32}P, the sensitivity observed in these two compounds resembles that observed in ionizing radiations (table 11).

The fact that the cells are most resistant when radioactive atoms are located on only one of the two DNA strands poses an interesting problem. In this case, $\alpha = 1/8000$ (table 11); the lethal modification which corresponds to this event is particularly rare. This cannot be a repair "error" in the most probable modifications, since such an error is not related to the location of ^{32}P on only one or on both of the DNA strands. It has been demonstrated that each transmutation brings about a great number of modifications located in the sugar molecules, and on the bases around the disintegrating atom, thereby causing modifications in the opposite strand (see Part I). A single event concerns exclusively the site where ^{32}P disintegrates — the presence of the new sulphur atom. We think that the *rare* lethal modification, which depends on ^{32}P location on a single or on the two DNA strands, is an "in situ" consequence of the sulphur atom itself, which, with a probability of 1/8000, appears in a particular form, incompatible with the structure of DNA. Mn will be this non-reparable modification; its probability of being lethal seems to be 1/8000. When the bromide analogue is located on the same strand as ^{32}P, this probability is unchanged (exp. 2, table 11); such an analogue cannot, in effect, modify a purely physical event.

When the ^{32}P atoms are equally divided between the two DNA strands, the value of α doubles: $\alpha = 1/4000$. This result does not depend on the specific activity of ^{32}P. For the specific activities studied (0.25 and 0.50 mCi/mg P), there is one ^{32}P atom for (2 or 4) $\times 10^7$ P atoms, and no interaction between the disintegrations of two ^{32}P atoms can take place: the nature of the Mn event cannot be modified.

In order to explain the doubling observed in the value of α, we suggest that the probability of the Mn event is not 1/8000, but $\epsilon_n = 1/4000$. In the precise case where ^{32}P exists on only one strand, the opposite strand contains no S atom, but simply reparable modifications; in this case, the strand seems capable of recombining, when the radioactive strand contains one, but not two, Mn events. When ^{32}P exists on each strand, the first Mn modification occurs with a probability of 1/4000; it is then lethal. In this case, the opposite strand contains concurrently both reparable modifications and S atoms, and the DNA strand containing only one Mn modification does not appear to be capable of recombination. The same is true when analogues are found in the strand opposite that containing the ^{32}P atoms (exp. 4, table 11). It therefore seems that, *with a low probability of 1/4000, a lethal Mn modification exists, which is non-reparable, and which results from a particular state of sulphur incompatible with the natural DNA structure;* when the opposite strand is "normal", without halogenated analogues and with all its phosphorus atoms, the recombination can lead to the replication of DNA, on condition that the ^{32}P labelled strand contains only one of these Mn modifications. When the halogenated analogues and the ^{32}P atoms are randomly distributed between the two DNA strands, one again observes the usual sensitizing which these analogues produce. By favoring charge transfers, these analogues determine specific lethal modification, whose probability is greater than that of the Mn modification.

6. Discussion

Each of the local modifications, M_i, defined in Part II, is the consequence of a secondary event which, in DNA, reveals the sudden substitution of a stable sulphur atom for a radioactive phosphorus atom. The primary events of this transmutation end, by means of several possible mechanisms, in various secondary events which are summarized in table 2 (Part I). An examination of this table shows that an identical local modification (for example, the single- or double-strand break in DNA) can be the result of various secondary events. There is thus no unequivocal relationship between a given localized modification and a precise secondary event, except for the localized modification, Mn; this one seems to be the result of the multiple ionization of sulphur (Auger effect).

At each disintegration of a ^{32}P atom, one sulphur atom appears. In the DNA molecule, it is quite probable that a sulphate structure replaces the initial phosphate structure (Part I, page 67). The biological consequences just analyzed indicate that the lethal efficiency of this transmutation is equal to 1 in the case of ϕX174 and S13 bacteriophages having single-stranded DNA (table 6). In addition, the lethal effect seems, in this case, to be a consequence of a single-strand break in such DNA (table 6) — a break which is essentially due to the dissipation of the high recoil energy of ^{32}S. When the recoil energy is low, as is true in the case of ^{33}P, such a break is produced with a lower probability, and as a result of the reorganization of the sulphate structure following the reorganization of the sulphur structure (see Part I). The break of a sulphatedeoxyribose bond is thus the only consequence of the substitution of a phosphate configuration. For cells with double-stranded DNA, the value of α is always less than 1. We have seen that the lethal modification in this case seems to be the result of charge transfers, of the dissipation of recoil and excitation energies, and of the radicals' activity, all of which phenomena correspond to the reorganization of the sulphur and the sulphate. *The substitution "per se" of sulphate for phosphate does not seem to cause serious damage.* It is probable that the dimensions and similar conformations of these two structures (fig. 3, Part I) make this a non-lethal substitution.

In the case of human cells, $\alpha = 1/4000$ (table 11). We assumed that the lethal modification, Mn, which is not easily repaired and which is characterized by this probability, was due to a particular sulphur state. It was seen that the sulphur in the transmutation $^{32}P \rightarrow {}^{32}S + \beta^-$ was ionized several times, with a probability of about 10^{-3}, as a result of secondary ionization phenomena (Part I, page 67). The inactivation probability $\alpha = 1/4000$ of mammalian cells might be related to some of the ^{32}P atoms, which were involved in this multiple ionization phenomenon (Auger effect). The biological consequence must then be analogous to those which take place upon disintegration by electronic capture of ^{125}I, whose inactivation probability α is about $1/40$ (table 3). If one considers only the Auger effect, the probabilities α_{125_I} and α_{32_P} become of the same order. Thus, it seems that, for mammalian cells, the lethal effect is the consequence of a non-easily repaired modification, following the multi-ionization of a newly formed atom. This atom, after reorganization and neutralization must, of necessity, have very different bonding from the initial sulphur atom. The DNA replication seems incompatible with the existence of only some of these atoms or molecules.

In all ^{32}P labelled DNA, the radioactive atoms were randomly distributed. It was seen that, for bacteriophages having single-stranded DNA, the lethal effect is characterized by a probability $\alpha = 1$. The disintegration of any of the ^{32}P atoms in DNA is thus a lethal event; *each of the ^{32}P atoms thus seems to have the same importance,* the integrity of the strand insuring the survival of the bacteriophages. For more complex cells, it seems that the same is true, since, for mammalian cells, the modification by the Auger effect of only some atoms, randomly localized in DNA, affects its replication. The necessity of DNA integrity for indefinite cellular division has already been stated regarding ^{32}P labelled bacteria [Cairns and Davern, 1966; Davern, 1968]. It would now appear that this is necessary for all cellular types.

The double-stranded DNA bacteriophages studied (λ, T1, T7, T4 and SP82G) have the same sensitivity of ^{32}P at $-196°C$ (average $\alpha = 1/9$), but are classed according to two groups at $4°C$. At this temperature, in fact, average $\alpha = 1/7$ for λ, T1 and T7, but average $\alpha = 1/13$ for T4 and SP82G (table 6).

At $-196°C$, the lethal modification is the double-strand break, M_{db}; at $4°C$, this break is lethal for only T4 and SP82G. For λ, T1 and T7, the lethal modification M_{r1} is more probable than the double-strand break (tables 6 and 8), and follows the attack of one radical. In order to explain why the action of this radical is not lethal for the two other bacteriophages T4 and SP83G, we envisage the possible influence of a different DNA environment. This different environment would favor the interaction of the radical with a component of a molecule other than DNA, but very near — a protein, for example. We still know very little about the "in situ" conformation of DNA in bacteriophages. Nonetheless, it seems that in viral particles, particular interactions do exist between DNA and proteins, interactions which depend on the type of bacteriophages studied [Tikchonenko, 1969]. It is also known that energy transfer between proteins and DNA are possible in bacteriophages [Lillicrap et al., 1974]. The radical which leads to the M_{r1} modification, lethal for λ, T1 and T7, must have a noticeably diminished reactivity at $-196°C$, and that is why, at that temperature, the lethal modification is the double-strand break, for all the bacteriophages examined.

All the results thus far discussed demonstrate that the lethal efficiency α depends on the complexity of the cells studied. We assumed that a particular modification would correspond to each value of α efficiency. We have seen (table 8) that the single-strand break, M_{sb}, is lethal only for single-stranded DNA bacteriophages, and the double-strand break, M_{db} for double-stranded DNA bacteriophages. Certain modifications are lethal for certain bacteriophages and bacteria, while others are lethal for bacteria and yeasts. Only the rare and non-easily repaired modifications, Mn, are lethal for mammalian cells alone. The lethal modifications in bacteria are the same as those in double-stranded DNA bacteriophages only when their growth medium contains two sources of carbon, glycerol and sodium lactate, or glycerol and glucose (table 9); it is probable that after growth in these media, the bacteria lose their ability to repair. This division of different cellular types into "groups" is concomitant with that already described by Kaplan and Moses [1964] and Terzi [1965]. If the first group is related to organisms whose DNA is single-stranded, the other groups correspond to those organisms whose DNA is double-stranded. These latter differentiate among themselves by their repair capacity.

In 1962, Riley and Pardee came to a conclusion about the complexity of the disintegration consequence of ^{32}P incorporated in bacteria. Harriman and Stent [1964] suggested the possibility of several types of lethal modifications. These authors supposed the existence in DNA of either double-strand breaks (a lesion they called: "long range hit") or of modifications of base or sugar molecules ("short range hit"). This classification was taken up by Krisch [1974]. Our study develops these notions, and shows that a dozen lethal modifications are probable, in relation to various and different means of sulphur and sulphate reorganization (table 2, Part I). The choice between these different possible means depends, in part, on the DNA environment itself, and in part, on chance, since the phenomena which accompany the transmutation of a radioactive atom do not take place in a privileged spacial direction. Given the position of phosphorus atoms in DNA, these phenomena are directed, randomly, toward the interior or the exterior of the double helix. The DNA environment itself depends on the cellular type studied, and on the storing conditions of cells, such as the presence of radical scavengers, of oxygen or of nitrogen, and the storage temperature of cells, which conditions the state (liquid or solid) of the medium.

One of the lethal modifications is the double-strand break in DNA. This break is the result of two distinct events. The first is the break in the sulphate-deoxyribose bond, provoked by recoil energy or the reorganization of the sulphate molecule. The other is the opposite strand-break in the phosphate-deoxyribose bond, provoked by the action of a radical or by a charge transfer. These two distinct events depend on the disintegration of a same radiophosphorus atom, and are separated by a very short time interval of about 10^{-13} s. This is why they seem to be brought on by a single event [Miller, 1970].

The reversible character of lethal modifications resulting from ^{32}P disintegration has been suggested by various authors [Hill, 1960; Riley and Pardee, 1962; Hartman and Kozinski, 1962; Tomizawa and Ogawa, 1967; Rosenthal and Fox, 1970a, b]. Presently, there are still only indirect proofs of possible reparations, such as the modifications of the lethal efficiency, α, with genetic factors [Tomizawa and Ogawa, 1968; Hatzfeld and Moustacchi, 1969], with growth conditions [Apelgot, 1972], or plating conditions [Hill, 1960]; in addition, microscopic examination of the development, in a solid nutritrive medium, of bacteria having accumulated ^{32}P transmutations was performed [Apelgot and Thiery, 1974].

Important analogies exist between the lethal effect produced by disintegrations of incorporated ^{32}P and by ionizing radiations. With these two agents, the activity of oxygen and of the radical scavengers is of the same order [Van Dyke, 1960; Matheson and Thomas, 1960], and the double-strand breaks in DNA cannot be the only lethal modification (table 8; [Stringini et al., 1963; Ginoza, 1967; Latarjet, 1973]). Cells which have been damaged by one or the other of these two agents (^{32}P or ionizing radiation) can effect mitoses before cellular division stops definitely; this is the phenomenon of "deferred death" [Holweck and Lacassagne, 1934; Apelgot and Thiery, 1974; Roots and Smith, 1975; Moustacchi, personal communication]. This phenomenon infers defective repair and leaves one to assume that the bacterial, yeast and mammalian cells are able to repair a great number of the lesions produced by ^{32}P disintegrations, as well as numerous lesions produced by ionizing radiations. This result is not surprising, because, on the one hand, with these two agents, the lethal modifications must be of the same type, since they result from phenomena which are also of the same type, e.g. charge transfer, dissipation of energy, radicals; on the other hand, the repair systems do not seem to be specific for a precise lesion, but rather for a group of similar lesions [Waldstein et al., 1971; Latarjet, 1974]. The only specific event in ^{32}P transmutation is the appearance of a sulphur atom, which transforms a phosphate molecule into sulphate. The results obtained seem to indicate that this substitution is non-lethal, except when the sulphur appears in a multi-ionized form. However, this possibility occurs with low probability, about 10^{-3}, and seems to affect only mammalian cells. We have called the corresponding lethal modification Mn; it has every chance of being "non-reparable".

7. Conclusions

An analysis of the localized events which occur after the reorganization, in DNA, of sulphur and of sulphate and which appear after ^{32}P or ^{33}P transmutation, suggests the simultaneity of several mechanisms having different probabilities (table 2, Part I). The existence of these different mechanisms justifies the different lethal modifications, M_i, which we have postulated, and which, subsequently, are characterized by a probability ϵ_i (table 8). Following a single-break in DNA, the most probable lethal modification corresponds to the action of one radical, and the least probable to the destruction of the sulphate structure produced when the sulphur atom is multi-ionized. The double-strand break in DNA is only one of the possible modifications, and seems to be lethal only in certain particular cases (table 8). These double-strand breaks, like the single-strand breaks, can be produced by several different mechanisms (table 2, Part I).

The probability which characterizes each of the possible lethal modifications depends on experimental conditions *and* on the probability of the event which gives birth to it. This is why the lethal modifications can depend on different evens when, in the same experiment, the cells are stored at different temperatures (ex. bacteriophages stored at 4° or at $-196°$C, table 6).

This report shows that, for probably all kinds of cells, each of the phosphorus atoms seems to have equal importance, and that the integrity of nearly all the DNA seems necessary to insure the ability to replicate indefinitely. However, the repair phenomena, when capable of acting, hide this fact. Our study also shows that the bacteriophages examined, seemingly divided into two groups (λ, T1, and T7, on the one hand, and T4 and SP82G on the other) must probably differ in terms of their in situ DNA conformation, and the energy transfer which might thereby be produced.

Acknowledgements

We wish to thank Miss E. Moustacchi, Mrs. A. Golde, Mr. M. Frilley and Mr. R. Latarjet for the remarks and criticisms which helped so much in the final elaboration of Part II of this report.

References

Apelgot, S. and Latarjet, R. 1962. Biochim. Biophys. Acta 55, 40.
Apelgot, S. and Latarjet, R. 1966. Int. J. Radiat. Biol. 10, 165.
Apelgot, S. 1972. Int. J. Radiat. Biol. 22, 557.
Apelgot, S. and Thiery, J.P. 1974. Radiat. Res. 59, 283 and unpublished results.
Burki, H.J., Roots, R., Feinendegen, L.E. and Bond, V.P. 1973. Int. J. Radiat. Biol. 24, 363.
Cairns, J. and Davern, C.I. 1966. J. Mol. Biol. 17, 418.
Carlson, T.A., Nestor Jr. C.W. and Tucker, T.C. 1968. Phys. Rev. 169, 27.
Davern, C.I. 1968. J. Mol. Biol. 32, 161.
Davison, P.F., Freifelder, D., Hede, R. and Lewinthal, C. 1961. Proc. Natl. Acad. Sci. USA 47, 1123.
Den Hollander, W., Van der Jagt, P.J. and Van Zanten, B. 1975. Radiochim. Acta 22, 101.
Domon, M., Barton, B., Porte, A. and Rauth, A.M. 1970. Int. J. Radiat. Biol. 17, 395.
Drobnik, J. 1964. Radiat. Res. 21, 339.
Feinendegen, L.E., Henneberg, P. and Tisljar-Lentulis, G. 1977. In: Molecular and Microdistribution of Radioisotopes and Biological Consequences. Int. Conf. Jülich 1975. Curr. Top. Radiat. Res. 12 (1978).
Fox, M.S. 1963. J. Mol. Biol. 6, 85.
Fuerst, C.R. and Stent, G.S. 1956. J. Gen. Physiol. 40, 73.
Gabrilovich, I.M. and Romanovkaja, L.N. 1970. Biochim. Biophys. Acta, 213, 231.
Ginoza, W. 1967. Annu. Rev. Nucl. Sc. 17, 469.
Grigg, G.W. 1968. Mol. Gen. Genet. 102, 316.
Harriman, P.D. and Stent, G.S. 1964. J. Mol. Biol. 10, 488.
Hart, E.J. and Anbar, M. 1970. The Hydrated Electron. (John Wiley, New York).
Hartman, P.E. and Kozinski, A.W. 1962. Virology 17, 233.
Hatzfeld, J. and Moustacchi, E. 1969. Int. J. Radiat. Biol. 15, 101.
Haynes, H.R., 1975. In: Molecular Mechanisms for Repair of DNA, part B. Eds. Hanawalt, P.C. and Setlow, R.B. (Plenum Publ. Corp., New York) p. 529.
Hershey, A.D., Kamen, M.D., Kennedy, Y.W. and Gest, H. 1951. J. Gen. Physiol. 31, 305.
Hill, R.F. 1960. Nature 188, 412.
Ho, K.S.Y. 1975. Mutat. Res. 30, 327.
Holweck, F. and Lacassagne, A. 1934. Radiophysiologie 3, 81 and 215.
Hot Atom Chemistry Status Report, 1975. IAEA, Vienna.
Jiang, V.W., Krohn, K.A. and Welch, M.J. 1975. J. Am. Chem. Soc. 97, 6551.
Kaplan, H.S. and Moses, L.E. 1964. Science 145, 21.
Krisch, R.E. 1970. Int. J. Radiat. Biol. 18, 259.
Krisch, R.E. 1972. Int. J. Radiat. Biol. 21, 167.
Krisch, R.E. 1974. Int. J. Radiat. Biol. 25, 261.
Latarjet, R., Ekert, B., Apelgot, S. and Rebeyrotte, N. 1961. J. Chim. Phys. 58, 1046.

Latarjet, R. 1974. In: Current Topics in Radiation Research, Vol. 8. Eds. Ebert, M. and Howard, A. (North-Holland, Amsterdam) p. 1.
Lederer, C.M., Hollander, J.M. and Perlman, I. 1967. Tables of Isotopes, 6th ed. (John Wiley, New York).
Ley, R.D. and Krisch, R.E. 1974. Int. J. Radiat. Biol. 25, 531.
Lillicrap, S.C., Fielden, E.M. and Dewey, D.L. 1974. Int. J. Radiat. Biol. 26, 65.
Manning, P.G. and Monk, C.B. 1962. J. Chem. Soc. 2573.
Matheson, A.T. and Thomas, C.A. 1960. Virology 11, 289.
Miller, R.C.J.R. 1970. Virology 5, 536.
Moustacchi, E. 1959. Biochim. Biophys. Acta 36, 577.
Moustacchi, E. 1968. In: Biological Effects of Transmutation and Decay of Incorporated Radioisotopes. Panel Proc. Ser. (IAEA, Vienna) p. 17.
Mouton, R.F. and Tremeau, O. 1969. Int. J. Radiat. Biol. 15, 369.
Nefedov, V.D., Zaitsev, V.M. and Toropova, M.A. 1963. Russ. Chem. Rev. 32, 604.
Raadschelders-Buijze, C., Roos, C.L. and Ros, P. 1973. Chem. Phys. 1, 468.
Ragni, G. and Szybalski, W. 1962. J. Mol. Biol. 4, 338.
Riley, M. and Pardee, A.B. 1962. J. Mol. Biol. 5, 63.
Roots, R. and Smith, K.C. 1975. Int. J. Radiat. Biol. 27, 595.
Rörsch, A. and Van der Kamp, C. 1961. Biochim. Biophys. Acta 46, 401.
Rosenthal, P.N. and Fox, M.S. 1970a. J. Mol. Biol. 50, 573.
Rosenthal, P.N. and Fox, M.S. 1970b. J. Mol. Biol. 54, 441.
Schmidt, A. and Hotz, G. 1973. Int. J. Radiat. Biol. 24, 307.
Scholes, G. 1963. Prog. Biophys. Mol. Biol. 13, 59.
Serber, R. and Snyder, H.S. 1952. Phys. Rev. 87, 152.
Sinsheimer, R.L. 1959. J. Mol. Biol. 1, 43.
Skorobogatov, G.A. and Nefedov, V.D. 1966. Zh. obsch. Khim. 36, 995.
Steinberg, C.M. and Von Borstel, R.C. 1966. Science 154, 429.
Stent, G.S. and Fuerst, C.R. 1955. J. Gen. Physiol. 38, 441.
Strigini, P., Rossi, C. and Sermonti, G. 1963. J. Mol. Biol. 7, 683.
Szybalski, W. 1962. The Mol. Basis of Neoplasia, 147.
Teoule, R., Bomicel, A., Bert, C., Cadet, J. and Polverelli, M. 1974. Radiat. Res. 57, 46.
Terzi, M. 1965. J. Theor. Biol. 8, 233.
Tessman, J. 1959. Virology 7, 263.
Tikchonenko, T.I. 1969. Adv. Virus Res. 15, 201.
Tomizawa, J.I. and Ogawa, H. 1967. J. Mol. Biol. 30, 7.
Tomizawa, J.I. and Ogawa, H. 1968. Cold Spring Harb. Symp. Quant. Biol. 33, 243.
Town, C.D., Smith, K.C. and Kaplan, H.S. 1974. In: Current Topics in Radiation Research, Vol. 8. Eds. Ebert, M. and Howard, A. (North-Holland, Amsterdam) p. 351.
Ullrich, M. and Hagen, U. 1971. Int. J. Radiat. Biol. 19, 507.
Van der Jagt, P.J., Den Hollander, W. and Van Zanten, B. 1975. Radiochim. Acta 22, 162.
Van Dyke, J.G. 1960. Biochim. Biophys. Res. Commun. 3, 190.
Vizdalová, M., Jonoská, E. and Zhestjanikov, V.D. 1971. Int. J. Radiat. Biol. 20, 49.
Waldstein, E.A., Zhestjanikov, V.D. and Tomilin, N. 1971. In: First European Biophysics Congress, Vol. II. Eds. Broda, E., Locker, A. and Springer-Lederer, H. (Verlag der Wiener Med. Akademie) p. 189.
Waters, R. and Moustacchi, E. 1974. Biochim. Biophys. Acta 353, 407.
Weiner, R.M. 1966. Phys. Rev. 144, 127.
Wexler, S. 1965. In: Actions chimiques et biologiques des radiations, VIIIe Série. Ed. M. Haissïnsky (Masson, Paris) p. 105.
Witkin, E.M. and Farquharson, E.L. 1969. In: Mutation as Cellular Process. Eds. Wolshenholme, G.E. and O'Connor, M. (Ciba Foundation Symposium, London) p. 36.
Wolfgang, R., Anderson, R.C. and Dodson, R.W. 1956. J. Chem. Phys. 24, 16.

DOSE-TIME RELATIONSHIPS IN IRRADIATED WEEVILS AND THEIR RELEVANCE TO MAMMALIAN SYSTEMS

W.E. LIVERSAGE * and R.G. DALE

Charing Cross Hospital, London, England

The investigation had two objectives:
1. To study, in an organised normal tissue, the form of relationship between the iso-effect dose and the overall irradiation time of fractionated regimes of radiation.
2. To test the General Formula for equating protracted and acute regimes of radiation [Liversage, 1969]. This formula had previously been shown to be approximately true for two mammalian tissues.

The LD $50_{(30)}$ for grain weevils has been determined for 63 different irradiation regimes ranging from single doses delivered in 24 min or continuously at low dose-rate over periods up to 106 h to fractionated acute regimes involving 1–20 fractions delivered in overall times varying from 0 to 28 days. The use of 68,200 weevils has enabled iso-effect data to be established which are probably more comprehensive and accurate than those obtained for any other tissue.

Weevils were housed and irradiated at constant temperature and at constant humidity. A ^{60}Co γ-ray beam unit was used for all irradiations.

Over the whole range investigated, the General Formula for equating protracted and acute regimes of radiation has certainly been found to be valid, to the degree of accuracy (±10%) claimed by Liversage [1969].

For fractionated regimes, the relationship between the LD 50 for weevils and overall time is not consistent with published iso-effect formulae but a new relationship has been established which fits the experimental data, provided the radiation is delivered in equally spaced fractions of equal magnitude. Unexpected results were obtained using unequally spaced fractions or doses of unequal magnitude. These results suggest that the rate of cellular repopulation at any instant of time is a function of the damage present.

The implications of these results in radiobiology and radiotherapy are discussed. It is suggested that the effects of cellular repopulation in rapidly proliferating normal tissues, such as skin and mucosa, may have been underemphasised and those of intracellular recovery overemphasised in recent years.

---

* Present address: Radiation Physics Department, St. Bartholomew's Hospital, London.
The experimental work reported in this article formed part of the thesis submitted to London University by one of the Authors (W.E.L.) for the award of a Ph.D. degree.

CONTENTS

1. INTRODUCTION 100
   1.1 The effect of repair processes on iso-effect dose 100
   1.2. The choice of end-point for the present experiments 102
   1.3 The care and custody of grain weevils 104
   1.4. The irradiation of weevils 107
   1.5. Pilot experiments 109

2. THE MAIN EXPERIMENTS 113
   2.1. General considerations 113
   2.2. The groups of experiments 114

3. THE RESULTS 120
   3.1. Graphical estimation of the LD 50 120
   3.2. The estimation of the LD 50 values by probit analysis 120
   3.3. Comparison of graphical results with those obtained by computer . . 122

4. A GUIDE TO THE ANALYSIS OF THE RESULTS 123

5. THE RELIABILITY OF THE OBSERVED RESULTS 124
   5.1. Variation of susceptibility between groups of experiments 124
   5.2. Estimated error in observed LD $50_{(30)}$ due to choice of day 0 124
   5.3. Variation of LD $50_{(30)}$ with age and pre-storage in plastic pots . . . 125
   5.4. Variation of LD $50_{(30)}$ with period of post-storage in plastic pots . . 125

6. THE SPLIT-DOSE RECOVERY CURVE FOR WEEVILS 125
   6.1. The observed shape 125
   6.2. Factors governing the shape of split-dose recovery curves 126
   6.3. Analysis of the split-dose recovery curve 127
   6.4. Estimation of τ (the half-life of the dose-equivalent of sub-lethal damage) from the split-dose recovery curve 127

7. THE RELATIONSHIP BETWEEN ISO-EFFECT DOSE AND OVERALL TIME WHEN USING A CONSTANT NUMBER OF EQUALLY SPACED FRACTIONS 128
   7.1. Existing iso-effect formulae 128
   7.2. Attempt to fit results to an Ellis-type formula 128
   7.3. Attempt to fit results to the Cohen-type formula 129
   7.4. Attempt to fit results to the Liversage-type formula 130
   7.5. The search for the true relationship 131
   7.6. Discussion 138
   7.7. Conclusion 139

8. THE VARIATION IN LD 50 WITH FRACTION NUMBER 140
   8.1. The difficulty of separating effects due to fraction number from those due to overall time 140
   8.2. Variation in the LD 50 with the number of daily fractions 141
   8.3. The increase in LD 50 due to intracellular recovery between fractions 142

8.4.	The effect of reducing the irradiation time per fraction	144
8.5.	The derivation of a dose-response curve for the critical cells of the weevil	146
8.6.	Conclusions	147

9. THE RELATIONSHIP BETWEEN THE RATE OF REPOPULATION AND THE DAMAGE PRESENT 148

9.1.	Experiments using two doses of unequal magnitude	148
9.2.	Predicted results assuming the rate of repopulation to be proportional to the damage present	150
9.3.	Comparison of observed and predicted results	151
9.4.	The anomalous result for 6 fractions in 7 days	154
9.5.	Conclusions	155

10. TESTING THE GENERAL FORMULA FOR EQUATING PROTRACTED AND ACUTE REGIMES OF RADIATION 155

10.1.	Variation in LD 50 with duration of low dose-rate continuous irradiation	155
10.2.	The correction for repopulation	156
10.3.	The application of the general formula to the above data	157
10.4.	Conclusions	159

11. DISCUSSION OF THE RESULTS AND THEIR IMPLICATIONS IN RADIOBIOLOGY AND RADIOTHERAPY 160

11.1.	Comparison of results with mammalian data	160
11.2.	The relationship between the rate of repopulation and the damage present	160
11.3.	The relationship between iso-effect dose and overall time of a fractionated regime	163
11.4.	The relative importance of intracellular recovery and repopulation	164
11.5.	Existing dose-time relationships	167
11.6.	The general formula for equating protracted and acute regimes of radiation	178

12. SUMMARY OF THE CONCLUSIONS 180

12.1.	The factors influencing iso-effect relationships in weevils	180
12.2.	The value of τ	181
12.3.	The validity of the General Formula	181
12.4.	The variation in LD 50 with overall time	182
12.5.	The relationship between repopulation and damage	182
12.6.	The relative importance of repopulation and intracellular recovery	183
12.7.	Revision of the estimated repopulation rate for skin	183
12.8.	A revised interpretation of the increase in iso-effect dose when fraction number only is varied	184
12.9.	The design of future radiobiological experiments	184
12.10.	The possible exploitation of repopulation effects	184

ACKNOWLEDGMENTS 184

REFERENCES 185

1. Introduction

There is much interest at present in determining iso-effect relationships in radiotherapy. The available data in this field are somewhat limited. Research into the form of these relationships in patients is quite rightly hampered by ethical considerations.

The work to be described is a parallel investigation of the effects of overall time, fraction number and dose-rate in a normal organised tissue in which the LD 50 of grain weevils was determined for 63 different irradiation regimes. Because very large numbers of specimens could be used, it was possible to obtain iso-effect data which are probably more comprehensive and more accurate than those available for any other organised tissue. Thus the shapes of these relationships could be studied in greater detail than has previously been possible. The shapes of the relationships observed in weevils suggest that we may need to revise our ideas regarding dose-time relationships in radiotherapy. It has been shown by Riemann and Flint [1967] that weevils given whole body doses of irradiation die due to the killing of the regenerative cells of the mid-gut and the results of the present series of experiments suggest that the critical cells in the weevil are exhibiting similar repair processes to those found in mammalian tissues. These repair processes include recovery from sub-lethal damage and repopulation governed by homeostatic control and radiation-induced mitotic delay.

The extent of the analogy between results in weevils and results in mammalian tissues has yet to be proven, but the results obtained in weevils will later be compared and contrasted with those observed in mammalian systems including man.

1.1. The effect of repair processes on iso-effect dose

During the course of a fractionated high dose-rate irradiation it is generally considered that the dose D_{NT} necessary to produce a given effect, increases with the number of fractions N and the overall irradiation time T days for two reasons:

(a) Intracellular sub-lethal damage is repaired between fractions and a quantity of radiation, which we will call ^RD_N, is used up in re-creating this sub-lethal damage.

(b) Cell proliferation increases the surviving fraction of cells during the irradiation time T and an additional quantity of radiation ^rD_T needs to be given to compensate for this repopulation effect.

These effects may be expressed mathematically,

$$D_{NT} = D_1 + {}^RD_N + {}^rD_T, \tag{1}$$

where D_1 is the single dose (at high dose-rate) necessary to produce the given effect. In the above context, high dose-rate means a dose-rate such that the repair of sub-lethal damage is negligible during the course of any individual fraction of irradiation.

The protracted dose D_t necessary to produce the same effect, if delivered continuously over a period of t hours may be represented mathematically by the following relationship

$$D_t = D_1 + {}^R D_t + {}^r D_t, \qquad (2)$$

where ${}^R D_t$ is the additional dose needed to compensate for repair of sub-lethal damage occurring during t hours and ${}^r D_t$ is the additional dose needed to compensate for any repopulation of cells occurring during the protracted irradiation.

Liversage [1969] published a General Formula for equating protracted and acute regimes of radiation. He claimed that ${}^R D_N$ would be equal to ${}^R D_t$ provided N and t were related by the following expression

$$N = \frac{\mu t}{2\left[1 - \frac{1}{\mu t}(1 - \exp(-\mu t))\right]}, \qquad (3)$$

where $\mu = 0.693/\tau$ and τ is the half life of the dose-equivalent of sub-lethal damage as defined by Lajtha and Oliver [1961] who deduced that τ was approximately equal to 1.5 h in mammalian cells irradiated at normal body temperature.

The value of τ may be estimated from the shape of a split-dose recovery curve but the exact determination of τ is difficult due to the recovery curve being affected by radiation-induced synchrony [Liversage, 1969a].

Using a value of $\tau = 1.5$ h, Liversage [1969] had shown that the General Formula was consistent with available data for skin reaction in patients and LD 50 measurements in mice. The present experiments were designed to test this formula more rigorously over the range 0.5–100 h for a different endpoint.

These ranges were chosen because of their relevance to the clinical situation involved on changing from a low dose-rate radium treatment of carcinoma of the cervix to a fractionated high dose-rate remotely controlled afterloading technique [O'Connell et al., 1967].

Examination of the above equations shows that in order to test the validity of the General Formula, it is necessary to be able to estimate fairly accurately the effect of repopulation of cells during the course of any fractionated acute regime. It was therefore decided that experiments should be included to determine how ${}^r D_T$ varied with overall time when using a constant number of fractions. It was this group of experiments which produced the most interesting and unexpected results.

1.2. The choice of end-point for the present experiments

To test and investigate the above relationships over the required ranges, experiments were designed to study, for one specific end-point, how the iso-effect dose varies with

(a) split-dose interval (11 regimes, 0–25 h)
(b) continuous irradiation time (13 regimes, 1–106 h)
(c) number of "daily" fractions (7 regimes, 1–20 fractions)
(d) overall irradiation time using 4 fractions (6 regimes, 3–28 days)
(e) overall irradiation time using 9 fractions (3 regimes, 10–28 days)

Thus it was estimated initially, that the iso-effect dose would need to be determined for 40 different irradiation regimes. As the work progressed, the need and opportunity to obtain additional data arose. The iso-effect dose was in fact determined for 63 different irradiation regimes. Most of the extra regimes were needed to provide information regarding the long-term repair process which was found to be more important and more complicated than originally envisaged. The remainder were due to repeating the daily fractionation regimes at a higher dose-rate made possible by the installation of a new ^{60}Co unit.

When comparing the effects of continuous irradiation with those of fractionated irradiation it is clearly advantageous if the geometrical dose distribution can be kept constant. The only practical methods of continuously irradiating part of an animal for 100 h involve the use of interstitial, intracavitary or surface applicators. The geometry of such devices is extremely critical and difficult to simulate with fractionated high dose-rate irradiation. With whole-body irradiations these problems do not arise. Both low and high dose-rate irradiations may be delivered by the same ^{60}Co γ-ray unit, using similar geometry but with or without the insertion of a lead absorber.

The experiments were intended to show small variations in iso-effect dose with changing conditions and it was, therefore, essential that for each regime the iso-effect dose should be determined with a high degree of accuracy. It was decided to aim at a somewhat arbitrary accuracy of ±3%. To obtain this degree of accuracy, using whole-body irradiation and lethality as an end-point, it was estimated that several hundred animals would be required for each of the 40 or more irradiation regimes. Thus the total number of animals required would be of the order of tens of thousands. (The number actually used was 68,200.) Housing such large numbers of small mammals and irradiating many of them for continuous periods of up to 100 h would require resources which are beyond most of our research institutes. These problems are greatly reduced if small animals such as insects are used.

The General Formula to be tested had already been shown to be valid to a limited degree of accuracy for two mammalian tissues [Liversage, 1969]. Thus experiments with insects would help in testing the general nature of the formula more effectively than the use of yet another mammalian tissue.

Grain weevils were chosen for these experiments because considerable data regarding the susceptibility of these insects to γ-radiation were available.

Grain weevils breed in grain and in experimental work are normally housed on grain. The female lays her eggs in the grain and fig. 1 taken from Cornwell and Morris [1959] shows the developmental stages existing between oviposition and the emergence of the adult weevil some 33 days later, if housed at 26°C.

The average life of a grain weevil housed at 70% R.H. and 25°C is of adequate duration, being about 110 days if kept on wheat and about 190 days if kept on flour [Coombs and Woodroffe, 1964]. Grain weevils do not fly and can only climb a vertical glass wall with difficulty. They are thus easily handled.

Jefferies and Banham [1961], using ^{60}Co γ-rays found the LD $50_{(28)}$ for

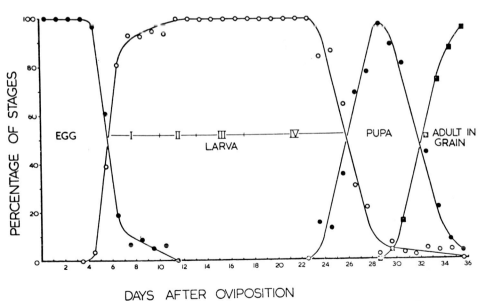

Fig. 1. Percentages of development stages in grain infested by grain weevils at 26°C and 77% R.H. Taken from Cornwell and Morris [1959].

grain weevils to be 4470r * using single doses and 6780r using 4 fractions in 3 days at dose-rates of 7300 to 10,000 r/h. Their weevils were housed at 26°C and 76% R.H. but were irradiated at temperatures which varied between 25° and 35°C. These workers found that grain weevils given two doses of 4,000r 10 days apart, showed the same percentage mortality as grain weevils given a single dose of 4,000r. They claimed that this demonstrated that complete recovery from 4,000r occurred in a period of ten days.

In organised tissues, the long-term repair of radiation injury is generally considered to be influenced by homeostatic forces, but the rate of repair is much slower, and hence more difficult to measure in mammals than in weevils. The latter appeared to be good experimental subjects in which to study the form of relationship, if any, between the rate of repair and the amount of damage present at any instant of time.

Martin et al. [1962] studied the effect of culture density on susceptibility of grain weevils reared and housed in a room kept at 26°C, 75% R.H. They showed that if grain weevils were cultured on wheat housed in glass jars, the single dose LD $50_{(42)}$ for the emerged adult decreased from 6,750r to 5,000r as the culture density was increased from 10 to 1000 parents/800 g of wheat. These authors showed that during the developmental stages, the temperature of the dense cultures rose by as much as 6°C above room temperature due to metabolic activity. They suggested that the LD 50 for grain weevils was very dependent on the temperature at which they had been kept during their development from egg to adult.

Further work by Pendlebury et al. [1962] investigated how the LD 50 varied depending on whether weevils were kept at 15°C or 30°C before, during or after irradiation. They found that keeping the weevils at 30°C instead of 15°C before irradiation increased their sensitivity by 50%, during irradiation decreased their sensitivity by 20% and after irradiation, had little effect on the ultimate percentage kill but decreased the time for radiation deaths to reach completion from 56 to 14 days.

The above information suggested that the LD 50 of grain weevils would be an excellent end-point for this study, but indicated that the temperature at which the weevils were cultured, housed and irradiated would need to be carefully controlled.

1.3. The care and custody of grain weevils

In all the irradiation experiments referred to above, the weevils had been reared, irradiated and subsequently housed on grain. In order to count the

---

* The unit of dose used by these workers was the 'roentgen-equivalent-physical' denoted by the symbol 'r'. One 'r' unit is equal to 0.97 rad and 100 rad equals 1.00 Gy.

surviving weevils in any batch, it is first necessary to sieve the grain to separate the weevils from it. This process is very time-consuming as the weevils burrow into the grain to feed and even after sieving for 15 min, some weevils may still lie hidden in the grain.

In prolonged experiments the grain needs to be changed at least every 28 days to prevent the emergence of offspring. Between changes, the metabolic activity of the developmental stages of these offspring may well affect the temperature of the batch under test. To reduce these difficulties, it was decided to keep the weevils on flour at 70% R.H. Sieving is then much easier and offspring do not interfere with the experiment for the simple reason that the female weevil does not normally lay eggs in flour. Weevils require a moist atmosphere but flour tends to cake and grow mould if stored at a relative humidity above 70% R.H.

The published data referred to above clearly indicated that the temperature needed to be very carefully controlled during all stages of the investigation and the building of a constant temperature, constant humidity, irradiation cabinet and storage room, were essential.

In all the experiments to be described, weevils were cultured on wheat but housed and irradiated on flour; both kept at 27°C, 70% R.H. Standard stock cultures were made by placing 120 adult weevils on 200 g of Argentine Plate Wheat which had been disinfested by heating to 60°C for 4 h and then conditioned by allowing it to stand in the Constant-Temperature (C.T.) Room for at least one week. The wheat was contained in glass jam jars 7.8 cm diameter sealed with filter paper fixed by wax and two layers of cotton cloth held by rubber bands.

Throughout these experiments, any jar containing a culture, flour, grain or weevils was stored on its own individual platform in a tray of oil on one of the shelves in the C.T. Room. The oil prevented any weevils which might escape from reaching other specimens, and helped to reduce the risk of specimens becoming infested with other species of insects attracted by the food and the environment. No new weevils would emerge from the stock culture for 33 days, but by 36 days, the rate of emergence exceeded 100 weevils per day. Young weevils could be separated by sieving the culture at frequent intervals during the next three or four weeks. Weevils pass through a British Standard No. 12 sieve but grain does not. Stock cultures were prepared every six weeks.

Special cultures were prepared to produce weevils for experiments. The grain was prepared as for stock cultures but only 100 weevils (obtained from standard stock cultures) were placed to oviposit on each jar of grain. The parents were removed by sieving 15 ± 1 days later. Between 10 and 20 such cultures were prepared approximately 60 days before the first irradiation in any group of experiments. The cultures were sieved again 24 days and 14 days before the first irradiation. The weevils obtained from the second sieving were used in the experiments and were thus all 14–24 days old on the first irradiation day. Each group of experiments, designed to determine the LD 50 for a number (about 10) of different irradiation regimes, required a total of about 300 plastic pots of 40 weevils. The plastic pots were 2.7 cm high and 2.6 cm in diameter. Two weeks before their first irradiation, these pots were loaded with the required amount (approximately 4.7 g) of disinfested flour and stored in the C.T. Room. The 30 or so pots to be used for a given irradiation regime were allocated consecutive numbers. Loading the 300 or so pots with weevils occupied two full days work, and this task was usually started 5 days before the first irradiation. Pots with numbers ending in zero were loaded first, and those ending in nine, last.

The problem of handling and counting the thousands of weevils involved, was solved surprisingly easily. If a thousand or more weevils are placed in a large upturned beaker, they tend to walk out of the spout, more or less in a single file, and are easily picked up by a "pouter". The latter is basically a miniature vacuum cleaner comprising a glass or perspex vessel 10 cm long and 2.5 cm diameter which is sealed by a rubber bung. Inlet and outlet glass tubes pass through the rubber burng. The outlet tube, whose entrance is covered by a wire gauze, is connected to a vacuum pump. The inlet tube is bent through a right angle and has a small hole in one side. When this is covered with a finger and the end of the inlet tube is placed near a weevil, the latter is sucked up into the "pouter". In this way, 40 weevils may be counted and collected in as many seconds, and then transferred to one of the plastic pots which are fitted with perforated polythene lids.

Before and after irradiation, the pots containing the weevils were housed in the C.T. Room in a definite sequence which depended primarily on the last digit of their allocated reference number. In this way, any variation in irradiation susceptibility produced by positional factors or loading sequence would be reflected in the results observed for any given irradiation regime, but would affect all regimes equally.

The early experiments revealed that weevils were capable of enlarging the perforations in the polythene lids sufficiently to escape. To prevent this happening, each plastic pot was stored before and after irradiation in a somewhat larger plastic container also fitted with a plastic perforated lid. The lids of the plastic pots were inspected daily and replaced when necessary. Weevils receiving single doses of irradiation were transferred to glass honey jars 6.6 cm in diameter 7 days after irradiation. These jars had been filled with 35 g of disinfested, sieved, whole wheat flour and housed in the C.T. Room for at least 7 days. Weevils receiving protracted or fractionated regimes of treatment were similarly transferred 7 days after Day 0, defined below.

Preliminary experiments showed that weevils killed by single doses usually died about 16 days after irradiation, and that weevils surviving more than 23 days, suffered a mortality similar to that of unirradiated weevils. Thus the majority of deaths occurring within 30 days could be ascribed to the effects of radiation, whilst those occurring after 30 days, were probably due to other causes. The LD $50_{(30)}$ was, therefore, considered to be a good yardstick by which to measure the susceptibility of weevils to single doses of radiation. When comparing the effects of fractionated irradiations with those produced by single doses, the question arose as to how one should define Day 30. It could be argued that the 30 day period could be counted from either half way through the irradiation regime or alternatively from the end of the irradiation regime. In view of this uncertainty it was decided to define Day 0 as the day which was three-quarters through the irradiation regime and Day 30 as being 30 days later. This was a somewhat arbitrary choice which had to be made in order to design the experiments. The error in the estimation of the iso-effect doses due to defining Day 0 in this way is discussed later (see section 5.2).

The number of weevils surviving was counted 5 or 6 times between Day 0 and Day 37. Counts were invariably made on Day 7, when the weevils were transferred from plastic pots to honey jars and on Day 30. At each count the weevils in any specimen were separated from the flour by sieving with a British Standard Sieve No. 25. The weevils were then placed on a white plastic table top. Those which moved were picked up by the "pouter"; those which did not move were placed on a sheet of metal which was then gently heated. Weevils which failed to respond to this stimulus were considered dead. Both the numbers of dead and alive weevils in any sample were counted in order to detect counting errors, which were found to be very rare. Dead weevils were discarded: live weevils were resealed in their respective jars.

1.4. The irradiation of weevils

Weevils were irradiated on flour in batches of 40 in the small plastic pots described above (section 1.3). During irradiation the weevils were housed in a constant temperature, constant humidity cabinet (fig. 2).

The cabinet was made of wood lined with 2.5 cm thick expanded polystyrene. An insulated partition, having a vent at top and bottom, divided the cabinet into a specimen compartment at one end and an air-conditioning compartment at the other.

The specimen compartment housed a perspex nest of 23 removable 12.7 mm thick perspex shelves. Each shelf contained 37 symmetrically arranged holes which during irradiation were filled with plastic pots containing either 40 weevils plus flour or flour only. In either case, the amount of flour in the pot was adjusted so that the weight of the pot and its contents was equal to the weight of perspex displaced. When exposed to a vertically downward

Fig. 2. The specimen compartment of the irradiation cabinet with the door open to show the nest of shelves housing the plastic pots containing weevils for irradiation.

^{60}Co γ-ray beam the dose-rate received by the weevils was determined by their position in the nest. Thus several hundred pots of 40 weevils could be irradiated to different doses simultaneously.

In the majority of the experiments (Groups A to F, see section 2.2), the vertically downward beam of a Cobalt-60 Unit giving a 22 cm square field at a source-cabinet distance of 60 cm was used for the irradiations. Each acute irradiation lasted between 71 and 79 min, but the doses delivered to the weevils varied between 700 and 7500 rad depending upon their position within the cabinet. During protracted continuous irradiation, lasting between 6 and 106 h, the dose-rates in the cabinet were reduced by introducing a

2 cm thick lead absorber into the beam at a distance of 47 cm from the source.

The original Cobalt-60 Unit was replaced by a more powerful and versatile Cobalt-60 Unit which allowed the high dose-rate acute irradiation experiments to be repeated using shorter irradiation times (24–26 min). This unit was used for the experiments in Groups G and H.

The dose-rate on any shelf in the irradiation cabinet was expected to be approximately 15% less than that on the shelf immediately above, and it was anticipated that there would be a gradual decrease in dose-rate toward the periphery of each shelf.

Details of the dosimetry involved, are described elsewhere [Liversage, 1973]. Essentially, the dose-rate received on each shelf and the variation across the shelf were measured using an ionization chamber in conjunction with an Ionex Mark III Dosemeter. These measurements were confirmed using T.L.D. (Lithium Borate) powder in small capsules half embedded in the surface of the flour.

1.5. Pilot experiments

After carrying out experiments to determine the effects of the environment on the life expectancy of the unirradiated weevil [Liversage, 1973], it was decided that pilot experiments should be carried out on weevils which had been cultured, housed and irradiated under the conditions to be employed.

Each of the groups of main experiments would involve the use of approximately 10,000 weevils and a considerable amount of experimental work spread over a period of approximately 5 months.

If the true LD 50 lay outside the range of dose levels chosen, or if too large a range of dose levels was employed, much time and effort would be wasted. It was therefore essential that pilot experiments should be carried out which would enable appropriate ranges of dose levels to be chosen for the main experiments.

The pilot experiments were therefore designed to obtain approximate values of the LD 50 for weevils for a small number of important conditions to be employed. These experiments would serve the additional objective of testing, in practice, the proposed experimental technique. It had been decided that the age of weevils employed in the main experiments should be 14–24 days. The pilot experiments were therefore also designed to see whether there was any gross difference in susceptibility of weevils of different ages within this range. Thus, in the pilot experiments, each batch of weevils employed was sub-divided into groups aged 14–18, 18–21 and 21–24 days.

For the pilot experiment it was decided to find the approximate LD 50 for the following regimes: (a) Single doses; (b) 4 fractions in 3 days; (c) 4 fractions in 21 days.

The results of (a) and (b) should enable reasonable estimates to be made of how the LD 50 would increase with the number of daily fractions whilst the results of (b) and (c) should enable estimates to be made of how the LD 50 for a given number of fractions would increase with overall time.

For each of the above regimes, the expected LD 50 was predicted from the results of Jefferies and Banham [1961] and dose levels chosen which were considered to span the predicted LD 50 by an adequate range of values. Three pots (one for each age group) each containing 40 weevils were irradiated at each dose level and three kept unirradiated to act as controls for each regime.

The dose levels were as follows:

Single doses, 8 dose levels   (2360–6580 rad)
4 fractions in 3 days, 7 dose levels   (4370–9440 rad)
4 fractions in 21 days, 7 dose levels   (6840–17580 rad)

Figs. 3 and 4 show the number of survivors plotted against time for two dose levels for the single dose and the 4 fraction in 21 days regimes. Fig. 5 shows the number of survivors plotted against dose for each of the three regimes.

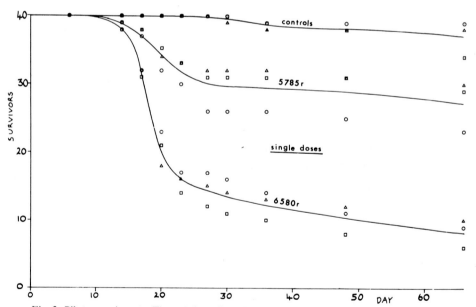

Fig. 3. Pilot experiments. The number of survivors plotted against time after single doses of 5,785 or 6,580 rad. Circles, 40 weevils 14–18 days old; Squares, 40 weevils 18–21 days old; Triangles, 40 weevils 21–24 days old.

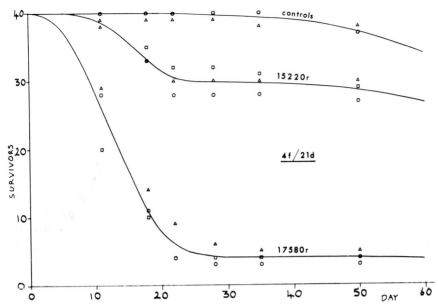

Fig. 4. Pilot experiments. The number of survivors plotted against time after total doses of 15,220 or 17,580 rad delivered in 4 fractions in 21 days. Day 0 was taken as the day which was three-quarters through the irradiation regime, as previously discussed (section 1.3). Circles, 40 weevils 14–18 days old; Squares, 40 weevils 18–21 days old; Triangles, 40 weevils 21–24 days old.

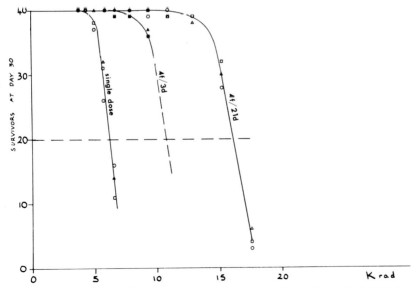

Fig. 5. Pilot experiments. The number of survivors at day 30 plotted against total dose for regimes involving either single doses, 4 fractions in 3 days or 4 fractions in 21 days. Circles, 40 weevils 14–18 days old; Squares, 40 weevils 18–21 days old; Triangles, 40 weevils 21–24 days old.

From these graphs it was estimated that the LD $50_{(30)}$ for each regime was an follows:

Single dose,	6200 ± 200 rad
4 fractions in 3 days,	10800 ± 800 rad
4 fractions in 21 days,	16100 ± 500 rad

The LD 50's observed for single doses and 4 fractions in 3 days were much higher than the corresponding values of 4470r and 6780r found by Jefferies and Banham [1961]. The latter workers had housed and irradiated their weevils on wheat rather than flour, had employed much higher culture densities and had not controlled their irradiation temperature. The differences in the observed results are therefore not surprising, especially when the changes in susceptibility known to be produced by changes in culture density or temperature are considered (see section 1.3.)

The present authors are deeply grateful to the research workers who discovered these facts. We choose to work with weevils rather than any other insect because their pioneering work provided us with a sound basic knowledge of the behaviour of this particular insect and how its sensitivity varied with environmental conditions.

It will be observed that the LD 50 for 4 fractions in 3 days lies outside the range of dose levels used for this regime. The value of the pilot experiments may be judged from the fact that this state of affairs never arose in any of the 63 different irradiation regimes for which the LD 50 was measured in subsequent experiments.

Analysis of the results of the pilot experiments failed to show any significant or consistent variation in susceptibility with age over the range studied.

The pilot experiments suggested that the LD 50 increased with fractionation at the rate of approximately 1,500 rad per daily fraction and this provided a useful guide to the dose levels to be given in the main daily fractionation experiments.

The pilot experiments also suggested that, for a constant number of fractions, the LD 50 increased with overall irradiation time at a rate of approximately 300 rad per day. This provided a useful guide to the dose levels to be given in the main fractionation experiments employing long intervals (up to 9 days) between fractions.

The ideal system for testing the General Formula would be one exhibiting a large effect due to recovery from sub-lethal damage and no effect due to repopulation. The pilot experiments suggested that the proposed system exhibited a large effect due to fractionation (1,500 rad per daily fraction) and a relatively small but not insignificant effect due to repopulation (300 rad per day). Thus, the proposed system seemed likely to provide a practical and suitable means of testing the General Formula provided appropriate corrections for repopulation could be accurately determined.

The pilot experiments employing four fractions showed that increasing the overall time from 3 to 21 days, caused an increase in the LD 50 or 49%. This is a much larger increase in iso-effect dose than that found for skin reactions by Fowler et al. [1963]. These workers found an increase of only 17% on changing from 5 fractions in 4 days to 5 fractions in 28 days. The proposed system therefore appeared to be an excellent one in which to study the form of relationship between iso-effect dose and overall time.

In the light of the above findings, it was decided that the main experimental work should continue as planned, but additional experiments to elucidate the exact nature of the repopulation correction should be included.

2. The main experiments

2.1. General considerations

To determine the LD 50 with the required degree of accuracy for one irradiation regime, it was estimated that approximately 1,000 weevils (25 pots of 40) would be required. The constant temperature room would house up to 330 jars (13,200 weevils), so that the number of regimes which could be investigated at any one time was approximately 13. The experiments were therefore split into a number of groups detailed below (section 2.2).

The special cultures required to produce weevils for irradiation needed to be prepared approximately 60 days before irradiation; irradiation regimes lasted up to 28 days and counting continued for at least another 35 days. Thus, more than 4 months elapsed between initiating a group of experiments and obtaining their results.

Because the susceptibility of weevils is so critically dependent on temperature before and during irradiation, it was decided that every experimental specimen should be accompanied by specimens taken from the same culture to act as controls. Some controls would be unirradiated, whilst others would receive single doses of varying magnitude to enable the single dose LD 50 to be determined for each culture. These weevils required single doses of the order of 6000rad and required to be placed on the upper shelves of the irradiation cabinet. During multiple fraction experiments, these weevils were subdivided; some being irradiated during each irradiation period.

To ensure that the ^{60}Co Unit and the requisite space in the irradiation cabinet would be available at the required time, detailed advance planning of these experiments was essential. Irradiation schedules, showing the data, exposure time and the position to be occupied by each specimen pot, were prepared two or three months in advance, (i.e., before preparing the necessary cultures).

2.2. The groups of experiments

The regimes used in each group are tabulated in the third column of table 1 (section 3).

Group A. This group of experiments involving 11,320 weevils had a single aim — namely the production of a split-dose recovery curve. This was necessary in order to be able to estimate the rate of recovery from sub-lethal damage and involved finding the LD 50 for weevils given two equal doses of radiation separated by intervals which varied in 2.5 h increments, from 0 to 25 h. The number of weevils involved was too large to permit the first of these two doses being delivered simultaneously to all weevils. By carefully planning the irradiation schedules, it was found that all weevils could receive their required doses by placing them in the cabinet during two out of a total of six irradiation periods.

Group B. This group of experiments involving 13,160 weevils was designed to study the three effects referred to below.
(a) How the LD 50 varied with the number of fractions when these were given at the rate of 5 fractions per week.
(b) How the LD 50 for either 4 or 9 fractions varied with overall time.
(c) How the single dose (S.D.) LD 50 varied with pre-storage in plastic pots for times up to 28 days prior to irradiation. N.B. It was realised that some weevils involved in experiments (a) and (b) above, would be 28 days older and would have been stored in plastic pots for 28 days longer when they received their last irradiation than weevils which received all their dose on the first day.

Group C. This group of experiments involving 7,800 weevils was designed primarily as a preliminary study of how the LD 50 varied with the duration of continuous irradiation. Their main objective was to ensure the success of the experiments which were to be performed in Group D when the required radiotherapy Cobalt-60 Unit would be available from Maunday Thursday to Easter Tuesday (106 h, instead of the 58 h applicable to a normal week-end).

It had also been realised that because of the way Day 0 was defined, (section 1.3), weevils receiving irradiation regimes of 28 days duration, could not be transferred from cramped plastic pots to more spacious glass jars until Day 7. The weevils in Group A experiments had been transferred earlier than this. Thus, the Group C experiments included a large number of single-dose con-

Table 1

Group	Ref. No.	Regime	LD 50(30) (rad) Eye-line	LD 50(30) (rad) Computed	95% F.I (%)	Normalised LD 50(30) and 95% F.L. (rad)	Remarks
A	1	S.D.	6,280	6,270	0.8	5,900	The weevils in this group had 1–4 days post-storage in plastic pots instead of the usual 7 days (see Ref. Nos. 27–30).
	2	2.5 h split	7,740	7,740	2.3	7,280 ± 170	
	3	5.0 h split	8,430	8,400	3.0	7,900 ± 240	
	4	7.5 h split	8,430	8,430	2.8	7,970 ± 230	
	5	10.0 h split	8,320	8,290	1.5	7,800 ± 130	
	6	12.5 h split	7,910	7,830	1.3	7,410 ± 110	
	7	15.0 h split	8,090	8,150	1.2	7,660 ± 110	
	8	17.5 h split	7,870	7,850	3.1	7,380 ± 240	
	9	20.0 h split	8,020	8,030	1.3	7,550 ± 110	
	10	22.5 h split	8,150	8,140	2.8	7,650 ± 220	
	11	25.0 h split	8,340	8,390	3.7	7,890 ± 300	
B	12	S.D.	5,920	5,910	1.3	5,900	Irrad. 1st week
	13	S.D.	5,750	5,730	2.7	5,900	Irrad. 4th week
	14	2F/3d	8,730	8,750	3.5	8,740 ± 320	
	15	4F/3d	10,420	10,450	1.4	10,440 ± 200	
	16	6F/7d	13,460	13,420	3.1	13,410 ± 460	
	17	9F/10d	17,420	17,530	3.1	17,680 ± 730	3d gap at end
	18	14F/17d	23,990	24,040	1.2	24,240 ± 730	
	19	20F/25d	31,260	31,530	3.2	32,130 ± 1350	
	20	4F/7d	12,710	12,570	1.3	12,660 ± 230	
	21	4F/10d	13,870	13,810	1.4	13,920 ± 420	
	22	4F/14d	15,100	15,150	2.4	15,280 ± 550	
	23	4F/21d	16,330	16,310	4.0	16,620 ± 800	
	24	4F/28d	16,870	16,850	4.4	17,160 ± 890	
	25	9F/18d	22,810	22,900	4.5	23,330 ± 1210	
	26	9F/28d	24,660	24,620	3.6	25,080 ± 1130	

Table 1 (continued)

Group	Ref. No.	Regime	LD 50(30)(rad) Eye-line	LD 50(30)(rad) Computed	95% F.L. (%)	Normalised LD 50(30) and 95% F.L. (rad)	Remarks
C	27	S.D.	5,860	5,890	1.1	6,110 ± 200	1d post-storage
	28	S.D.	5,920	5,930	4.0	6,150 ± 310	4d post-storage
	29	S.D.	5,710	5,690	3.1	5,900	7d post-storage
	30	S.D.	5,500	5,490	4.8	5,690 ± 320	14d post-storage
	31	7h cont.	7,930	7,920	3.1	8,210 ± 360	
	32	15h cont.	9,590	9,550	2.6	9,910 ± 400	
	33	36h cont.	11,330	11,350	5.4	11,760 ± 730	
	34	58h cont.	13,340	13,330	4.4	13,820 ± 750	
D	35	S.D.	6,140	6,130	2.4	5,900	
	36	6h cont.	7,910	7,920	4.5	7,620 ± 390	
	37	14h cont.	9,860	9,870	2.3	9,490 ± 310	
	38	22h cont.	10,500	10,480	3.2	10,080 ± 400	
	39	38h cont.	12,680	12,670	5.9	12,190 ± 780	
	40	54h cont.	13,680	13,720	3.6	13,200 ± 570	
	41	68h cont.	14,960	14,920	4.5	14,360 ± 730	
	42	84h cont.	14,100	15,090	2.6	14,520 ± 510	
	43	106h cont.	16,710	16,700	3.1	16,120 ± 630	
E	44	S.D.	5,860	5,850	1.7	5,900	
	45	4F/1d	9,027	9,260	2.1	9,340 ± 250	
	46	4F/3d	10,490	10,520	2.4	10,600 ± 310	
	47	6F/7d	14,490	14,480	3.1	14,600 ± 510	3d gap in middle
	48	9F/8d	16,790	16,830	1.0	16,970 ± 340	
	49	6F/5d	13,030	13,040	1.3	13,140 ± 280	
F	50	S.D.	5,740	5,730	3.5	5,900	
	51	4F/9h	9,250	9,260	2.9	9,540 ± 430	

#						
	2F/1.025d					Normalised values
52	$D_A = 1,010$	6,520	6,500	2.1	6,690 ± 270	$D_A = 1,040$
53	$D_A = 2,070$	7,410	7,420	2.4	7,640 ± 320	$D_A = 2,130$
54	$D_A = 3,380$	8,050	8,040	1.8	8,280 ± 320	$D_A = 3,480$
55	$D_A = 4,520$	7,890	7,950	2.4	8,190 ± 340	$D_A = 4,660$
56	$D_A = 5,090$	7,230	7,220	2.9	7,430 ± 330	$D_A = 5,240$
	2F/3.83d					Normalised values
57	$D_A = 1,000$	6,650	6,550	1.9	6,850 ± 270	$D_A = 1,040$
58	$D_A = 2,070$	7,690	7,510	3.2	7,840 ± 370	$D_A = 2,130$
59	$D_A = 3,300$	8,750	8,730	1.9	8,990 ± 360	$D_A = 3,400$
60	$D_A = 4,480$	9,040	9,040	2.3	9,310 ± 390	$D_A = 4,610$
61	$D_A = 5,090$	8,510	8,550	2.7	8,800 ± 390	$D_A = 5,240$
	2F/8.14d					Normalised values
62	$D_A = 1,010$	6,905	5,960	2.6	7,170 ± 320	$D_A = 1,040$
63	$D_A = 2,070$	7,870	7,860	1.0	8,090 ± 290	$D_A = 2,130$
64	$D_A = 3,350$	9,020	9,040	1.4	9,130 ± 350	$D_A = 3,450$
65	$D_A = 4,450$	9,460	9,410	1.7	9,690 ± 380	$D_A = 4,580$
66	$D_A = 4,980$	8,320	8,360	3.6	8,610 ± 430	$D_A = 5,130$
G						
67	S.D. (72')	5,610	5,620	1.3	5,900	Irrad. 1st day
68	S.D. (24')	5,060	5,070	3.3	5,320 ± 190	Irrad. 1st day
69	S.D. (25.5')	5,140	5,140	2.0	5,400 ± 130	Irrad. 28th day
70	9F/10d	16,220	16,380	5.0	17,190 ± 890	26' exp.
71	9F/28d	22,390	22,470	3.7	23,580 ± 920	26' exp.
72	14F/17d	21,280	21,340	2.4	22,400 ± 600	26' exp.
73	20F/25d	27,670	27,680	3.5	29,050 ± 1070	26' exp.
H						
74	S.D. (72')	5,805	5,840	2.9	5,900	Irrad. 1st day
75	S.D. (24')	5,180	5,180	3.7	5,230 ± 250	Irrad. 1st day
76	4F/3d	9,860	9,970	3.7	10,070 ± 470	26' exp.
77	6F/7d	13,390	13,390	3.9	13,530 ± 660	26' exp.

trols to study how the LD 50 varied with the duration of post-irradiation storage in plastic pots.

Group D. This group of experiments involving 6,000 weevils was designed to determine how the LD 50 varied with the duration of continuous irradiation over the range 1–106 h. Weevils were sub-divided into groups and irradiated to appropriate dose levels for 1.25, 6, 14, 22, 38, 54, 68, 84 or 106 h. These experiments were crucial in that they could not be repeated until a year later when the irradiation facility would next be available for such a long period. The excitement of these 106 h ($4\frac{1}{2}$ days), was enhanced by the possibility of an electrical power strike disrupting experiments. Fortunately, the only power disruption during those 106 h lasted only a few minutes and one of the workers involved happened to be awake and available to switch on the Cobalt Unit as soon as power was restored. This disruption of a few minutes is not considered important in irradiations lasting many hours, but nevertheless, should be reported.

Group E. This group of experiments involving 4,920 weevils was designed to answer questions raised by the results of earlier experiments. The experiments (Group B), to determine how the LD 50, using a constant number of fractions, increased with overall time, were designed with the object of obtaining information which would enable the daily fractionation data to be corrected for the effect of repopulation. These experiments showed that the rate of the repopulation effect (expressed as rad per day), decreased appreciably as the interval between fractions was increased. To determine the correction applicable to daily fractionation, the observed data needed to be extrapolated backwards to zero time interval and, when this was done, suggested that the initial rate of the repopulation effect was equivalent to about 800 rad/day. This was a larger correction than originally envisaged and unreliable in that the extrapolation had to be extended to a region likely to be greatly affected by radiation-induced mitotic delay. To reduce the amount of extrapolation, it was decided to extend the experiments to include fraction intervals of 8 h. This was done by determining the LD 50 for 4 fractions in 1 day and comparing this directly with the LD 50 for 4 fractions in 3 days. The results of earlier experiments also suggested that the position of a 3-day gap in irradiation carried out at 5 fractions per week, could influence the value of the LD 50. Anomalous results had occurred when 6 fractions were given with a week-end gap between the fifth and sixth fractions. Experiments were therefore included to find the LD 50 when 6 fractions were given in 7 days with a 3-day gap between the 3rd and 4th fraction and when 6 fractions were given in 5 days without a gap.

These experiments involved irradiating specimens every day for eight consecutive days. The opportunity was therefore taken to determine the LD 50 for 9 truly daily fractions (i.e., 9 fractions in 8 days).

Group F. This group of experiments involved 12,800 weevils. Additional information was required on the effect of repopulation occurring during very short fractionation intervals, and therefore included an experiment to determine the LD 50 for 4 fractions in 9 h.

The main object of this group of experiments was to test a new hypothesis developed from studying the results of earlier work to be discussed in section 7. This analysis suggested that during a fractionated regime of acute irradiation, the rate of the repopulation effect (expressed as rad/day), decreased exponentially with time. The magnitude of the total effect occurring during a multi-fraction regime and the rate of the exponential decrease, suggested indirectly that the instantaneous rate of repair (rad/day), might be approximately proportional to the damage present (expressed in rad). To test this simple hypothesis and to discover whether the rate of repair was a function of the damage present, experiments were designed to measure the rate of repair occurring after initial doses of different magnitudes.

Weevils were irradiated with an initial dose D_A rad and t_s days later subjected to a second dose D_B. By keeping D_A and t_s constant, and varying D_B, the total dose $(D_A + D_B)$ necessary to kill 50% of the weevils in 30 days was determined. In this way, the LD 50_{30} was determined for five values of D_A ranging from 1,000 to 5,000 rad and three values of t_s, namely 1.03, 3.83 and 8.14 days. The change in LD 50 with time interval enabled estimates to be made of the rate of repair occurring after each initial dose. From the above it will be seen that the LD 50 needed to be determined for 15 different irradiation regimes and this required a total of 12,000 weevils.

Groups G and H. These groups of experiments involving a total of 8,600 weevils were intended to be conducted as one group and were designed to study the effect of shortening the irradiation time during acute fractionated irradiation. The times of 71–79 min used in all the acute irradiation regimes referred to above, were longer than desirable as it was considered that an appreciable amount of sub-lethal damage recovery could have occurred during the course of each fraction. The installation of a new Cobalt-60 Unit enabled LD 50 single doses to be given in irradiation times of the order of 25 min. It was therefore decided, that all the daily fractionation experiments referred to in Group B should be repeated using the shorter irradiation time. The above regimes included 9 fractions in 10 days and, by also including 9 fractions in 28 days, it would be possible to assess the change that such a shortening would have on corrections for repopulation effects. Irradiated controls would be irradiated for both the long and the short irradiation times, thus enabling the single dose LD 50 to be determined under both conditions. Unfortunately, "The best laid schemes of mice (or weevils) and men gang aft agley and leave us nought but grief and pain for promised joy" [Burns, 1785]. The supply of Argentine Plate Wheat which had been used for all cultures up to this time was running low and fresh supplies were no longer available. It was therefore decided that Australian Wheat would need to be used for some of the experiments, and these would require their own irradiated and unirradiated controls. The regimes were therefore sub-divided into two batches — one to use Argentine Plate Wheat, the other Australian Wheat. These experiments were planned to run concurrently, but twelve days after being placed to oviposit on the cultures made from Australian Wheat, the weevils involved were found to be dead — presumably due to insecticide contamination. Because of this, the second batch of experiments (Group H), had to be delayed for one month. Fresh cultures were prepared for these experiments using Manitoba Wheat kindly supplied by the Pest Infestation Laboratory at Slough. The regimes employed in each batch are listed in table 1.

3. The results

3.1. Graphical estimation of the LD 50

The survival at Day 30 in each pot was corrected for natural mortality using the results of the appropriate unirradiated controls and the formula of Abbott [1925].

For each irradiation regime, the average corrected percentage survival at each dose level was plotted against the logarithm of dose on probability graph paper. The best fitting straight line (as judged by eye), was then drawn through the plotted points. The dose corresponding to 50% corrected survival on the eye-line will be referred to as the eye-line $LD\ 50_{(30)}$ and the values obtained are compared in table 1 with those obtained using a computer programme described below.

3.2. The estimation of the LD 50 values by probit analysis

For each irradiation regime, the probit regression line and the 95% fiducial limits of the LD 50 and slope were estimated using the method of maximum likelihood (Finney, 1952). This is an iterative method which starts by accepting a provisional regression line (usually the eye-line) as the basis of the calculations. This provisional line determines the weighting factors to be allocated to the individual results.

The first cycle of calculations enables a more accurate regression line to be found and this is used as the starting point for the next cycle. Provided the probits of the observed data plotted against the logarithm of dose lie close to a straight line, there is little difficulty in deciding where the regrsssion line should be drawn. Little change in the position of the regression line is produced by the first cycle of calculations and even less by the second.

The introduction of computer techniques enables a much larger number of cycles to be carried out without additional effort and eliminates any subjective errors which might otherwise be caused by the investigator's choice of eye-line.

The computer programme written in Basic Language consists of three parts. The first part determines the parameters of the most likely probit regression line, the second part performs a χ^2 test to determine how well the observed data fit this line and calculates the heterogeneity factor [Finney, 1952]. The third part calculates the 95% fiducial limits on both the LD 50 and the slope of the probit regression line taking into account the heterogeneity factor when significant.

The programme assumes originally that the provisional regression line is a horizontal straight line of probit value 5. By repeating the cycle of calculations the computed regression line is moved closer to the line of maximum likelihood. Ten such cycles were used routinely, but it was found from experience that four or five would probably have been sufficient as little or no change in the parameters of the regression line normally occurred after the fourth of fifth cycle.

All the observed data, including those for pots having zero or 100% survival, were used in obtaining the regression line. The χ^2 test is unreliable when applied to test the discrepancy between observation and prediction if, in any pot, the number of specimens expected to respond (or not to respond) is less than about five [Finney, 1952]. Thus in part 2 of the programme the results obtained with such pots were pooled with those of others irradiated to the same dose level, before applying the χ^2 test. If pooling all the data for the given dose level failed to raise the number to above 5, the programme ignored the data for that dose level in compiling the value of χ^2 applicable to that regime and a warning was printed stating that this was so. All such data were then examined to see whether the discrepancy between predicted and observed survival was abnormally large. The greatest difference found was for the 20 fraction in 25 days regime in which, at one of the six dose levels used, 24 weevils out of 200 died instead of the predicted 16. The latter number consisted of 12 which were expected to die from natural causes (by comparison with the unirradiated control group) and 4 which were expected to die from the radiation. The value of χ^2 for this individual result is 4.4. On statistical ground, a discrepancy of this magnitude would be expected to occur in 4% of the observations.

The 20 or 30 pots involved in each treatment regime had been uniformly distributed in the constant-temperature room which was known to have a temperature gradient from one side to the other. In calculating the 95% fiducial limits, it was essential to take into account any heterogeneity arising from this or other causes. Wherever χ^2 was found to be significant at the 95% confidence level, all variances were therefore multiplied by the heterogeneity factor as suggested by Finney [1952]. The third part of the computer programme performed the appropriate calculation and printed out the requisite 95% fiducial limits.

The programme was carried out on the Leasco Time Sharing System employing a Hewlett Packard 2000A Computer operated on-line from a Westrex terminal. The data were fed in using punched tape.

3.3. Comparison of graphical results with those obtained by computer

At least one of the present authors refuses to accept results provided by a computer unless they agree with those obtainable by himself using alternative and perhaps more old-fashioned methods.

The results of all the experiments are presented in table 1. Examination of this table shows that the maximum discrepancy between the eye-line LD $50_{(30)}$ and the computed LD $50_{(30)}$ for any regime is 1.1% (experiment No. 76). In general it will be seen that the spreads of the computed LD $50_{(30)}$ values (as specified by the 95% confidence limits), are less than ±4%.

In the penultimate column of table 1, the normalised values, which are referred to in all subsequent analyses, were obtained by multiplying the computed LD $50_{(30)}$ by 5,900 and dividing by the single dose LD $50_{(30)}$ obtained in the same group of experiments. When experiments involved prolonged fractionation, the single dose LD 50 was determined at the beginning and end of the group of experiments and an intermediate value (interpolated to Day '0' of the relevant fractionated regime) was used for the purpose of normalisation. Any error caused by this procedure was likely to be small as the variation observed in the single dose LD 50 between the beginning and end of a group of experiments was never more than 3%.

The 95% fiducial limits quoted in the penultimate column were obtained by taking the square roots of the sum of the squares of the limits of the computed LD 50 for the relevant regime and those of the corresponding single dose LD 50.

The results obtained by graphical analysis were usually obtained earlier than those obtained by the computer and were used to verify that the computer was producing sensible results. The computer on the other hand, could determine the 95% confidence limits for a regime in a matter of seconds, whereas both of the present authors required two hours and a desk calculating machine to determine these limits by alternative means.

4. A guide to the analysis of the results

Some readers may be interested in some of the weevil results but not have time to study all of them. This section is designed to give information to enable such readers to find the section of interest.

In section 5, the observed data are examined to check whether their reliability had been adversely affected by extraneous factors such as environmental changes during the two years of the experiments. Other factors such as the rather arbitrary choice of Day 0 and some of the technical factors involved, are also discussed.

In section 6, the split-dose recovery curve and the estimation of the half-life of the dose-equivalent of sub-lethal damage are discussed.

In section 7, the shape of the relationship between iso-effect dose and the overall irradiation time when a constant number of fractions is given in varying overall times, is examined. The results presented in this section are perhaps the most important and certainly the most interesting results emerging from these experiments.

In section 8 the variation in LD 50 with fraction number is examined both before and after correction for the effects of repopulation occurring during the course of a fractionated regime.

In section 9 the results of the experiments in Group F are discussed. These experiments involved determining the LD 50 for weevils given two doses of unequal magnitude, separated by different intervals of time. They were designed to test a new hypothesis (section 7), which suggested that the rate of repopulation (expressed in rad/day) is proportional to the damage present (expressed in rad).

In section 10, the results obtained in the experiments designed to determine how the LD 50 varied with the duration of low dose-rate irradiation, are examined. They are then compared with the variation in LD 50 (corrected for the effects of repopulation) with fraction number using acute irradiation. These results are then used to test the validity of the General Formula for Equating Protracted and Acute Regimes of Radiation which was the original aim of these experiments.

In section 11, the large body of accurate iso-effect data obtained by using large numbers of weevils are compared with those obtained (perhaps in less detail or with less certainty) using systems which are more relevant to the clinical radiotherapy situation. The similarity of the results in weevils and those in more relevant tissues, suggests that there are lessons to be learnt from the weevil experiments which could be important to radiobiologists designing experiments employing more relevant systems.

Whilst the weevil experiments cannot give quantitative guidance in a clinical situation, they may be of interest to the radiotherapist who wishes to query some of the iso-effect relationships frequently used in radiotherapy.

In section 12, an attempt is made to summarise the findings which appear to the authors to be of importance.

5. The reliability of the observed results

5.1. Variation in sensitivity between groups of experiments

The irradiated controls were used to determine the single dose LD $50_{(30)}$ for the weevils involved in each group of experiments and this was found to be 5900 rad ±5% for every group except those in Group A which had an LD $50_{(30)}$ of 6270 rad, (i.e., 6% high). The reason for this group being less sensitive than the average was considered to be mainly due to these weevils having been cultured on a shelf where the temperature had been 0.5–1°C higher than that on which all subsequent groups of weevils had been cultured.

Variation in the sensitivity of weevils between batches was corrected for by the normalising procedure referred to above (section 3.3).

5.2. Estimated error in observed LD $50_{(30)}$ due to choice of Day 0

Throughout the experiments, Day 0 had been defined (section 1.3) as the day which was three-quarters through the irradiation regime. If this somewhat arbitrary definition had been well chosen, then the interval between Day 0 and radiation death might be expected to be similar for all irradiation regimes. The survivals observed in 38 experiments (Ref. Nos 12–49) were analysed to see whether this was so. The survivors had been counted at approximately 5 day intervals between Day 7 and Day 37. For each regime, for the dose levels immediately above and immediately below the LD 50, graphs were drawn plotting survival against time. The interval between Day 0 (as defined above) and the day on which the radiation mortality was 50% of that occurring by Day 30, was determined from each of the 76 graphs. The intervals so determined were all found to lie between 6 and 18 days with 90% between 9.5 and 16.5 days. These results were examined to see whether there was any trend suggesting that this interval was dependent on the percentage surviving or the overall irradiation time. No consistent variation with percentage survival could be detected, but it was found that this interval tended to decrease from 16 days to 10 days as the overall irradiation time increased from 0 to 28 days. This suggested that the effective Day 0 and Day 30 for prolonged fractionated regimes may have occurred up to 6 days earlier than had been assumed, and that the effective Day 0 occurred closer to half through rather than three-quarters through the irradiation regime.

All the above regimes having overall irradiation times in excess of 10 days were examined to determine the difference between their corrected percentage survivals at 24 days and at 30 days. The maximum difference found was 2.7%. The regimes showing the greatest changes were then further examined to determine the difference between their LD $50_{(30)}$ and LD $50_{(24)}$. The maximum difference so found was 0.8%. It was therefore considered that no appreciable error in the determination of the LD $50_{(30)}$ had resulted from the choice of Day 0.

5.3. Variation of LD $50_{(30)}$ with age and pre-storage in plastic pots

Experiments Nos 12 and 13 and Nos 68 and 69 show that there is less than 3% difference in sensitivity between weevils which have been housed in small plastic post for approximately 4 days and those which have been so housed for 28 days longer and were 28 days older at the time of irradiation. Thus LD $50_{(30)}$ observed for prolonged fractionation regimes should be little affected by changes in sensitivity due to ageing and housing during the course of the regime. The method used to normalise the observed LD 50's (section 3.2) should, however, compensate for any slight changes due to this cause.

5.4. Variation of LD 50 with period of post-storage in plastic pots

Experiments Nos 27–30 show that as the period of post-irradiation storage in plastic pots is increased from 1 to 14 days, the observed LD $50_{(30)}$ decreases by approximately 400 rad (i.e., an average of 30 rad per day).

To eliminate this source of variation, weevils in subsequent experiments (Nos 31–77) were transferred from plastic pots to glass jars after the same period of post-irradiation storage as that employed in the main fractionation experiments (Nos 12–26), i.e., the weevils were transferred on Day 7. Thus in all the experiments except the split-dose experiments (group A), this possible source of error was eliminated. Any error occurring in the Group A experiments due to this cause should have been eliminated or greatly reduced by the normalising procedure (section 3.3).

6. The split-dose recovery curve for weevils

6.1. The observed shape

In experiments Nos 1–11, weevils were irradiated with two equal doses separated by intervals of 0–25 h. The LD $50_{(30)}$ and corresponding 95% fiducial limits determined for each of the split-dose intervals are presented in table 1.

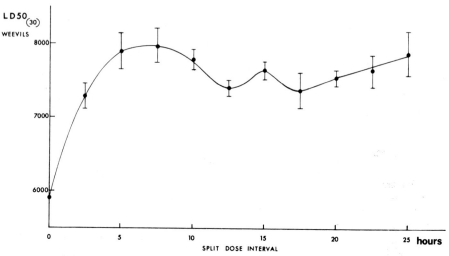

Fig. 6. The observed split-dose recovery curve. The total LD $50_{(30)}$ for weevils irradiated with two equal doses is plotted against the interval between doses. The limits shown are the 95% fiducial limits estimated by probit analysis as previously discussed (section 3.2).

The observed results, normalised in accordance with their appropriate single dose controls as previously described (section 3.3), are plotted against interval in Fig. 6. The resultant split-dose recovery curve exhibits similar characteristics to those obtained with mammalian cells and tissues. Before attempting to interpret this curve, the factors which influence the shape of such curves will be considered.

6.2. Factors governing the shape of split-dose recovery curves

Young and Fowler [1969] suggest that there are three factors which influence the shape of a recovery curve: intracellular recovery, progression and repopulation. Such curves generally show an initial rapid rise in the surviving fraction of cells which reaches a peak in a few hours and is considered to be due mainly to the repair of intracellular sub-lethal damage. The recovery curve may then show a number of peaks and troughs believed to be due to the progression through the division cycle of a partially synchronised population of cells surviving the first dose.

When the split-dose interval exceeds the mitotic delay induced by the first dose, the curves show a gradual rise due to repopulation. The first and second effects and the second and third effects overlap, and it is consequently difficult to analyse the relative effects of the three components.

Young and Fowler [1969] used a model incorporating experimental data for synchronous and asynchronous Chinese hamster cells [Sinclair, 1967, 1968], to investigate the effects of progression and repopulation on split-dose experiments. This model assumed that full intracellular recovery occurs before the second dose is given. Thus their predicted split-dose recovery curves showed how the surviving fraction of cells should vary with interval due to progression and repopulation, but could not show the rise due to intracellular recovery.

By applying the theory of Lajtha and Oliver [1961], Liversage [1969a] predicted how the surviving fraction of cells in Sinclair's split-dose experiment should vary due to the repair of intracellular sub-lethal damage in the absence of any other effect. He then combined the curve obtained by Young and Fowler [1969], which showed the effect of progression and repopulation with his own which showed the effects of repair of sub-lethal damage. This gave a curve demonstrating all three effects and which was in reasonably close agreement with the curve obtained experimentally by Sinclair.

6.3. Analysis of the split-dose recovery curve

Liversage [1973] used the model of Young and Fowler as a guide in interpreting the factors responsible for the shape of the split-dose recovery curve for weevils. He deduced that progression and radiation induced synchrony had little effect in weevils, causing a variation in LD 50 with split-dose interval of no more than ±4% as shown in fig. 6. He also deduced that radiation-induced mitotic delay inhibited repopulation from occurring during split-dose intervals of less than 17 h. Despite these advantageous circumstances, he found that a precise determination of the value of τ, the half-life of the dose-equivalent of sub-lethal damage was not possible.

6.4. Estimation of τ (the half-life of the dose-equivalent of sub-lethal damage) from the split-dose recovery curve

As discussed by Liversage [1973], a minimum value of τ may be found from a knowledge of the time taken to reach the first peak of the curve, whilst a maximum value may be obtained by considering the length of time required for the apparent recovery to be half-complete.

Immediately after the first irradiation, the surviving cells are in a resistant phase of the cycle. As they progress into a more sensitive phase, they will tend to reduce the split-dose LD 50 observed. The first peak in the LD 50 curve therefore occurs when the rate of decrease due to progression becomes equal to the rate of increase in the LD 50 due to the repair of sub-lethal damage. Thus the first peak occurs before sub-lethal damage repair is complete. Now, the repair of sub-lethal damage is virtually (99%) complete in a period of 7τ hours and the first peak in the split-dose LD 50 curve for weevils occurs at 7 h. It follows that τ must be more than 1 h.

Having estimated a lower limit for τ, it is now necessary to estimate an upper limit.

The first peak of the weevil recovery curve is 2,070 rad higher than the single dose LD $50_{(30)}$ of 5,900 rad suggesting that the total amount of recovery (expressed in rad) is 2,070 rad. However, as the repair of sub-lethal damage is not complete by 7τ hours, and as progression above is causing a fall in the LD 50 by 7 h, the true rise due to complete repair of sub-lethal damage is likely to be greater than 2,070 rad, perhaps even 200 or 300 rad more. The observed split-dose recovery curve rises by 1,035 rad, (i.e., half of 2,070 rad) in 1.8 h and by 1,170 rad, (i.e., half of 2,340 rad) in 2.0 h. Thus, as judged from the split-dose recovery curve, the repair of sub-lethal damage appears to be half complete in approximately 2 h. Oliver [1964] and Liversage [1969a] have shown that because radiation effect and dose are not linearly related, the half-life τ of the dose equivalent of sub-lethal damage is less than the time required for the recovery to be half complete as judged from an observed split-dose recovery curve. On the above reasoning, τ must be less than 2 h.

The split-dose recovery curve for LD $50_{(30)}$ for weevils suggests therefore that the half-life τ of the dose-equivalent of sub-lethal damage, for the critical cells in weevils, lies between 1 and 2 h when these are irradiated and housed at 27°C 70% R.H.

7. The relationship between iso-effect dose and overall time when using a constant number of equally spaced fractions

7.1. Existing iso-effect formulae

Several formulae which claim to relate iso-effect dose and overall treatment time in radiotherapy have been published [Ellis, 1967; Cohen, 1968; Liversage, 1971]. The shapes of the relationships described by these formulae differ from one another and it should be stressed that the exact form of the true relationship is not known. Clearly one would not expect that the numerical values of the parameters or constants used in any of the above formulae would be applicable to weevils but the weevil results are examined below to compare the shapes of the relationship predicted by each type of formula with that found in weevils.

7.2. Attempt to fit results to an Ellis-type formula

Ellis [1967] claims that iso-effect doses are proportional to N^A and T^B where N is the number of fractions, T is the overall time and A and B are constants.

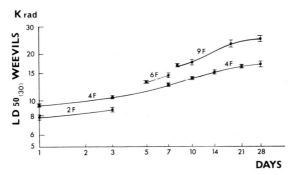

Fig. 7. The LD $50_{(30)}$ (with 95% fiducial limits) for weevils plotted on a double logarithmic scale against overall irradiation time for fractionated regimes involving 2, 4, 6 or 9 fractions.

If the LD 50 for weevils varies with overall time T in this way, then for any constant number of fractions N, the LD 50 values plotted against overall time T on a double logarithmic scale should lie on a straight line of slope B.

With the exception of experiment No. 16, the results of all the experiments involving 2, 4, 6 or 9 fractions, each of 71–79 min duration, together with their 95% fiducial limits are plotted in this way (fig. 7). Experiment No. 16 (6 fractions in 7 days) was excluded as it involved a 3-day gap between the 5th and 6th fraction. The LD 50 observed in this experiment is believed to be greatly influenced by the position of the gap and will be discussed more fully later (section 9).

Examination of fig. 7 shows that the exponent B of time is not constant but depends on both N and T. Between 10 days and 28 days, the 9-fraction data have an average slope of 0.34 ± 0.07 whilst the 4-fraction data have an average slope of 0.205 ± 0.055. Between 1 day and 3 days, the 4-fraction data have an even smaller slope of 0.115 ± 0.04.

It may be concluded that the Ellis [1967] type relationship does not apply to LD 50 observations on weevils.

7.3. Attempt to fit results to the Cohen-type formula

The formula of Cohen [1968] implies that keeping the number of fractions constant but increasing the overall time, increases the iso-effect dose by C times the increment in overall time, where C is considered to be a constant.

The data plotted in fig. 7 are replotted on a linear-linear scale in fig. 8. Examination of this diagram shows that for a constant number of fractions, the LD 50 for weevils is not linearly related to overall time, e.g., from the 4-fraction data it will be seen that the LD 50 increases by 630 rad per day as

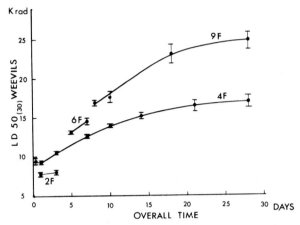

Fig. 8. The LD $50_{(30)}$ (with 95% fiducial limits) for weevils plotted on a linear-linear scale against overall time of irradiation for fractionated regimes involving 2, 4, 6 or 9 fractions.

the overall time is increased from one to three days but only by 77 rad per day as the overall time is increased from 21 to 28 days. Clearly the value of C is not a constant but decreases as the overall irradiation time increases.

It may be concluded that the Cohen [1968] type relationship does not apply to LD 50 observations on weevils.

7.4. Attempt to fit results to the Liversage-type formula

Liversage [1971] using published data for skin reactions, suggested that for a constant number of fractions, the rate of increase in iso-effect dose would decrease as the irradiation regime was prolonged. He suggested that the rate of increase C was a variable which decreased as $(T - T_F)$ was increased and could be considered as a function of $(T - T_F)$ where T is the overall time used and T_F is the time required to give the same number of fractions at five fractions per week.

In the present experiments, data were obtained for 2, 4, 6 and 9 fractions given strictly daily and the fomula of Liversage may be re-written

$$D_{NT} - D_{NT_D} = C(T - T_D) \qquad (4)$$

where D_{NT} is the LD 50 for weevils irradiated with N fractions in T days and T_D is the time in days required to give N strictly daily fractions, i.e., $T_D = (N - 1)$ days. The parameter C should now be a variable which decreases as $(T - T_D)$ increases.

To test whether the weevil data fitted such a relationship, all the data

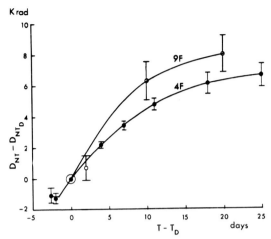

Fig. 9. The increase in LD $50_{(30)}$ on changing from N daily fractions to N fractions in T days plotted against the corresponding increase in overall time for fractionated regimes involving 4 or 9 fractions. (D_{NT} is the LD 50 when N fractions are delivered in T days and T_D is the time required to give N daily fractions.)

included in figs. 7 and 8 were analysed and values of $(D_{NT} - D_{NT_D})$ were plotted against $(T - T_D)$ in fig. 9.

The results show that the value of C does decrease as $(T - T_D)$ is increased. Unfortunately, it will also be observed that C is a function of the number of fractions N. For example, on changing from 9 fractions in 8 days, (i.e., daily fractions) to 9 fractions in 18 days, the LD 50 increases by 6,360 rad, i.e., 636 rad per day. On changing from 4 fractions in 3 days to 4 fractions in 13 days, fig. 9 shows that the LD 50 increases by only 4,500 rad, i.e., only 450 rad per day even though $(T - T_D)$ is 10 days in both cases.

It may be concluded that the formula of Liversage [1971] (eq. 4) does not apply to LD 50 observations on weevils as the value of C varies not only with the time increment $(T - T_D)$, but also with the number of fractions N. The problem is discussed further below.

7.5. The search for the true relationship

7.5.1. The importance of fraction interval

Examination of fig. 9 suggests that the increase in LD 50 with increase in overall time is not just a function of the increment in overall time, but is also some function of fraction interval t_s.

The average amount of repopulation (expressed in rad) occurring between consecutive fractions t_s days apart, minus that between consecutive fractions

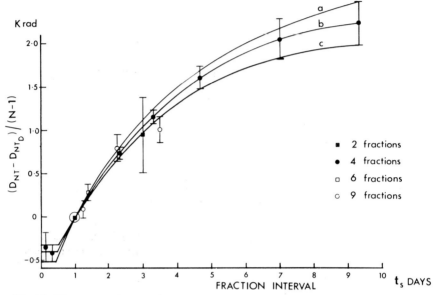

Fig. 10. The average increase in iso-effect dose per fraction interval due to cellular repopulation between consecutive fractions appears to be principally a function of fraction interval t_s. The most likely relationship is represented by curve (b) and its upper and lower limits by curves (a) and (c).

one day apart will be $(D_{NT} - D_{NT_D})/(N-1)$. The numerical values of this expression are plotted against fraction interval t_s (fig. 10) for the same experimental data as those used in the previous three diagrams.

Irrespective of the number of fractions employed, it will be seen from fig. 10 that all the plotted data lie close to a single curve (b). The maximum and minimum limits of the relationship are represented by curves (a) and (c). It may be concluded that for any irradiation regime involving 2–9 fractions and killing 50% of the weevils, the average amount of repopulation per fraction interval is principally a function of that interval and is practically independent of N and T.

Before attempting to use the data in fig. 10 to estimate the effect of repopulation during irradiation regimes in general, a close scrutiny of the data is called for.

A quantity of radiation $^r D_T$ has previously been defined (section 1.1) as that quantity of radiation necessary to compensate for the effects of repopulation occurring during a fractionated regime involving N fractions delivered in an overall time of T days. Thus $^r D_T$ is a measure of the amount of repopulation (expressed in rad) occurring during such a regime.

Using this terminology, the amount of repopulation per fraction interval will be $^rD_T/(N-1)$ and this should be equal to the increase in $(D_{NT} - D_{NT_D})/(N-1)$ as the fraction interval is increased from 0 to t_s days, as shown in fig. 10. The curves in this diagram have been drawn with a discontinuity occurring at approximately 0.5 days. The only data enabling the position of this discontinuity to be located, are data obtained for regimes involving 4 fractions. For regimes involving different numbers of fractions, the discontinuity could occur earlier or later than shown in fig. 10.

The discontinuity is believed to result from radiation-induced mitotic delay preventing repopulation occurring for a period t_m days after each fraction of radiation. Using Chinese hamster cells, Elkind et al. [1963], showed that mitotic delay was directly proportional to the dose delivered. Thus, t_m is likely to vary with fraction size.

It may be concluded that if the data in fig. 10 are to be used to estimate the total repopulation occurring during any fractionated regime, due allowance must be made for the influence of mitotic delay and how this is likely to vary with fraction size. These effects are discussed below.

7.5.2. The effects of mitotic delay

In an attempt to estimate the effects of mitotic delay the LD $50_{(30)}$ was determined for 4 fractions in 3 days, 4 fractions in 1 day and 4 fractions in 9 h; i.e., for fraction intervals of 1, 0.33 and 0.125 days (Experiments Nos 46, 45 and 51). The computed LD 50's observed were 1,0520 ± 2.4%, 9,260 + 2.1% and 9,260 ± 2.9%. Thus, no difference was observed in the LD 50 for intervals of 0.33 and 0.125 days. (After normalising in accordance with their respective single dose controls, the LD 50 for $t_s = 0.125$ days was found to be 2% higher than that for $t_s = 0.33$ days but this difference is not significant.)

The LD 50's for $t_s = 1$ day and for $t_s = 0.33$ days were determined in the same group of experiments and may therefore be compared without normalising. The difference between them is 1,260 ± 330 rad, i.e., 420 ± 110 rad per interval. Thus, on average, this amount of repopulation occurs after each daily fraction of 2,630 rad. By extrapolating curve (b), as drawn in fig. 10, it will be observed that repopulation is required only for the last 0.52 days of each daily period in order to obtain 420 rad per day. It is therefore estimated that a dose of 2,630 rad produces a mitotic delay t_m of 0.48 days, i.e., 0.18 days/1000 rad.

As the mitotic delay t_m will vary with the dose delivered per fraction, the total amount of repopulation occurring during any fractionated regime is more likely to be $(N-1)$ times some function of $(t_s - t_m)$ rather than $(N-1)$ times some function of t_s. The experimental data presented in fig. 10 will therefore be reconsidered below.

7.5.3. The shape of the relationship between repopulation and fraction interval

The mitotic delay t_m for each of the 4-fraction regimes has been calculated assuming the delay to be 0.18 days/1,000 rad. The values of $(D_{NT} - D_{NT_D})/(N-1)$ together with their 95% confidence limits have been replotted against $(t_s - t_m)$ in fig. 11. It is considered that, at the 95% level of confidence, the curve relating these two factors is likely to lie between curves (a) and (c). Curve (b) was calculated from a mathematical relationship derived below.

In fig. 11, the slope of the curve passing through the experimental points at time $(t_s - t_m)$ is equal to the average instantaneous rate of repopulation M (expressed in rad/day) occurring t_s days after each of the N fractions except the last. (N.B. It is necessary to insert the word "average" here as the rate of repopulation may vary from one fraction interval to the next.)

The slopes of curves (a) and (c) in fig. 11 have been plotted against $(t_s - t_m)$ on a logarithmic-linear scale in fig. 12 to obtain upper and lower limits to the average instantaneous rates of repopulation and to show how this decreases with time. It will be observed that, to within the limits of

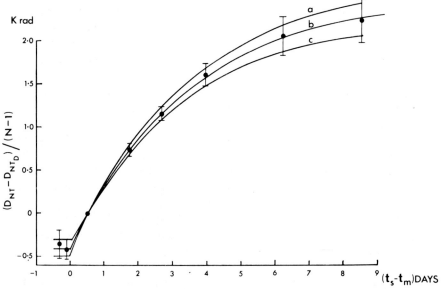

Fig. 11. The average increase in iso-effect dose per fraction interval as the interval between fractions is increased from 1 day to t_s days plotted against $(t_s - t_m)$ where t_s is the fraction inverval and t_m is the mitotic delay after each fraction. Curves (a) and (c) represent the upper and lower limits of the experimental data. Curve (b) was calculated from eq. (6) and is in excellent agreement with the plotted 4-fraction experimental data.

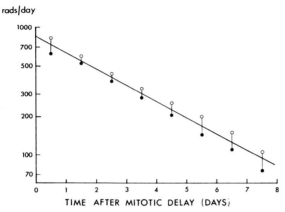

Fig. 12. The average instantaneous rate of repopulation (expressed in rad/day equivalent) during the interval between fractions plotted on a logarithmic-linear scale against time after mitotic delay. The upper limits (open circles) and the lower limits (closed circles) were obtained by measuring respectively the slopes of curve (a) and curve (c) in fig. 11 at $(t_s - t_m)$ days.

experimental accuracy, the average instantaneous rate of repopulation M, t days after the mitotic delay, decreases exponentially with time and may be represented by the equation

$$M = M_0 \exp(-\lambda t) \qquad (5)$$

where M_0 is the average instantaneous rate of repopulation (expressed in rad/day) immediately following the mitotic delay period and λ is the exponential decay constant.

It follows that the total repopulation ^rD_T (expressed in rad) occurring during a fractionated regime delivering a total dose D_{NT} rad in N equally spaced fractions in an overall time of T days will be given by:

$$^rD_T = (N-1) \int_0^{(t_s - t_m)} M \, dt$$

$$= (N-1) \frac{M_0}{\lambda} [1 - \exp(-\lambda(t_s - t_m))], \qquad (6)$$

where t_s is the interval between fractions (i.e., $T/(N-1)$ days) and t_m is the mitotic delay estimated to be $0.18 \, D_{NT}/1{,}000 \, N$ days for the critical cells of the weevil.

From fig. 12 it is estimated that $M_0 = 850$ rad/day and $\lambda = 0.29$ days^{-1}.

(The time required for the rate of repopulation to fall to half its initial value is thus 2.4 days.)

In order to verify that the above relationship does in fact fit the shape of the experimental data from which it is derived, eq. (6) was used to calculate how the expression $(D_{NT} - D_{NT_D})/(N - 1)$ should vary with $(t_s - t_m)$. The results of these calculations are plotted (curve b) in fig. 11. The precise agreement at all points on the curve shows that the above mathematical relationship certainly has the correct shape. Had the best smooth curve passing through the experimental points been drawn on this diagram, it would have been practically indistinguishable from the curve calculated from eq. (6).

The total repopulation $^rD_T/(N - 1)$ per fraction interval t_s is equal to the increase in $(D_{NT} - D_{NT_D})/(N - 1)$ as the fraction interval is increased from t_m to t_s days. For any irradiation regime this may be estimated from curve (b) in fig. 11. The 95% confidence limits of this estimate may be obtained by considering the corresponding increases in curves (a) and (c). For convenience, the increases in these curves have been replotted against $(t_s - t_m)$ in

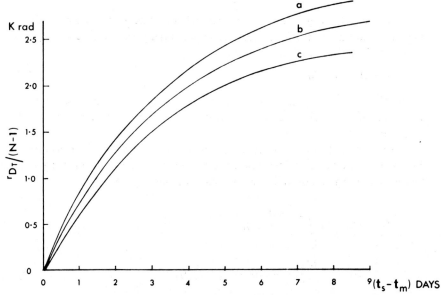

Fig. 13. The total repopulation (rad-equivalent) per fraction interval plotted against $(t_s - t_m)$ where t_s is the fraction interval and t_m is the mitotic delay after each fraction. The curves were obtained from fig. 11 by transforming the origin of the co-ordinates. Thus curve (b) represents the estimated repopulation as calculated from eq. (6) whilst curves (a) and (c) are the experimentally derived 95% confidence limits of this estimate.

Table 2
Values of $(D_{NT} - {}^rD_T)$ i.e., the LD 50 for N fractions corrected for the effects of repopulation occurring during the overall irradiation time T. The method of correction is described in section 7.5.3

N	T days	$(D_{NT} - {}^rD_T)$ rad
2	0.104	7,280 ± 170
	0.208	7,900 ± 240
	0.313	7,970 ± 230
	0.417	7,800 ± 130
	0.521	7,410 ± 110
	0.625	7,660 ± 110
	0.729	7,340 ± 240
	0.833	7,430 ± 110
	0.938	7,450 ± 230
	1.042	7,630 ± 310
	3.0	7,350 ± 360
4	0.375	9,540 ± 430
	1.0	9,340 ± 250
	3.0	9,370 ± 400
	3.0	9,210 ± 320
	7.0	9,160 ± 490
	10.0	9,160 ± 650
	14.0	9,280 ± 780
	21.0	9,280 ± 1,080
	28.0	9,120 ± 1,210
6	5.0	10,790 ± 530
	7.0	11,080 ± 770
9	8.0	12,950 ± 800
	10.0	12,350 ± 1,140
	18.0	13,890 ± 1,680
	28.0	11,500 ± 1,740
14	17.0	14,750 ± 1,660
20	25.0	17,750 ± 2,570

fig. 13, which has been used to estimate the total repopulation correction ${}^rD_T$ for each of the fractionated regimes investigated.

Values of $(D_{NT} - {}^rD_T)$ i.e., the LD 50 for N fractions if the repopulation effect is eliminated are tabulated in table 2. The consistency of the values obtained for a given number of fractions but for different overall times, provides supporting evidence for the validity of the derived mathematical relationship (eq. 6).

7.6. Discussion

It should be stressed that the above relationship was derived from experiments involving fractions of equal magnitude. In general, these were as equally spaced as a five day week would allow. The most unequal spacing occurred in regimes involving 5 fractions per week as a 3-day gap occurred at each weekend. For this reason, the 6-fraction and 9-fraction regimes were repeated using strictly daily fractions.

The 3-day gaps occurring in the 5-fraction per week experiments were distributed symmetrically about the mid-point of the regimes employed except for one experiment (No. 16). In this experiment, involving 6 fractions in 7 days, a 3-day gap occurred between the 5th and 6th fractions and the LD $50_{(30)}$ was found to be 13,410 ± 460 rad. When the experiment was repeated but with the 3-day gap between the 3rd and 4th fraction (experiment No. 47), the LD $50_{(30)}$ was found to be 1,200 rad higher (i.e., 14,600 ± 510 rad).

In deriving the above relationship, experiment No. 16 was ignored on the grounds that this result was greatly influenced by the presence of a gap so close to the end of the regime. This result will be further discussed in section 9 in the light of additional experimental evidence presented in the same section.

Although eq. (6) is more complex than any of the empirical iso-effect formulae previously referred to, it is based on the simple eq. (5) which shows that during each interval, the instantaneous rate of repopulation decreases exponentially with time. This experimental finding suggests that the rate of repopulation may be directly proportional to the damage present, provided both entities are expressed in terms of their dose equivalents.

To satisfy such a hypothesis, the repopulation occurring during an infinitely long interval must be equal to the damage present at the start of that interval. For this to be true, the numerical value of M_0/λ in eq. (6) must be equal to the average damage (expressed in rad) present at the start of each fraction interval of any regime which kills 50% of the weevils. For any such regime, irrespective of the number of fractions or the overall time employed, the ultimate damage is equivalent to that produced by a single dose of 5,900 rad. Thus, during the whole of any such regime, the average damage present will be approximately 2,950 rad which is in close agreement with the value of 2,920 rad found for M_0/λ (M_0 = 850 rad/day; λ = 0.29 days^{-1}).

The above analysis is full of approximations and is complicated by the fact that damage is not linearly related to dose. Nevertheless, the agreement is encouraging and suggests that the hypothesis is worthy of further investigation and consideration. To develop the concept more fully, a dose-response curve for the critical cells of the weevil is required. This is derived in section 8. The hypothesis is discussed further in section 9, together with the results of experiments designed to investigate the relationship between the rate of repopulation and the damage present.

7.7. Conclusion

The LD $50_{(30)}$ for weevils varies with overall irradiation time in a way which is not consistent with published iso-effect formulae purporting to estimate such variation for normal tissues in patients or skin reactions in animals.

The total repopulation ^rD_T (expressed in rad) occurring during a fractionated regime delivering a total dose D_{NT} rad in N equally spaced fractions in an overall time of T days has been found to be consistent with the following relationship:

$$^rD_T = (N-1)\frac{M_0}{\lambda}[1 - \exp(-\lambda(t_s - t_m))], \qquad (6)$$

where t_s is the interval between fractions and t_m is the mitotic delay estimated to be equal to $0.18\, D_{NT}/1{,}000\, N$ days. It is estimated that $M_0 = 850$ rad/day and $\lambda = 0.29$ days^{-1}.

For any regime killing 50% of the weevils, the average repopulation occurring per fraction interval will be $^rD_T/(N-1)$ and may be calculated from the above formula or may be determined, together with appropriate 95% confidence limits, from the experimentally derived curves presented in fig. 13. Curve (b) on this diagram is the curve obtained by calculation using the above formula and is centrally situated between the upper and lower limits of the experimental data (curves a and c) thus demonstrating the excellent agreement between the formula and the experimental data.

In section 1.1, the dose "wasted" due to intracellular recovery ^RD_N and the dose necessary to compensate for repopulation ^rD_T were mathematically related by the simple eq. (1):

$$D_{NT} = D_1 + {^RD_N} + {^rD_T}. \qquad (1)$$

In order to test the general formula for equating protracted and acute regimes of radiation, the original objective of these experiments, it is necessary to find how the dose "wasted" due to intracellular recovery varies with fraction number. This may now be determined by applying eqs. (1) and (6) to the fractionated LD 50 results. The appropriate analysis is presented in section 8.

The finding that the rate of repopulation decreases exponentially with time between consecutive fractions, suggests that a simple relationship may exist between the rate of repopulation and the damage present. The results of experiments designed to discover whether such a relationship exists, are presented and discussed in section 9.

8. The variation in LD 50 with fraction number

8.1. The difficulty of separating effects due to fraction numbers from those due to overall time

Fowler [1971] reviewed animal experiments designed to determine how iso-effect doses vary with the number of fractions and overall time. He pointed out that few experiments have been carried out in which the effects of overall time and fraction number were investigated separately.

The present investigation demonstrates that it is not always simple to design experiments which will do this completely. The results obtained with weevils show that the increase in LD $50_{(30)}$ with fraction number using a constant overall time is, in fact, a function of the overall time chosen.

It has been shown (section 7) that in any irradiation regime, the additional dose necessary to compensate for the repopulation occurring during the regime is a function not only of the overall irradiation time, but also of the interval between fractions. Thus, irradiation regimes employing a constant overall irradiation time but different numbers of fractions, require unequal additional doses to compensate for repopulation. The reason for this is that, when a small number of large fractions are delivered in a given overall time, the interval between fractions is large, and from fig. 12 it will be seen that the rate of repopulation decreases as the interval increases. Thus, the total amount of repopulation (expressed in rad) during the overall time is less than that occurring when a large number of small fractions are used. In the case of weevils, both may be estimated from eq. (6).

The Ellis [1967] Formula is of the form

$$D_{NT} = D_1 \times N^A \times T^B \tag{7}$$

and implies that if the overall time of irradiation is kept constant and N varied from 4 fractions to 9 fractions, the iso-effect dose should increase by a factor of $(9/4)^A$. This factor should be independent of the overall time. Results of experiments Nos 17, 21, 24 and 26 (table 1), show that this factor changes from 1.27 ± 0.04 to 1.46 ± 0.08 as the overall time is increased from 10 to 28 days. (The limits quoted are the 95% fiducial limits.)

It may be concluded that in iso-effect experiments, designed to study the effect of fraction number only, the influence of overall time is not necessarily eliminated simply by keeping it constant. The implications of this finding are discussed further in section 11.5.

8.2. Variation in the LD 50 with the number of daily fractions

The LD 50 was determined for 2, 4, 6 and 9 fractions using strictly daily fractions and for 6, 9, 14, and 20 fractions delivered at the rate of five fractions per week. The results (after normalising in accordance with their single dose controls) are plotted on a linear scale against fraction number (fig. 14). The results of all the experiments in which fractions were of 71–79 min duration are included except the result of experiment No. 16. The latter was excluded because of the influence of the 3-day gap between the last 2 fractions as previously discussed (section 7.6).

It will be observed from fig. 14 that from 1 to 2 daily fractions, the LD 50 increases by 1,900 rad (i.e., 32%), but from 2 to 20 "daily" fractions the increase per fraction is practically constant at 1,350 rad per "daily" fraction. (The term "daily" in this context includes regimes in which five fractions were given per week.)

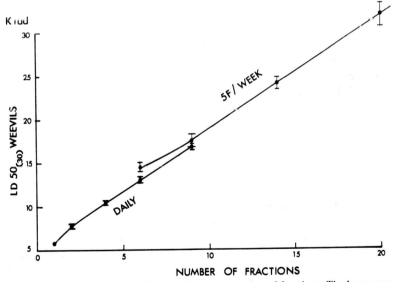

Fig. 14. The LD $50_{(30)}$ for weevils plotted against number of fractions. The lower curve is for strictly daily fractionation, (e.g. 9 fractions in 8 days) whilst the upper curve is for 5 fractions per week (e.g. 9 fractions in 10 days).

8.3. The increase in LD 50 due to intracellular recovery between fractions

As previously discussed (section 7.7), eq. (6) may be applied to any irradiation regime to estimate the magnitude of the repopulation component ^rD_T. The intracellular recovery component ^RD_N may then be determined by substituting the estimated value of ^rD_T into eq. (1).

$$D_{NT} = D_1 + {}^RD_N + {}^rD_T, \qquad (1)$$

where D_{NT} is the LD $50_{(30)}$ for weevils irradiated with N fractions in an overall time of T days and D_1 is the single dose LD $50_{(30)}$.

The intracellular recovery component ^RD_N is a function of N but is assumed to be independent of T provided the interval between fractions is sufficiently large to allow full intracellular recovery between fractions. The split-dose recovery curve (fig. 6) indicates that such recovery is essentially complete in an interval of 0.3 days.

The value of ^RD_N for any regime may be estimated from eq. (1) by using experimental values of D_{NT} and D_1 and estimated values of ^rD_T calculated from eq. (6). The most reliable estimates of ^RD_N are obtained when ^rD_T is small as the 95% confidence limits on the value of ^rD_T will then also be small (fig. 13). Thus, to obtain reliable estimates of the effects of complete intracellular recovery between N fractions, the fraction interval should be small but greater than 0.3 days. Five fractions per week data are available for all the different numbers of fractions investigated and these employ average fraction intervals of 1.4 days or less. To estimate how the LD 50 varies with fraction number, it is necessary to include the 5 fraction per week data as these are the only data available for 14 and 20 fractions.

In view of the above considerations, the estimated values of $(D_{NT} - {}^rD_T)$ obtained for all regimes employing fraction intervals of between 0.3 and 1.5 days, were abstracted from table 2 (section 7.6). For each fraction number, the average of these values is given in table 3 and plotted against fraction number in fig. 15. These values are in fact, the same, to within ±0.5%, as would have been obtained using all the data presented in table 2.

Fig. 15 shows how the LD 50, corrected for repopulation, increases with the number of fractions whilst fig. 14 shows how the LD 50 uncorrected for repopulation increases with the number of daily fractions.

The increase in LD 50 with the number of daily fractions is mainly due to two effects; intracellular recovery and repopulation. As previously discussed (section 6.2), synchrony may also affect the observed LD 50, especially when the number of fractions is small. However, examination of the split-dose recovery curve (fig. 6) reveals that even when only two fractions are given, synchrony only produces a variation of ±4% in the observed LD 50. It is believed that the effects of synchrony on the two-fraction data have been

Table 3
Average values of $D_{NT} - {}^rD_T$ for each fraction number

Number of fractions N	LD 50 − Repopulation correction $(D_{NT} - {}^rD_T)$ rad
1	5,900
2	7,590 ± 180
4	9,310 ± 320
6	10,940 ± 650
9	12,650 ± 970
14	14,750 ± 1,660
20	17,750 ± 2,570

Each of the above fractions were delivered in a period of 71–79 min. The derivation of this table is described in section 8.3 and the repopulation correction in section 7.5.3.

largely eliminated by averaging the eight values of $(D_{NT} - {}^rD_T)$ presented in table 2 for time intervals varying from 0.3 to 1.04 days. When more than two fractions are employed, the effects of synchrony are likely to be appreciably less than ±4% and have been neglected in this analysis. The averaging proce-

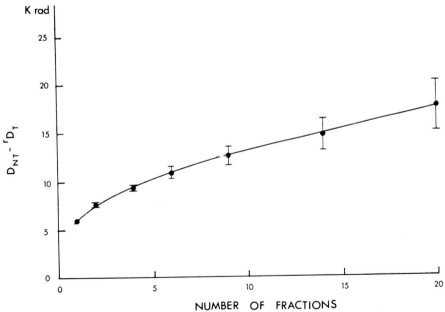

Fig. 15. The LD $50_{(30)}$ for weevils, corrected for repopulation between fractions plotted against the number of fractions. This curve shows how the LD $50_{(30)}$ increases with the number of fractions if complete intracellular recovery but no repopulation occurs between fractions. The plotted data are taken from table 3.

dure, though less effective in these cases, has probably reduced any small error due to this cause.

By comparing figs. 14 and 15, the relative importance of intracellular recovery and repopulation may be estimated. Fig. 14 shows that from 2 to 6 fractions the combined effects of intracellular recovery and repopulation increase the LD 50 by an average of 1,340 rad per daily fraction. From fig. 15, it will be seen that of this 1,340 rad, 840 rad result from intracellular recovery and the remaining 500 rad from repopulation. From 9 to 20 fractions the total increase averages 1,310 rad per daily fraction of which 460 are due to intracellular recovery and 850 rad due to repopulation. It may be concluded that in this biological system, when the number of "daily" fractions is less than six, intracellular recovery has more effect than repopulation, but that the relative importance of these two effects is reversed when the number of daily fractions is greater than nine.

8.4. The effect of reducing the irradiation time per fraction

All the fractionated results analysed previously in this and earlier sections were obtained from experiments in which each fraction of irradiation had been delivered in a period of 71–79 min.

The main objective of this investigation is to test the validity of the general formula [Liversage, 1969] for equating protracted and acute regimes of radiation discussed earlier (section 1.1). In deriving this formula Liversage [1969] assumed that the acute irradiation was delivered at a sufficiently high dose-rate for its effects to be dose-rate independent, i.e. he assumed that the amount of sub-lethal damage repaired during each fraction was negligible compared with the amount produced.

It has been estimated (section 6.4), that the half-life of the dose equivalent of sub-lethal damage for the critical cells of the weevil is between 1 and 2 h. Thus an irradiation time of 71–79 min would allow appreciable repair of sub-lethal damage to occur during each fraction. To test the validity of the general formula, fractionated experiments should therefore be done using shorter irradiation times.

As previously mentioned, the installation of a new cobalt unit enabled the irradiation times to be shortened to between 24 and 26 min. The regimes listed in table 4 were therefore repeated using these shorter irradiation times. Single dose controls using an irradiation time of 72 min were included in each batch of experiments and this enabled the LD 50's observed to be normalised in the usual manner.

The observed and normalised LD $50_{(30)}$'s for each regime are presented in the comprehensive list of results (Experiment Nos 67–77) in table 1. The results obtained with 24–26 min exposures are compared with those ob-

Table 4
The reduction in the LD 50 due to reducing the irradiation time per fraction from approximately 75 min to approximately 25 min.

Regime	Experiment No.	Normalised LD 50 71–79 min irradiations	Experiment No.	Normalised LD 50 24–26 min irradiations	Ratio of LD 50's
Single dose	67	5,900	68	5,320 ± 190	1.11
Single dose	67	5,900	69	5,400 ± 130	1.09
Single dose	74	5,900	75	5,230 ± 250	1.13
4F/3d	15	10,440 ± 200	76	10,070 ± 470	1.04
4F/3d	46	10,600 ± 310	76	10,070 ± 470	1.05
6F/7d	47	14,600 ± 510	77	13,530 ± 660	1.08
9F/10d	17	17,680 ± 730	70	17,190 ± 890	1.03
14F/17d	18	24,240 ± 730	72	22,400 ± 600	1.08
20F/25d	19	32,130 ± 1,350	73	29,050 ± 1,070	1.11
9F/28d	26	25,080 ± 1,130	71	23,580 ± 920	1.06
Mean ratio					1.08

Table 5

Number of fractions N	LD 50 − Repopulation correction ($D_{NT} - {}^rD_T$) rad
1	5,320 ± 190
4	8,890 ± 540
6	10,150 ± 870
9	12,210 ± 1,250
14	13,360 ± 1,560
20	15,490 ± 2,460

Each of the above fractions were delivered in a period of 24−26 min. The derivation of this table is described in section 8.4.

tained with 71−79 min exposures in table 4 from which it will be seen that the latter are on average 8% higher than the former. The single dose controls were irradiated at approximately 230 rad/min and 78 rad/min respectively. The results show that dose-rate effects can be appreciable at dose-rates in excess of 78 rad/min when the effect being studied is produced by a fairly large single dose of 5,000 to 6,000 rad which consequently requires an irradiation time in excess of one hour.

To test the validity of the general formula referred to above, it was necessary to determine how the LD 50 corrected for repopulation varied with fraction number. Thus, for each of the regimes employing 24−26 min exposures, it was necessary to estimate the repopulation component ${}^rD_T$. This was done as described previously (section 8.3), but because the efficiency of the radiation was 8% higher than with 71−79 min irradiations, M_0 in eq. (6) was assumed to be 850/1.08; i.e., 787 rad/day. The values of $(D_{NT} - {}^rD_T)$ so obtained for different numbers of fractions, are tabulated in table 5. In section 10 these results are compared with predictions based on the application of the general formula to data obtained using continuous low dose-rate irradiation.

8.5. The derivation of a dose-response curve for the critical cells of the weevil

Following the method of Fowler and Stern [1963], let it be assumed that a weevil dies within 30 days of irradiation due to some unknown critical population of cells being reduced to a surviving fraction of b^{-20}, where b is some unknown but constant number.

The fraction of cells surviving a single dose of $\frac{1}{N}(D_{NT} - {}^rD_T)$ should therefore be $b^{-20/N}$ as N such doses delivered at intervals too small to permit any repopulation, but sufficiently large to allow full intracellular recovery between fractions, would kill 50% of the weevils. The data tabulated in table 3 may thus be used to obtain a theoretical cell survival curve (fig. 16) for doses delivered in 71−79 min. The absolute level of survival is not ob-

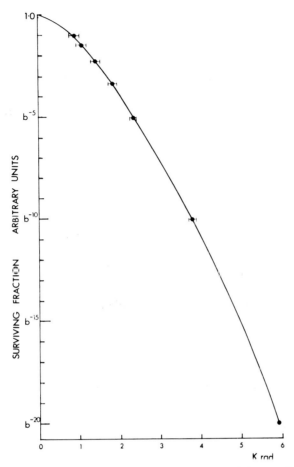

Fig. 16. Theoretical cell survival curve for the critical cells of the weevil derived as described (section 8.5) from the results of fractionation experiments.

tainable from this curve, as the value of b is unknown.

Liversage [1973], using the slopes of the probit mortality curves for weevils and the model of Lange and Gilbert [1968], estimated that the D_0 value for this survival curve is less than 535 rad and that to kill 50% of the weevils, their critical cell populations need to be reduced by a factor in excess of 1,000.

8.6. Conclusions

The LD $50_{(30)}$ for weevils increases with the number of "daily" fractions, the increase is practically constant at 1,350 rad per fraction. It is estimated that from 2 to 6 fractions, 63% of this increase is due to intracellular recovery and 37% to repopulation. From 9 to 20 fractions, 35% of the increase is estimated to be due to intracellular recovery and 65% to repopulation.

When weevils are irradiated with 1 to 20 fractions, shortening the irradiation time per fraction from 75 min to 25 min, decreases the LD 50 by an average of 8%. For single dose irradiations the average decrease observed was 11% showing that appreciable dose-rate effects can occur at dose-rates in excess of 78 rad/min when the effect being studied requires an irradiation time in excess of one hour.

The fractionated data, corrected for repopulation have been used to derive the shape of the cell survival curve (fig. 16) for the critical cells of the weevil which are believed to be the regenerative cells of the mid-gut.

9. The relationship between the rate of repopulation and the damage present

9.1. Experiments using two doses of unequal magnitude

The relationship (eq. 6) derived previously (section 7.5.3), provides a means of estimating the total repopulation occurring during a fractionated

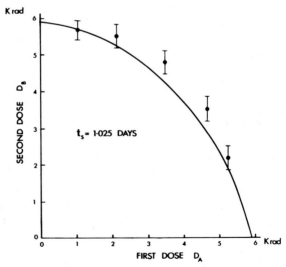

Fig. 17. The dose D_B necessary to kill 50% of the weevils in 30 days if delivered 1.025 days after an initial dose D_A plotted against initial dose D_A. The observed values of D_B may be compared with the curve predicted by the hypothesis discussed in section 9.2.

regime which kills 50% of the weevils, provided the regime consists of equally spaced fractions of equal magnitude.

As already discussed (section 7.6), the data from which this relationship was derived suggest that the rate of repopulation (expressed in rad/day) may be directly proportional to the damage present (expressed in rad). If such a simple relationship exists, it could be used to estimate the repopulation occurring during regimes involving fractions of unequal magnitude and unequal spacing. Experiments Nos 52–66 were designed to find whether a simple relationship existed or not.

In these experiment, weevils were irradiated with an initial dose D_A and t_s days later subjected to a second dose. By keeping D_A and t_s constant and varying the second dose, the total dose $(D_A + D_B)$ necessary to kill 50% of the weevils in 30 days was determined. In this way, the LD $50_{(30)}$ was found for five values of D_A ranging from 1,040 to 5,240 rad and three values of t_s namely; 1.025, 3.83 and 8.14 days. The results of these experiments are presented in the comprehensive results (table 1, section 3.3). The doses D_B (normalised in accordance with their appropriate single dose controls), are plotted against D_A (similarly normalised) in figs. 17, 18 and 19 for each of the three time intervals. For purposes of comparison, theoretical curves predicted as described below, are also shown.

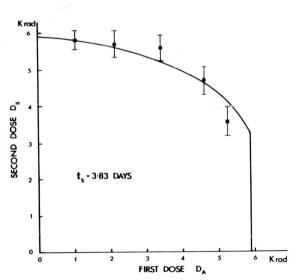

Fig. 18. The dose D_B necessary to kill 50% of the weevils in 30 days if delivered 3.83 days after an initial dose D_A plotted against initial dose D_A. The observed values of D_B may be compared with the curve predicted by the hypothesis discussed in section 9.2.

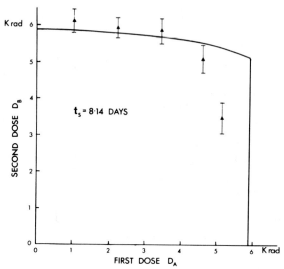

Fig. 19. The dose D_B necessary to kill 50% of the weevils in 30 days if delivered 8.14 days after an initial dose D_A plotted against initial dose D_A. The observed values of D_B may be compared with the curve predicted by the hypothesis discussed in section 9.2.

9.2. Predicted results assuming the rate of repopulation to be proportional to the damage present

The hypothesis to be tested is that at any instant of time (except during the mitotic delay), the instantaneous rate of repopulation (expressed in rad/day) is proportional to the damage A (expressed in rad) present at that instant of time.

The question arises as to how the dose equivalent of damage (A_0 rad) present immediately after a single dose of D_A rad should be defined. A_0 cannot be as large as D_A as part of the damage produced by D_A rad is repaired by intracellular recovery. The value of A_0 should represent the damage which remains after intracellular recovery is complete but before any repopulation has occurred. Let us therefore define the damage A_0 present immediately after a dose D_A by the following equation:

$$A_0 = D_1 - D'_B, \qquad (8)$$

where D_1 is the single dose necessary to kill 50% of unirrradiated weevils and D'_B is the dose necessary to kill 50% of the weevils if delivered shortly after the dose D_A, at a time when intracellular recovery is complete but before repopulation has commenced. For any value of D_A, the value of D'_B may be determined from the theoretical cell survival curve derived previously (fig. 16).

According to the above hypothesis, the total amount of repopulation (expressed in rad) occurring within t_s days of an initial dose D_A will be ^rD_T where

$$^rD_T = A_0 \left[1 - \exp(-\lambda(t_s - t_m))\right], \tag{9}$$

where λ is a constant previously estimated (section 7.5.3) to be 0.29 days^{-1} and t_m is the mitotic delay previously estimated (section 7.5.2) to be 0.18 days/1,000 rad. The dose D_B necessary to kill 50% of the weevils if delivered t_s days after an initial dose D_A is therefore given by:

$$D_B = D'_B + {}^rD_T. \tag{10}$$

Values of D_B calculated from the above equation have been used to plot the theoretical curves shown in figs. 17, 18 and 19.

9.3. Comparison of observed and predicted results

Fig. 17 shows that for initial doses of 3,000–5,000 rad, the predicted values of D_B at 1.025 days are 500–600 rad less than those found. The discrepancy may be partly due to experimental uncertainties (the 95% fiducial limits on the LD 50 of the irradiated controls for these experiments are unusually large at ±3.5%) and partly due to cell synchrony as the split-dose recovery curve (fig. 6) shows that the critical cells of the weevil are moving into a resistant phase 24 h after an initial dose of 4,000 rad.

In view of the above considerations, the agreement is reasonably satisfactory. However, as the amounts of repopulation occurring in one day are predicted to be no more than the equivalent of 200 rad, this comparison is more a test of the reliability of the derived cell survival curve (fig. 16) than the repopulation relationship.

For intervals of 3.83 days, fig. 18 shows excellent agreement between prediction and experiment except at D_A = 3,400 rad where the prediction coincides with the lower limit of the observed result and at D_A = 5,240 rad where the prediction is 600 rad higher than the observed result. If in these diagrams the observed results had not been normalised, they would all be 2.7% lower than plotted. Thus, whilst the normalising procedure may be blamed for part of the discrepancy in fig. 17 and at D_A = 3,400 rad, in fig. 18 it is likely to have minimised the discrepancy at D_A = 5,240 rad. Cell synchrony could still be having some effect at 3.83 days but it is doubtful whether it can account for the latter discrepancy.

For intervals of 8.14 days the suggested hypothesis predicts that the majority of the damage caused by the initial doses would be repaired by repopulation before the second dose was given. This is borne out by experiment for initial doses below 4,000 rad. Above 4,000 rad the prediction begins to deviate from observation and for an initial dose of 5,130 rad over-estimates the repopulation by as much as 1,950 rad.

In view of the above discrepancies, it was decided to plot the relationship between the repopulation observed, the initial damage A_0 and the initial dose D_A for each of the three time intervals used (fig. 20). The repopulation observed is assumed to be $(D_B - D'_B)$ and the initial damage is assumed to be $(D_1 - D'_B)$ as previously defined (section 9.2).

As previously discussed, the effects of cell synchrony are believed to have inflated the values of $(D_B - D'_B)$ observed for intervals of 1.025 days and thus the lower curve in fig. 20 is believed to overestimate the true repopulation occurring during this interval. Examination of fig. 20 shows that for all three intervals the repopulation increases as the initial dose D_A increases and at first is approximately proportional to the initial damage A_0. As D_A is increased beyond 3,500 rad, the repopulation increases less rapidly and reaches a peak when D_A is approximately 4,500 rad. As D_A is further increased, the amount of repopulation per interval falls dramatically. It may be argued that the dramatic decrease in the value of $(D_B - D'_B)$ for high values of D_A may not be due to a decrease in repopulation, but could be due to the first dose D_A killing a proportion of the weevils so that the second dose does not have to kill so many. Examination of the single dose mortality curves shows that 5,200 rad kills 10% of the weevils. Thus, to reduce the total surviving fraction of weevils to 50%, the second dose D_B is only required to kill 44.4% of the 90% surviving the first dose. The single dose mortality curves show that the dose required to kill 44.4% is only 70 rad less than that required to kill 50%. For intervals of 8.14 days, the decrease in the value of $(D_B - D'_B)$ which occurs as D_A is increased from 4,500 to 5,200 rad is 960 rad and thus cannot be explained by the killing effect of the first dose.

It may be concluded that although the total repopulation is approximately proportional to the initial damage for values of D_A less than 3,500 rad the subsequent rapid decrease in repopulation shows that there is no simple relationship between repopulation and damage which is applicable to the whole range considered.

The results suggest that homeostatic control mechanisms in the weevil are able to speed up the repopulation rate (rad/day) in approximate proportion to the damage present, but that after doses of 4,000 rad or more, the reproductive capacity of the surviving cells is severely impaired and they are thus unable to respond to the homeostatic control. An alternative explanation would be that the control mechanism itself is disrupted by such large doses (4,000 rad is 68% of the single dose LD 50).

The fact that no simple relationship has been found between repopulation

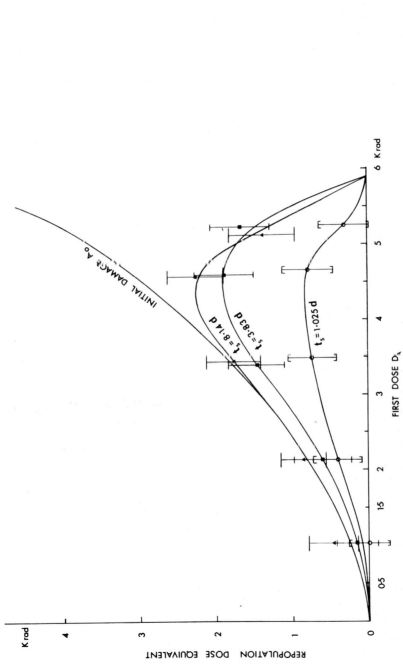

Fig. 20. The amount of repopulation (rad-equivalent) occurring in a period of t_s days after an initial dose D_A, plotted against D_A. The three lower curves show the repopulation dose-equivalents found by experiment for periods of 1.025, 3.83 and 8.14 days and how these vary with initial dose. The initial damage A_0 (upper curve) is defined by eq. (E) and represents the dose-equivalent of the damage present shortly after an initial dose D_A at a time when intracellular recovery is complete but before repopulation has commenced.

and damage in no way invalidates the relationship (eq. 6) established previously (section 7.5.3) for regimes employing equally spaced fractions of equal magnitude. The extension of this relationship to regimes involving unequal doses or unequal intervals is, however, made more difficult. In certain cases, guidance may be obtained by considering the shapes of the curves relating repopulation and damage (fig. 20) as is illustrated by considering the anomalous result discussed below.

9.4. The anomalous result for 6 fractions in 7 days

The LD 50, corrected for repopulation, observed for 6 fractions in 7 days (Experiment No. 16), was found to be 1,200 rad below the smooth curve drawn through the remainder of the daily fractionation data similarly corrected. This anomaly was thought to be associated with the 3-day gap between the 5th and 6th fraction as this was the only "daily" regime in which such a gap occurred so close to the end of the regime. Experiments Nos 47 and 49 were done to test this theory and their results are compared below.

	LD 50
No. 16, 6F/7d (gap between 5th and 6th fractions)	13,410 ± 460 rad
No. 47, 6F/7d (gap between 3rd and 4th fractions)	14,600 ± 510 rad
No. 49, 6F/5d (no gap)	13,140 ± 280 rad

Comparison of experiment Nos 47 and 49 shows that extending the gap between the 3rd and 4th fractions from 1 day to 3 days allows additional repopulation, equivalent to 1,460 rad, to occur. Comparison of experiments Nos 16 and 49 shows that a similar extension between the 5th and 6th fractions allows additional repopulation equivalent to only 270 rad to occur. This difference of 1,190 rad may be explained as follows:

In experiment No. 16, immediately before the 3-day gap, 11,170 rad were delivered in 5 fractions in 4 days. This dose is 95% of the LD 50 for 5F/4d (fig. 14). Thus the damage present at the start of this 3-day gap was probably similar to that caused by 95% of the single dose LD 50, (i.e., $5,900 \times 0.95 = 5,600$ rad). After such a single dose, fig. 20 shows that the amount of repopulation occurring in 3.83 days, should be approximately 700 rad. The exact amount will depend on how the curve is drawn in this rather critical region.

In experiment No. 47, on the other hand, the dose delivered immediately before the 3-day gap was 7,300 rad in 3 fractions in 2 days which is only 80% of the LD 50 for 3F/2d (fig. 14). Thus, the damage present at the start of the 3-day gap in this experiment, was probably similar to that caused by 80% of the single dose LD 50 (i.e., $5,900 \times 0.80 = 4,720$ rad). After such a single dose, fig. 20 shows that the amount of repopulation occurring in 3.83 days should be close to the peak value of 1,900 rad.

Thus the 3-day gap in experiment No. 47 occurred at a time when it would have maximum effect, whereas in experiment No. 16 it occurred at a time when it would have relatively little effect.

9.5. Conclusions

The relationship between the rate of repopulation (expressed in rad/day) and the damage present (expressed in rad), is not a simple one. When the damage present is less than that caused by a single dose of 3,500 rad, the amount of repopulation occurring in a given interval is approximately proportional to the damage present at the start of that interval. The majority of such damage is completely repaired by this process in a period of 8 days.

When the damage present is in excess of that caused by a single dose of 3,500 rad, the amount of repopulation is appreciably less than that predicted by a proportional relationship.

When the damage present is in excess of that caused by a single dose of 4,500 rad, the rate of repopulation decreases dramatically as the damage increases. This suggest that the critical surviving cells of the weevil may be sub-lethally damaged to such an extent that their proliferative capacity is greatly impaired or, alternatively, that the homeostatic control system of the weevil is disrupted by such high doses.

If, at some stage during a fractionated regime, a rest period is introduced, the amount of repopulation occurring during that period will be very dependent on the damage present at the start of that period. In weevils, maximum repopulation occurs when the initial damage is fairly large (equivalent to that produced by a single dose of 4,500 rad). If the rest period occurs too close to the end of the regime, the damage present may be sufficiently large to inhibit cell proliferation and little effect may be produced.

The implications of these findings and their relationship to results obtained by other workers will be discussed in section 11.

10. Testing the general formula for equating protracted and acute regimens of radiation

10.1. Variation in LD 50 with duration of low dose-rate continuous irradiation

The LD 50's observed with continuous low dose-rate irradiation are presented in the comprehensive table of results (Experiments Nos 31–43 in table

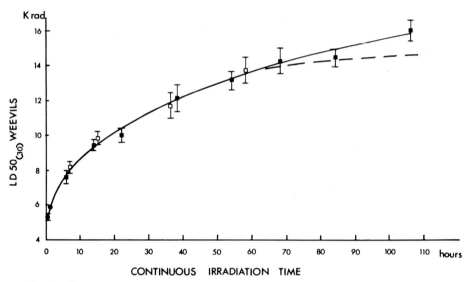

Fig. 21. The LD $50_{(30)}$ for weevils plotted against continuous irradiation time. The Group C results (open squares) are in close agreement with the Group D results (closed squares). It is estimated that the LD 50 corrected for repopulation lies between the full and dotted curves, as discussed in section 10.2.

1, section 3.3). These results (after normalising in accordance with their appropriate single dose controls), are plotted against irradiation time on a linear scale in fig. 21. It will be seen that the results obtained in Group C (open squares) are in close agreement with those obtained 3 months later in Group D (closed squares). The LD 50 for 0.4 h irradiation was obtained from the Group G and H experiments (Nos 68, 69 and 75).

It will be observed that the LD 50 increases by a factor of three as the irradiation time is increased from 0.4 to 106 h.

The derivation of the dotted curve in fig. 20 is discussed below.

10.2. The correction for repopulation

To test the general formula equating protracted and acute regimes of radiation [Liversage, 1969], it is necessary to determine how $(D_t - {}^rD_t)$ varies with irradiation time. As previously defined (section 1), D_t is the LD 50 for continuous irradiation lasting t hours and ${}^rD_t$ is the dose necessary to compensate for repopulation occurring during the irradiation.

Hall [1972] suggests that the dose-rate necessary to stop cell division occurring during irradiation is inversely proportional to cell cycle time and is equal to 19 rad/h for HeLa cells having a cell cycle time of 24 h. The large repopulation effect observed in weevils during fractionated regimes of radiation suggests that the cell cycle time for the critical cells of the weevil can be very short. The maximum rate of repopulation is estimated (section 7.5.3) to be 850 rad/day. The critical cells of the weevil are estimated to have a D_0 of 535 rad, although Liversage [1973] suggested that this could be somewhat in excess of the true value. However, combining these two estimates suggests that the cell cycle time for the critical cells of the weevil can be as short as 10.3 h. It is doubtful whether it is valid to extrapolate the data of Hall [1972] to weevils, but his data would suggest that a dose-rate of 44 rad/h should be adequate to stop cell division in the weevil. The LD 50 for the most protracted continuous irradiation regime was found to be 16,710 rad delivered in a period of 106 h at a dose-rate of 158 rad/h, (i.e., at a dose-rate 3.6 times higher than that necessary to stop cell division). It might therefore be concluded that repopulation by cell division was negligible during all the low dose-rate continuous irradiations, but the extrapolation from mammalian cells to insect cells throws doubt on this conclusion.

An alternative approach is to assume that radiation induces a mitotic delay of 0.18 days/1,000 rad, as found earlier (section 7.5.2). If the delay t_m, so calculated, exceeds the duration of the regime, it is assumed that no repopulation occurs. If, however, the delay is less than the duration of the regime, it is assumed that repopulation occurs at the maximum possible rate (equivalent to 850 rad/day) for a period of $(t - t_m)$ h

$$\therefore {}^rD_t = \frac{850}{24}(t - t_m), \tag{11}$$

where t and t_m are both expressed in hours.

The dotted curve in fig. 21 shows how $(D_t - {}^rD_t)$ varies with irradiation time if the amount of repopulation occurring is calculated from the above equation.

It will be observed that both methods of approach indicate that no repopulation occurs when the regime is of less than 60 h duration. For regimes of 60–106 h duration, eq. (11) suggests that a small repopulation component is present whereas the original approach suggests there is no repopulation effect in this range. It may be concluded that during the continuous irradiation regimes, repopulation effects were small, if not negligible, and that the true curve showing how $(D_t - {}^rD_t)$ varies with irradiation time is likely to lie between the dotted curve and the full curve in fig. 21.

10.3. The application of the general formula to the above data

According to the general formula for equating protracted and acute regimes of radiation discussed previously (section 1), $(D_{NT} - {}^rD_T)$ should be approximately equal to $(D_t - {}^rD_t)$ provided N and t are related by the following equation:

$$N = \frac{\mu t}{2\left[1 - \dfrac{1}{\mu t}(1 - \exp(-\mu t))\right]}, \qquad (3)$$

where D_{NT} is the LD 50 for N acute fractions delivered in T days, ${}^rD_T$ is the dose necessary to compensate for the repopulation occurring in T days, D_t is the LD 50 for continuous irradiation given over a period of t hours, ${}^rD_t$ is the dose necessary to compensate for repopulation occurring in t hours and μ is equal to $0.693/\tau$ where τ is the half-life of the dose equivalent of sub-lethal damage.

Fig. 21 shows how $(D_t - {}^rD_t)$ varies with irradiation time t and may there-

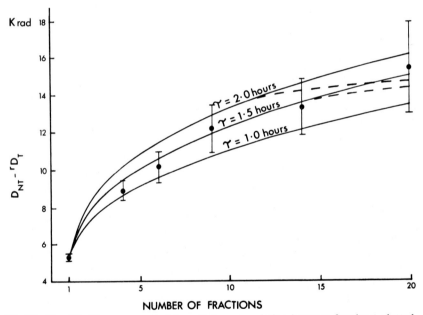

Fig. 22. The LD $50_{(30)}$, corrected for cellular repopulation between fractions, plotted against the number of fractions. The curves show the predicted values obtained by applying the general formula (eq. 3) to the low dose-rate continuous irradiation data, assuming the half-life of the dose-equivalent of sub-lethal damage to be 1.0, 1,5 or 2.0 h. The results obtained from fractionated high dose-rate experiments, corrected for repopulation as described in section 8.4, are also shown for purposes of comparison. The 95% confidence limits associated with these plotted points take into account the uncertainty in the observed Ld 50's and that involved in the repopulation correction.

fore be used in conjunction with the above relationship to predict how $(D_{NT} - {}^rD_T)$ should vary with fraction number.

Under the conditions used in these experiments, the half-life of the dose equivalent of sub-lethal damage for the critical cells of the weevil has been shown (section 6.4) to be greater than 1 h and less than 2 h. To allow for the uncertainty in the value of τ, the variation in $(D_{NT} - {}^rD_T)$ with fraction number has been predicted, as described above, using the values of τ equal to 1.0, 1.5 and 2.0 h. The resultant curves are plotted in fig. 2.2. The full and dotted curves in this diagram reflect the repopulation correction uncertainty inherited from fig. 21.

Values of $(D_{NT} - {}^rD_T)$, together with their 95% confidence limits, derived from the fractionation experiments (table 7, section 17.3) are also plotted in fig. 22 for comparison.

The values of 1.0 and 2.0 h are the lower and upper limits respectively of the value of τ estimated from the split-dose recovery curve. The true value of τ is probably closer to 1.5 h. However, any one of these three values of τ leads to predictions which are within ±12% of the experimental results obtained over the whole range investigated (i.e., from 1 to 20 fractions). Both the full curve and the dotted curve predicted using a value of τ equal to 1.5 h agree with experiment to within ±7% over the whole range investigated, and both lie within the 95% confidence limits of all the experimental points. Fig. 22 suggests, however, that predictions using the general formula and a value of τ equal to 1.5 h, slightly overestimate the LD 50 for fractionated regimes involving small numbers of fractions (less than 6) and possibly slightly underestimate the LD 50 for regimes involving larger number of fractions (about 20).

10.4. Conclusions

Liversage [1969] acknowledged that his formula for equating protracted and acute regimes of radiation was derived by making a number of assumptions and involved a number of approximations. In view of this, he claimed his formula would not be strictly true but should be accurate to ±10%.

The experimental results with weevils certainly substantiate this claim over the whole range investigated, (i.e., from 1 to 20 fractions) but suggest that from 6 to 20 fractions the formula is probably accurate to ±4%. The implications of these findings will be discussed in section 11.

11. Discussion of the results and their implications in radiobiology and radiotherapy

11.1. Comparison of results with mammalian data

Qualitatively, but not quantitatively, the LD 50 for weevils varies with split-dose interval, overall irradiation time and number of fractions in a similar manner to the variation in iso-effect dose observed in normal mammalian tissues and cells. Effects have been observed in weevils which in mammalian tissues would have been attributed to each of the following phenomena:

(1) Intracellular sub-lethal damage repair (section 6.2).
(2) Progression of a partially synchronised population of surviving cells through phases of varying sensitivity (section 6.2).
(3) Repopulation of damaged tissues by cell proliferation (section 7.0).
(4) Radiation induced mitotic delay (section 7.5.2).
(5) Homeostatic control of repopulation (section 9).

The system studied in the present investigation thus constitutes an excellent model for investigating precisely how these phenomena are likely to affect the shape of iso-effect dose relationships in normal tissues. The magnitude of the effect produced by any one of these phenomena is, of course, liable to be very different in weevils from that occurring in some other biological system.

The use of weevils has enabled a large body of accurate iso-effect data to be obtained. These iso-effect data are more comprehensive and more accurate than those obtained for any other biological system to date. They have been used to test a number of hypotheses and may be used to test future hypotheses as and when these are formulated.

The findings with weevils may be compared with those obtained (perhaps in less detail or with less certainty) using systems which are more relevant to the clinical situation. If the relationships observed in the more relevant tissues are consistent with those found in weevils, the latter may be used as a guide when interpreting, interpolating or extrapolating the more relevant data. Examples of this are discussed below.

11.2. The relationship between the rate of repopulation and the damage present

The weevil experiments involving fractions of unequal magnitude, previously analysed (section 9), show that when the damage present is less than that caused by a single dose of 3,500 rad, the rate of repopulation (expressed in "rad per day equivalent") is approximately proportional to the damage present (expressed in "rad equivalent"). As the damage is repaired, the rate of repopulation should decrease proportionately and should thus decrease exponentially with time. The multi-fraction experiments analysed previously (section 7), show that this is true and that the time required for the repopulation rate to fall to half its initial value is 2.4 days.

It is interesting to compare these findings with those obtained for mouse skin. Denekamp [1973], gave 4, 9 or 14 daily fractions of 300 rad to mouse skin followed by a series of test doses at either 0, 1, 4, 8 or 15 days later. By comparing the skin reactions, she was able to estimate the rate of repopulation (rad/day) occurring after each of the fractionated regimes.

After 4 doses of 300 rad, she found no significant repopulation occurred during the next 15 days. (The possibility that some repopulation was present but masked by synchrony, is discussed below.) After 9 doses of 300 rad, the average rates of repopulation during the first and second weeks were 58 and 22 rad/day respectively. After 14 doses of 300 rad, the corresponding rates were 130 and 18 rad/day. These skin reaction results show:

(a) the rate of repopulation increases with the damage present,

(b) as the damage is repaired, the rate of repopulation decreases,

(c) the time required for the rate of repopulation to fall to half its initial value is between 2.5 and 5.0 days (assuming it falls exponentially).

These findings have important implications in split-course therapy [Scanlon, 1960; Sambrook, 1962, 1963; Holsti and Taskinen, 1964] and suggest, as Denekamp points out, that in split-course therapy the interval between courses should not be allowed to be too long or the benefit of the faster rate of repopulation may be lost.

Denekamp's results also suggest that the most beneficial time to introduce a rest period would be towards the end of treatment when the damage present (and consequently the rate of repopulation) are likely to be near their maximum values. Denekamp refrained from drawing this conclusion and the weevil results suggest that her restraint was well founded.

The weevil results show the three effects (a), (b) and (c) referred to above but the fall in the repopulation rate is slightly faster than in mouse skin. (The time to fall to half the initial rate in weevils is 2.4 days.) The weevil results demonstrate that this fall does occur exponentially (fig. 12), but show that the rate of repopulation only increases with the damage present provided the damage present is less than that caused by a single dose of 4,000 rad. As the damage present increases above this level, the rate of repopulation falls dramatically (fig. 20). Whether such a decrease occurs in mouse skin or in the normal tissues of patients is not known, and requires further investigation. The weevil results however, provide a warning of what might be expected. The experiments using 6 fractions in 7 days, with a 3-day gap half-way or alternatively near the end of the irradiation regime, emphasise the point as previously discussed (section 9.4).

When considering the possible implications of these results it should, however, be remembered that during the weevil experiments, the whole body was irradiated whereas in radiotherapy, only a small portion of the body is normally irradiated.

The results observed in mouse skin are perhaps at variance with the weevil results in one respect. After 4 doses of 300 rad, Denekamp [1973] found no effect due to repopulation during the period 1–15 days. The repopulation effect only became appreciable after a further week of daily irradiation. Denekamp suggested that the delay in the onset of the repopulation effect probably reflected the necessity to accumulate an appreciable amount of damage which would be both dose dependent and time dependent if the damage is only expressed at mitosis. She pointed out that the normal turn-over time of the mouse epidermis (without mitotic delay), is at least 4.5 days and that the rate of expression of damage will be further slowed by radiation-induced mitotic delay from each dose.

The weevil results show that the rate of repopulation is certainly dose dependent but they do not exhibit any comparable time delay. Fig. 11 suggests that rapid repopulation is occurring in the critical cell population of the weevil 12 h after a dose of 2,630 rad. It has been estimated that the cell cycle time of these cells could be as short as 10.3 h, but if allowance is made for the mitotic delay, it would appear that homeostatic control has taken effect in the weevil before many cells could have suffered a mitotic death. This suggests that either interphase death is occurring in the weevil or the homeostatic control mechanism in the weevil can detect potentially lethal damage before the majority of that damage has been expressed in the form of cell death. Further evidence on this point is required.

The data of Denekamp regarding the rate of repopulation in skin after 4 doses of 300 rad are also worthy of further consideration. The test dose necessary to produce a given skin reaction, 8 days after the 4 doses, was 450 rad more than that required zero days after the 4 doses. According to Denekamp, 195 rad of this increase was due to intracellular recovery and this suggests that 255 rad was due to repopulation and synchrony. By 8 days the effects of synchrony are likely to be small and it may therefore be deduced that, during the first 8 days, repopulation occurred at an average rate of 32 rad/day. Denekamp's estimate of no repopulation (her table actually shows − 10 rad/day) during the period 1–8 days, may have been greatly influenced by the effects of synchrony inflating the dose necessary at one day, as she herself points out. It is thus possible that the homeostatic forces in mice are acting just as fast as those in weevils. The problem of separating the effects of synchrony from those of repopulation will make it difficult to obtain decisive evidence on this point.

11.3. The relationship between iso-effect dose and overall time of a fractionated regime

It has been shown (section 7), that none of the published relationships, purporting to predict how iso-effect doses for the normal tissues of patients vary with overall treatment time, are of the shape necessary to account for the variations observed in the LD 50 for weevils. The published formulae may be useful in predicting small changes in iso-effect dose for the normal tissues of patients but may be inaccurate when large changes are involved.

The form of the true relationship is more easily and accurately determined for weevils than for normal mammalian tissues, as previously discussed (section 11.1). Because normal mammalian tissues and the critical cells of the weevil both exhibit the same effects (section 11.1), it is probable that the form of relationship (eq. 6) found for weevils will be equally applicable to normal mammalian systems. The values of the parameters involved are of course likely to be very different. From the work of Denekamp, it appears that the repopulation rate M in mouse skin decreases to half its initial rate in 2.5–5 days. If eq. (5), as previously derived for weevils (section 7.5.3) applies to mouse skin,

$$M = M_0 \exp(-\lambda t) . \qquad (5)$$

The value of λ for mouse skin will therefore lie between 0.14 and 0.28 days^{-1}.

It follows that the average rate of repopulation occurring during the first 7 days after irradiation will be equal to $M_0 (1 - \exp(-7\lambda)/7\lambda$, where M_0 is the initial rate of repopulation. Using the above estimated value of λ, it follows that the initial rate of repopulation in mouse skin is 1.6 to 2.3 times the average rate occurring during the first 7 days.

To estimate the rate of repopulation occurring in normal tissues during fractionated regimes, it is common practice to compare the iso-effect dose for N fractions given daily with that for N fractions given, say, 7 days apart. Liversage [1969] applied this procedure to the pig skin results of Fowler et al. (1963), to obtain a repopulation rate in pig skin of 25 rad/day. The above calculation shows that the initial rate of repopulation could be twice as great as the 7-day average rate found by Liversage [1969]. Both Fowler [1971] and Liversage [1971] drew attention to the fact that repopulation rates were likely to be larger during short time intervals than those found during long time intervals, but neither could accurately predict the magnitude of such increases.

When daily fractions are used, the rate of repopulation in pig skin, according to the above reasoning, is likely to be approximately 50 rad/day, but the periods during which repopulation occurs are likely to be considerably less than the intervals between fractions due to radiation induced mitotic delay. The latter will depend upon the dose delivered per fraction and possibly on the cell cycle time of the epithelial cells at the time of irradiation [Hegazy, 1969]. Thus the amount of repopulation occurring in skin during a daily fractionation regime could actually be less than that estimated on the basis of 25 rad/day. The increase in repopulation occurring on changing from a fraction interval of 1 day to a fraction interval of 2 days could, however, be twice as great as that estimated on the basis of 25 rad/day.

It may be concluded that although eq. (6) is likely to apply to mammalian tissues, the values of the parameters involved are not known to a sufficient degree of accuracy at present to enable its application to be a feasible proposition.

11.4. The relative importance of intracellular recovery and repopulation

Before 1963, the importance of fraction number was not appreciated by the majority of workers in radiotherapy or radiobiology. Iso-effect doses were generally considered to be mainly determined by overall irradiation time and the number of fractions was considered to be relatively unimportant. The iso-effect data of Strandqvist [1944], for example, showed how iso-effect dose varied with overall time but generally no attention was paid to the number of fractions involved. Elkind and Sutton [1959] and others revealed the importance of the shoulder of the cell survival curve and the repair of sub-lethal damage. The crucial experiments demonstrating the importance of fraction number were undoubtedly the pig skin experiments of Fowler et al. [1963]. The iso-effect doses found by these workers are given in table 6. These results showed that when 5 fractions were given, increasing the overall time from 4 days to 28 days increased the iso-effect dose by only 600 rad. On changing from 5 "daily" fractions to 21 "daily" fractions the iso-effect dose increased by 1,900 rad but only 600 rad (i.e., 32% of this increase) could be attributed to the increase in overall time. The remaining 68% of the increase was there-

Table 6
Iso-effect doses observed in pig skin by Fowler et al. [1963]

Pig Skin Reaction

Regime	Iso-effect dose
Single dose	2,000 rad
5f/4d	3,600 rad
5f/28d	4,200 rad
21f/28d	5,500 rad

fore due to increasing the number of fractions. Increasing the overall time from 0 to 4 days is unlikely to have produced much effect in view of the increase observed in changing from 4 to 28 days. Thus practically the whole of the 1,600 rad increase accompanying the change from a single dose to 5 fractions in 4 days was probably due to the increase in the number of fractions rather than the increase in the overall time.

These results showed that the major factor which determines skin response is the size or number of fractions. This conclusion of Fowler and his co-workers has greatly influenced radiotherapy thinking during the last decade. Ellis [1968] developed the concept quantitatively by the introduction of a formula which implied that doubling the number of fractions would increase the iso-effect dose by 18% whilst doubling the overall time would only increase the iso-effect dose for normal tissue by 8%.

The constancy of these percentage variations is now being questioned. Recent work, [e.g., Dutreix et al., 1973; Denekamp, 1973; Berry et al., 1972; Douglas et al., 1975; Moulder and Fischer, 1976] suggests that the relative importance of fraction number decreases as the number of fractions is increased and that the relative importance of overall time increases as the overall time is increased. The analysis of published experimental and clinical data [Liversage 1971] also suggested that in order to fit the experimental data the exponent of time in the formula of Ellis [1968] would need to increase as the overall time is increased.

Evidence is thus accumulating which suggests that whilst fraction number is the major factor responsible for the increase in iso-effect dose when the number of fractions is small, the importance of fraction number may have been over-emphasised, and the importance of overall time under-emphasised, when the number of fractions is large.

The above discussion has concentrated on the relative importance of fraction number and overall time but has not yet discussed the relative importance of intracellular recovery and repopulation. The weevil results are important in this respect because they clearly demonstrate that the increase in iso-effect dose with increasing fraction number is not only due to intracellular recovery but is in part due to an increased rate of repopulation (at least when the latter is expressed as rad per day equivalent).

The effect of fraction number on the amount of repopulation occurring is illustrated by the weevil results presented in table 7. From these results it will be observed that as the number of fractions delivered in a given overall time is increased, the amount of repopulation occurring (expressed in rad) also increases. Thus part of the increase in iso-effect dose found by Fowler et al. [1963] on changing from 5 fractions in 28 days to 21 fractions in 28 days may have been due to an increase in the amount of repopulation. As previously discussed (section 11.3), lack of accurate data regarding mitotic delay

Table 7
The amount of repopulation occurring in 18 days in weevils varies with the number of fractions employed

Regime	LD 50	Dose equivalent of repopulation in 18 days
9F/10d	17,680 rad	7,400 rad
9F/28d	25,080 rad	
4F/10d	13,920	3,240 rad
4F/28d	17,160 rad	

and the rate of decrease in the rate of repopulation prevents an accurate estimation of the magnitude of the effect in the pig skin experiments.

For the LD 50 results for weevils the necessary factors have been determined and the repopulation component ^rD_T and the intracellular recovery component ^RD_N can be estimated with reasonable accuracy as previously described (section 7).

Analogous results to those shown for pig skin in table 6 are presented in table 8 together with their estimated repopulation and intracellular recovery components.

The results in table 8 show that nearly half the increase in LD 50 on changing from 4 fractions in 25 days to 20 fractions in 25 days is due to an increase in the repopulation component ^rD_T.

The changes in LD 50 with regime are of a similar nature to those found for skin reactions listed in table 6 and show that fraction number is more important than overall time. On changing from 4 fractions in 3 days to 20 fractions in 25 days the LD 50 increases by 21,610 rad. Using the same reasoning as that used for the pig skin results, it will be seen that 70% of this increase is

Table 8
The LD 50 for weevils given N fractions in T days is higher than the single dose LD 50 by an amount ^rD_T due to repopulation and by an amount ^RD_N due to intracellular recovery. The estimated values of these components are tabulated to show their relative importance

Regime	LD 50 (rad)	^rD_T (rad)	^RD_N (rad)
Single dose	5,900	0	0
4f/3d	10,520	1,210	3,410
4f/25d	16,930	7,620	3,410
20f/25d	32,130	14,380	11,850

due to the increase in fraction number and 30% to the increase in overall time.

Examination of the repopulation and recovery components on the other hand, reveals that 39% of the increase is due to intracellular recovery and 61% due to repopulation.

The above results thus confirm the conclusion of Fowler et al. [1963], in that they show that for the regimes under consideration, fraction number is more important than overall time. They do suggest, however, that repopulation may be more important than hitherto considered, as part of the increase with fraction number appears to be due to repopulation and not all associated with intracellular recovery, as has been generally accepted in the past.

As previously discussed (section 8.3), the weevil results also show that the relative importance of repopulation increases and that of intracellular recovery decreases as the number of daily fractions is increased. This, coupled with the work of Dutreix et al. [1973], Denekamp [1973], Berry et al. [1972] and Douglas et al., [1975], together with the data presented in table 8, suggests that repopulation effects in radiotherapy are probably more important than recently believed. More experimental work with mammalian tissues is needed to determine the magnitude of these effects.

11.5. Existing dose-time relationships

It is now ten years since Ellis [1967] introduced the N.S.D. concept. His basic formula attempts to describe the biological effectiveness of a given treatment regime in terms of a single number which is referred to as the RET dose. In simple terms the RET dose may be defined as the single dose of radiation of the same quality which, if applied to the same area of skin, would produce the same skin reaction as the given treatment retime. Unfortunately, due entirely to an inconsistent definition of overall time, this simple statement is not valid.

The basic concept of Ellis has been extended by the introduction of partial tolerance [Ellis et al., 1969], the T.D.F. factors of Orton and Ellis [1973] and the CRE system of Kirk et al. [1971]. The foundation of all these systems is the simple formula which Ellis proposed in 1967

$$D_{NT} = D_1 \times N^A \times T^B, \tag{12}$$

where D_1 is the single dose necessary to produce a given biological effect, D_{NT} is the fractionated dose delivered in N fractions in T days which would produce the same biological effect and A and B are constants, which Ellis [1968] suggested were equal to 0.24 and 0.11 respectively.

The validity of this basic formula will now be discussed. In 1944 Strandqvist showed that if the dose necessary to produce a given effect when using

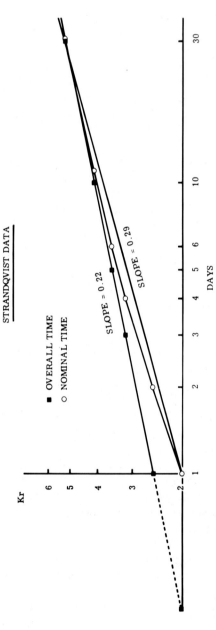

Fig. 23. Strandqvist's [1944] iso-effect doses plotted against treatment time on a log-log scale. Squares, using true overall time. Circles, using nominal time (i.e., true overall time plus one day). The slope of the curve is very dependent on the convention used for overall time.

5 or 6 fractions per week was plotted on a log scale against true overall treatment time, also on a log scale the clinical data could well be fitted by a straight line. Strandqvist found for skin reactions, that this straight line had a slope 0.22. Cohen [1960] analysed the clinical results of Paterson et al. [1952], Jolles and Mitchell [1947] and Ellis [1942]. He found that when plotted on a log-log plot, the iso-effect curve for skin tolerance had a slope of 0.33. This slope is very different from that (0.22) found by Strandqvist [1944].

However, this difference is entirely due to the definition of overall time. Cohen defined overall time T as being the true overall time T' plus one day, i.e., 5 fractions on 5 consecutive days were by Cohen defined as having an overall time of 5 days (not 4 days as defined by Strandqvist).

When comparing the slopes of such curves, the definition of time is vitally important as shown by fig. 23 (from Liversage [1973]). This shows the data of Strandqvist plotted using both conventions. The best straight line fitting the data changes dramatically, depending upon which convention is used. The conventions used by various workers are shown in table 9.

Ellis [1967], in deriving his formula, used the slope of 0.33 found by Cohen [1960] but unfortunately defined overall time in the same way as Strandqvist. But for this discrepancy, the RET dose would be equal to the equivalent single dose rather than the Nominal Standard Dose as defined by Ellis [1968].

The raw data used by Cohen [1960] are shown in fig. 24. These data are plotted using the same convention as that used by Cohen, namely that the nominal overall time T is equal to the true overall time T' plus one day. These data are replotted in fig. 25 after being normalised to 5,000 roentgens in 5 weeks. One point, (that for five fractions in five days) has been deleted from this diagram as Ellis [1970] kindly wrote to one of the present authors saying

Table 9
The different definitions of overall time used by various authors has led to confusion in dose-time relationships. In the above table T' is the true overall time in days, i.e., the number of days between the first and last treatment

Strandqvist [1944]	$T = T'$
Cohen [1960]	$T = T' + 1$
Ellis [1967]	$T = T'$
Kirk et al. [1971]	$T = T' + 1$
Orton and Ellis [1973]	$T = T' + T'/(N - 1)$

where T' is the true overall time in days i.e., the number of days between the first and last treatment.

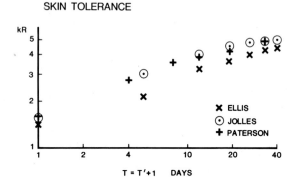

Fig. 24. The raw data (tabulated by Cohen [1960]) and used by Ellis [1967] in the derivation of his N.S.D. formula. Skin tolerance is plotted against nominal overall time (i.e., true overall time plus one day) on a log-log scale.

that he had some doubts about the dosimetry involved for this regime in 1942.

Examination of fig. 25 shows that the plotted data lie on a curved line of decreasing slope but the best straight line through these data has a slope of 0.33. This straight line is correct at one fraction and also at 20 fractions in 26 days, when using the Cohen convention regarding overall time. Using this convention the straight line of slope 0.33 fits the clinical data for five fractions per week from one to thirty fractions to within approximately ±10%. It will also be observed that there is good agreement between the Strandqvist curve and the data of Cohen [1960], except at one fraction (fig. 25).

The basic Ellis formula assumes that $A + B$ in eq. (12) is equal to 0.33 and Ellis deduced that A was equal to 0.22 and B to 0.11. In retrospect, it would

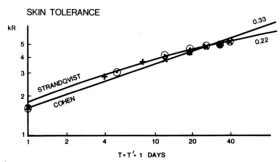

Fig. 25. The data of Cohen (after normalising to 5,000R in 5 weeks) compared with the curve of Strandqvist when both are plotted using the same convention for overall time.

seem that Ellis was correct in giving more weight to the exponent of N than to the exponent of T when the number of fractions is small but incorrect when the number of fractions is greater than 15 or 20, as is discussed below. In fig. 26, the data of Strandqvist [1944] and those of Fowler [1965] are plotted using the Orton and Ellis [1973] convention for nominal overall time. The agreement between Strandqvist and Cohen shown in fig. 25 and that between Strandqvist and Fowler (fig. 26) are remarkable.

The curve of Strandqvist and the plotted points of Fowler in fig. 26 show how skin tolerance varies with time when using five fractions per week. The straight line of slope equal to 0.35, labelled Orton and Ellis, shows how the skin tolerance predictions using the Ellis formula varies with time. Anyone who accepts the Ellis formula as Gospel, and makes a radical change from a five week treatment to a one week treatment is likely to find that the change in iso-effect dose differs from that predicted by as much as 24%.

The radiotherapist who makes a modest change in his fractionation regime will need to modify his prescribed dose. A modification based on the Ellis formula is probably better than no modification. On the other hand, a radiotherapist who changes his treatment from one "daily" fractionation regime to another "daily" fractionation regime would be well advised to modify his prescribed dose in accordance with the curve of Strandqvist or Fowler rather than use the Formula of Ellis. If he changes to a regime which involves a departure from a true five fractions a week regime, it may still be preferable to modify the dose in accordance with Fowler's Curve A [Fowler, 1965] and to make a small correction for the variation in overall time such as that suggested by Liversage [1971]. The weevil results (section 11.4) suggest that the correction advocated by Liversage could be erroneous.

However, the errors involved in the latter method could be less than those involved in applying the Ellis formula. The radiotherapist should perhaps use

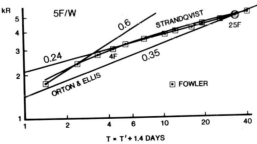

Fig. 26. The skin tolerance of Strandqvist [1944] and those of Fowler [1965] both plotted using the Orton and Ellis [1973] convention for overall time. The straight line of slope 0.35 shows the predictions of the Ellis Formula.

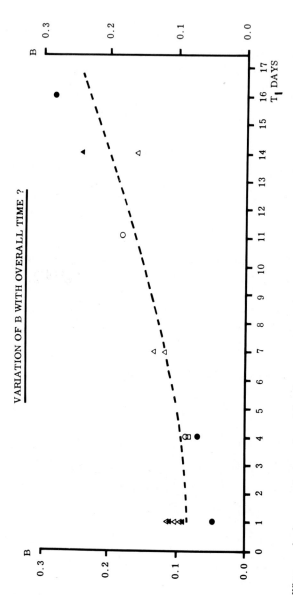

Fig. 27. When employing a constant number of fractions, the iso-effect dose should increase as the overall time is increased from T_1 to T_2 days. The results of such experiments on the skin of animals have been used to calculate the value of B in eq. (12). The above diagram (taken from Liversage [1971]) shows that the values of B so obtained, increases as the overall time T_1 increases.

both methods of prediction. The range of the predicted results should temper his clinical judgment but not over-rule it.

The introduction of the Ellis N.S.D. concept 10 years ago has stimulated research into the relative effects of fraction number and overall time. This research has been useful but today we need to re-assess the situation. An examination of the data on which the Ellis formula was based (figs. 25, 26), shows that the slope of the iso-effect curve (which should be equal to the sum of $A + B$), decreases as the number of daily fractions increases. Liversage [1971] showed that if the overall time is increased from T_1 days to T_2 days the value of B increases as the overall time T_1 is increased (fig. 27). His data are scanty but they imply that the value of A must decrease as the number of fractions increases. The work of Douglas et al. [1975] and Dutreix et al. [1973] demonstrated that the value of A does decrease as N increases (fig. 28). Thus A is not a constant but appears to decrease as N increases.

If T is kept constant and N varied, it should be possible to find the value of A (in eq. 12) for any given end-point. Experiments designed on these lines have shown that A is not a constant but is a function of T. Fig. 29 shows how the LD 50 for weevils increases as the number of fractions is increased from 4 to 9. The slope A of the curves obtained when these results (table 1, Ref. Nos 17, 21, 24, 26) are plotted on a log-log scale of dose versus number of fractions changes with overall time. When the overall time is 10 days A has a

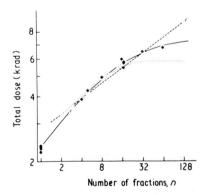

Fig. 28. Taken from Douglas et al. [1975]. The total dose required to produce a given skin reaction in mice, when proliferation is eliminated by keeping the overall time short, is plotted against the number of fractions on a log-log scale. The experimental results obtained by Douglas et al. using 4–64 fractions in an overall time of 8 days (closed diamonds), may be compared with the full curve described by the equation $N(\alpha D + \beta D^2) = \gamma$ where α, β and γ are constants, the dotted curve of Dutreix et al. [1973] and the broken curve having a constant slope of 0.24 as suggested by Ellis [1968]. The experimental results show that the slope of the iso-effect curve decreases as N increases.

value of 0.30 whereas when the overall time is 28 days, A has a value of 0.47.

Hopewell and Wiernik [1977] studied the effect of radiation on the kidney of the pig and also found that A in eq. (12) changed from 0.24 to 0.49 as the overall time was increased from 18 days to 39 days (fig. 30). These experiments show that A is not only a function of N but is also dependent on T.

If N is kept constant and T is varied, it should be possible to find the value of B in eq. (12) for any particular end-point. The weevil experiments illustrated in fig. 31 show that B is not a constant but is a function of N. The slope of the LD 50 versus overall time curve (plotted on a log-log plot), is 0.20 when 4 fractions are used, but is 0.34 when 9 fractions are used. Similarly, in their pig kidney experiments, Hopewell and Wiernik [1977] found the value of B in eq. (12) to be 0.11 when they used 6 fractions, but 0.38 when they used 14 fractions. The variation in B which occurs with variation in the overall time and also with the number of fractions in weevils, is perhaps best illustrated by the variations in the slopes of the curves shown in in fig. 7. All these results show that B is not a constant. The value of B is a function of T but is also dependent on N.

The above results also suggest that when T is small A decreases as N increases. They also suggest that when N is large, B increases as T increases.

In fig. 32 an attempt has been made to illustrate these effects graphically. The effects of overall time will be very dependent on the rate of repopulation of the tissues involved and therefore no attempt has been made at putting

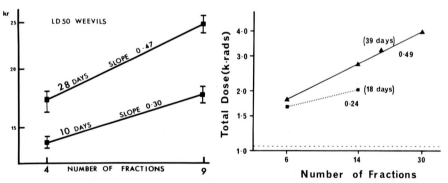

Fig. 29. The LD 50 for weevils plotted against number of fractions on a log-log scale. The exponent of N (in eq. 12) is dependent on the constant overall time employed in such experiments.

Fig. 30. Taken from Hopewell and Wiernik [1977]. The total renal tolerance dose in pigs plotted against number of fractions on a log-log scale. The exponent of N (in eq. 12) is dependent on the constant overall time employed in such experiments.

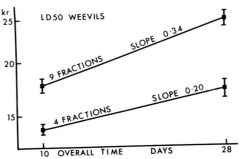

Fig. 31. The LD 50 for weevils plotted against overall time on a log-log scale. The exponent of T (in eq. 12) is dependent on the constant numbers of fractions employed in such experiments.

numerical values on the graphs displayed. These graphs are intended to explain the qualitative effects which could be occurring in a rapidly proliferating tissue such as skin.

If the effects of repopulation are kept small by keeping T small, it would appear that the iso-effect dose increases as N increases to 15 or 20 fractions, but after that a further increase in N produces very little increase in iso-effect dose (fig. 32, top left).

If, on the other hand, one is employing a large number of fractions delivered over a protracted period of time, one would expect repopulation to become more and more important as the regime is prolonged. The work of Denekamp [1973] has shown that towards the end of a treatment regime, the rate of repopulation in mouse skin can be equivalent to as much as 130 rad/day. Moulder and Fischer [1976] have reported a rate of 175 rad/day in rat skin. Fig. 20 also illustrates that the rate of repopulation in weevils is under some form of homeostatic control and increases as the damage increases, provided the latter is not too great. Thus, if one delivers a large number of small fractions over a prolonged period, one can envisage a situation arising where the rate of production of damage is no greater than the rate of repair, and one would expect there to be a steep increase in the iso-effect dose (fig. 32, top right). The lower graph in fig. 32 shows the combined effect of increasing both N and T, as in daily fractionation. This form of relationship is likely to apply to the early reactions observed in rapidly proliferating tissues such as skin and mucosa. The tissues exhibiting late reactions are probably proliferating more slowly and would be unlikely to show the up sweep in their iso-effect curves unless very protracted regimes are employed.

Paterson [1948] published a table showing how skin tolerance increased with overall time when treating at the rate of five fractions per week. His table included treatments involving 50 fractions delivered in 10 weeks, (i.e.,

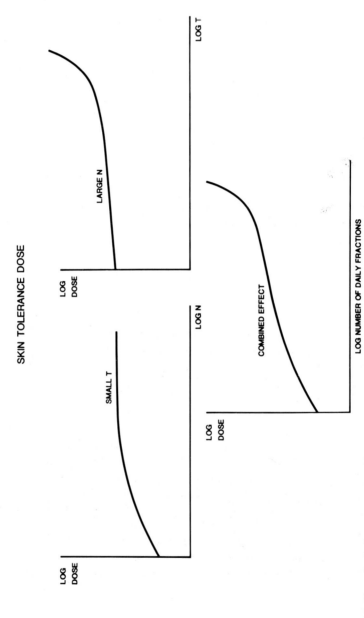

Fig. 32. Hypothetical curves to show how the acute skin tolerance should vary. (a) as the number of fractions increases but the overall time is kept short; (b) as the overall time increasses but the fraction interval is kept short; (c) as the number of daily fractions increases.

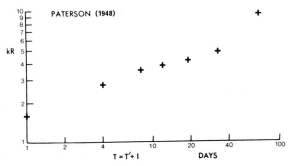

Fig. 33. Skin tolerance doses (5 fractions/week) in radiotherapy [Paterson, 1948] plotted against nominal overall time, (i.e., true overall time plus one day).

67 days) and are plotted in fig. 33. It may be significant that this particular regime which demonstrates the up-sweep, was not included in later editions of Paterson's book. One explanation of this exclusion could be that he subsequently found that late reactions differ from early reactions, particularly when treatment regimes are very protracted.

A similar curve to that obtained by Paterson, clearly demonstrating the up-sweep, has more recently been obtained by Moulder and Fischer [1976] for acute reactions in the skin of rats.

The initial rise in the iso-effect dose versus number of daily fractions curve (fig. 32) is due to intracellular recovery from sub-lethal damage. The ultimate rise is due to repopulation. Future experiments designed to study dose-time relationships in any biological system should be designed to separate these two effects. They should not be designed to evaluate the supposed constants A and B in eq. (12). The weevil experiments have clearly demonstrated that the effects of repopulation are not eliminated by keeping the overall time constant. Thus, if one wishes to study the effects of recovery from sub-lethal damage, one needs to vary the number of fractions, but the overall time needs not to be kept just constant but needs to be kept as small as possible and the interval between fractions should, where necessary, be reduced to 3 or 6 h, such as in the experiments of Douglas et al. [1975], who delivered 4–64 fractions in a constant overall time of 8 days.

Alternatively, one may employ the ingenious multiple split-dose technique of Dutreix et al. [1973]. The effects of intracellular recovery in patients may be assessed from the clinical results of Swoboda [1976] who delivered large numbers of fractions to patients in very short overall times by treating several times a day, thus minimising the effects of repopulation.

Experiments designed to study the effects of repopulation in an organised tissue should not be designed simply to show how the iso-effect dose, when employing a constant number of fractions, varies with overall time. The weevil experiments show that the amount of repopulation occurring in an organised tissue in any given time interval, is very dependent on the damage present at that time. Thus, a rest period three-quarters through treatment could have much more effect than a similar rest period one-quarter through treatment. The weevil experiments also suggest that the interval between fractions is a more important parameter than overall time (section 7.5) and that the amount of repopulation occurring in a given overall time is very dependent on fraction interval and hence on the number of fractions delivered during that overall time. Ideally, experiments designed to study the effects of repopulation should be designed with the objective of establishing, for their particular biological end-point, the value of the parameters involved in eq. (6) rather than those in eq. (12).

11.6. The general formula for equating protracted and acute regimes of radiation

The general formula for equating protracted and acute regimes of radiation was tested [Liversage, 1969], using data for skin reaction and LD 50 for mice which had been published by different workers. For both effects, the formula was shown to be true to an accuracy equal to that of the experimental and clinical data, estimated to be ±10%. Hall [1968] applied the formula to the in vitro cell survival data obtained for acute and protracted irradiation of HeLa cells by Hall and Bedford [1964] and by Bedford and Hall [1963]. He found excellent agreement between observations and predictions based on the general formula.

As discussed earlier (section 1), the derivation of the general formula was stimulated by the need to determine suitable fractionated acute regimes of radiation which could be used in clinical trials using a remotely controlled high dose-rate afterloading unit for the treatment of carcinoma of the cervix.

The clinical trial initiated by O'Connell et al. [1967] was designed to determine the acute dose, delivered in 6 fractions in 12 days, necessary to produce the same effect as a continuous low dose-rate regime delivering 4,000 rad to point A in 48 h.

O'Connell [1973] has established that a total dose of approximately 3,300 rad to point A, in 6 fractions in 12 days produces the same vaginal epithelium reaction as that produced by 4,000 rad to point A delivered continuously during 48 h. The geometrical arrangement of the sources is the same as far as is possible in both methods of treatment and the vaginal epithelium receives a dose which is approximately 2.5 times larger than that delivered to point A.

According to the general formula, 4,000 rad to point A (or 10,000 rad to the vaginal epithelium) delivered continuously in 48 h should produce the same effect as 4,000 rad to point A (or 10,000 rad to the vaginal epithelium) delivered in 12 fractions in a hypothetical time of 48 h. If the number of fractions is reduced from 12 to 6, the data of Fowler [1965] shows that the total dose should be reduced by 16%; i.e., to 3,360 rad to point A (and 8,400 rad to the vaginal epithelium). In the clinical trial, the six fractions were not given in 48 h but in 12 days, and therefore, the doses predicted from the general formula need to be increased to allow for repopulation occurring in the vaginal epithelium in a period of 10 days. As each fraction is approximately 1,400 rad, appreciable mitotic delay will reduce, if not prevent, repopulation occurring during the 2–3-day intervals between fractions. If and when repopulation is occurring, the rate could be as high as 50 rad/day if allowance is made for the factors previously discussed (section 11.3), and if vaginal epithelium is considered to repopulate at the same rate as skin. Thus, the increase in overall time from the hypothetical time of 48 h to the actual time of 12 days is liable to increase the vaginal epithelium iso-effect dose by between 0 and 500 rad (i.e., by 0–5%).

Thus, after making appropriate corrections for reducing the number of fractions from 12 to 6, and increasing the overall time to 12 days, the general formula predicts that the equivalent regime should be 3,360–3,530 rad delivered in 6 fractions in 12 days. This prediction is in good agreement with clinical observation.

The validity of the general formula has thus been tested for a number of different mammalian end-points, (skin reactions, LD 50 for mice, and HeLa cells in vitro) and is consistent with the clinical observations of O'Connell [1973] on the vaginal epithelium.

The present investigation using weevils has tested the formula against a completely different non-mammalian tissue and certainly shown it to be accurate to +12% and probably to ±7%.

Liversage [1969] claimed that, according to theory, the formula should be approximately true for all tissues but only to an accuracy of ±10%. The present study coupled with the above findings provides strong evidence supporting this claim.

This evidence strengthens the claim of Liversage [1969] that there is no inherent advantage in low dose-rate continuous irradiation compared with high dose-rate irradiation provided the latter is fractionated in accordance with eq. (3) (section 1.1). This claim is however, based on the assumption that intracellular recovery is more important than repopulation and that any small effects due to repopulation on increasing the overall time from t hours to T days are, in any case, likely to increase the iso-effect doses for normal tissues to an equal if not greater extent than those for tumours.

The present investigation, whilst agreeing with the general formula, has shown that repopulation effects in mammalian tissues are likely to be more important than recently considered. Thus, on changing from a low dose-rate continuous irradiation regime lasting t hours to a fractionated acute regime, in which N fractions are given in T days, the effects of intracellular recovery are likely to be the same, provided N and t are related in accordance with eq. (3), but the effects of repopulations are likely to be more important than previously realised.

Whether increasing the overall time from t hours to T days is likely to benefit the normal tissues or the tumour, is entirely a matter of speculation. The present position is, however, more optimistic than hitherto in that the present investigation (particularly when considered in conjunction with the recent results of Denekamp), has shown that the homeostatic factors which govern the rates of repopulation in normal tissues may be manipulated to obtain maximum effect, (e.g., by such devices as selecting the optimum time for the introduction of a rest period). Indeed, Ellis [1968] has stressed that the most important difference between normal tissues and tumours is that repopulation in the former is under homeostatic control, whereas in general, in the latter this is not so.

The fact that repopulation must now be considered to be more important than formerly envisaged, suggests that by manipulating the homeostatic influences to increase the rates of repopulation in normal tissues, these effects might be exploited to the benefit of the patient and the detriment of the tumour. However, before such benefit can be achieved in clinical practice at will, much more will need to be known about repopulation effects in both tumours and normal tissues. This suggests that more research should be directed toward this end.

12. Summary of the conclusions

12.1. The factors influencing iso-effect relationships in weevils

The manner in which the LD 50 for weevils varies with split-dose interval, overall irradiation time and number of fractions is similar to that in which iso-effect doses vary for normal mammalian tissues as explained in detail above. The LD 50 for weevils has been used as an end-point to establish iso-effect data which, because of the large number of weevils that could be used, are probably more comprehensive and accurate than those previously obtained for any tissue.

On changing from 1 to 20 fractions, the LD 50 (corrected for repopulation) increases by a factor of 3 due to intracellular recovery (section 8, table 3). Thus the magnitude of the effect due to intracellular recovery is only slightly larger than that observed for mammalian skin reactions; Fowler [1965] found the corresponding increase for skin reactions (uncorrected for repopulation) to be a factor of 2.76. The split-dose recovery curve (fig. 6), shows that radiation induced synchrony has relatively little effect in weevils, causing a variation in the LD 50 with split-dose interval of no more than ±4%.

For irradiation regimes involving a constant number of fractions, the percentage increase in iso-effect dose with increase in overall irradiation time is much greater for lethality of weevils than it is for mammalian skin reactions. Under certain comparative conditions, the percentage increase is approximately three times greater as shown by the pilot experiments (section 1.5). Thus, when the LD 50 for weevils is the chosen end-point, repopulation effects are relatively large and, therefore, conductive to investigation.

The iso-effect data for weevils may be, and have been here, used to investigate some of the fundamental relationships between iso-effect dose and radiation regime; relationships which are believed to be governed by intracellular recovery from sub-lethal damage and cellular repopulation influenced by mitotic delay and homeostatic control. In view of the above findings, the LD 50 for weevils is a particularly useful end-point for studying the effects of repopulation in an organised tissue.

12.2. The value of τ

When irradiated and housed at 27°C and 70% R.H , the half-life τ of the dose-equivalent of sub-lethal damage for the critical cells of the weevil is between 1 and 2 h and is therefore similar to that found for mammalian cells (section 6.4).

12.3. The validity of the General Formula

The General Formula for equating protracted and acute regimes of radiation [Liversage, 1969], should enable the dose necessary to produce a given effect using low dose-rate continuous irradiation to be converted to an approximately equivalent multi-fraction acute dose. Application of this formula requires a knowledge of the value of τ for the relevant cells but the equivalent multi-fraction dose so determined is only weakly dependent on the value assumed for τ (fig. 22). Uncertainty in the estimation of τ for the critical cells of the weevil unfortunately leads to some uncertainty in the predicted equivalent dose. However, over the whole range investigated, (i.e., from 1 to 20 fractions), predictions based on τ equal either to 1.0 or 2.0 h agree with observation to ±12%. The most likely value of τ is considered to be 1.5 h and predictions based on this value agree with observations to an accuracy of ±7% over the whole range investigated and ±4% over the range 6–20 fractions.

Considering the approximations involved in the derivation of the formula, the above agreement is considered to be satisfactory. The General Formula has been shown to be true to an accuracy of ±10% for several mammalian tissues. Now it has been tested against a very different type of tissue it may be generally applied with a greater degree of confidence than hitherto.

12.4. The variation in LD 50 with overall time

For fractionated regimes, the LD 50 for weevils does not vary with overall time in a way which is consistent with published formulae purporting to estimate such variation for the normal tissues of patients (section 7).

The dose ^rD_T necessary to compensate for cellular repopulation occurring during a fractionated regime which kills 50% of the weevils in 30 days is consistent with the following relationship:

$$^rD_T = (N-1)\frac{M_0}{\lambda}[(1 - \exp(-\lambda(t_s - t_m))], \qquad (6)$$

where N is the number of fractions; M_0 is the average rate of repopulation (expressed in rad/day) immediately after the mitotic delay t_m induced by each fraction, t_s is the interval between fractions; and λ is an exponential decay constant.

The above relationship only applies when the irradiation regime consists of equally spaced fractions of equal magnitude. For weevils, M_0 is estimated to be 850 rad/day, λ is estimated to be 0.29 days^{-1} and t_m is estimated to be 0.18 days/1,000 rad (section 7.5.3). The above relationship probably applies to other end-points in other tissues but insufficient data are available to enable reliable estimates to be made of the appropriate values of M_0, λ and t_m for any other tissue at present.

12.5. The relationship between repopulation and damage

When the damage present in the weevil is less than that caused by a single dose of 3,500 rad, the dose necessary to compensate for cellular repopulation occurring in a given interval is approximately proportional to the dose-equivalent of the damage present at the start of that interval. This suggests that the homeostatic control in the weevil does not simply switch on or off the repopulation mechanism but is capable of varying the rate of repopulation in accordance with the needs of the system. The rate of repopulation can be sufficiently high to repair in a period of 8 days, the majority of the damage caused by a single dose of 3,500 rad (fig. 19).

As the damage is increased to higher levels, the rate of repopulation (expressed in rad/day) fails to increase as rapidly as would be required by a proportional relationship and, in fact, decreases dramatically when the damage present is in excess of that caused by a single dose of 4,500 rad (fig. 20). This suggests that the critical surviving cells of the weevil may be sublethally damaged to such an extent that their proliferative capacity is greatly impaired or, alternatively, that the homeostatic control system of the weevil is disrupted by such high whole-body doses. (N.B. The single dose LD 50 for weevils is 5,900 rad.)

Because the rate of repopulation varies greatly with the damage present, it follows that the beneficial effect of a rest period during a fractionated regime should be dependent on the stage at which such a rest period is introduced. The weevil experiments show that this is so. For example, the LD 50 for 6 fractions in 7 days can vary by as much as 1,190 rad, (i.e., 9%) depending upon the position of a 3-day rest period. These results coupled with the work of Denekamp [1973] suggest that there could be an optimum time at which such a rest period should be introduced in split-course therapy and that further experiments using local irradiation of organised mammalian tissues should be done to determine whether this is so.

12.6. The relative importance of repopulation and intracellular recovery

The LD 50 for weevils increases with the number of daily fractions. As for mammalian tissues, the increase in iso-effect dose per daily fraction can be explained as being partly due to intracellular recovery and partly due to cellular repopulation. For the critical cells of the weevil, the former appears to be the major factor when the number of daily fractions is less than six and the latter the major factor when the number of fractions is greater than nine. Recent clinical and experimental work, discussed previously, (section 11.4) suggests that a similar change-over in the relative importance of these factors may occur in mammalian tissues as the number of daily fractions is increased. It may be anticipated, however, that the rapidly proliferating tissues of the weevil would be likely to exhibit this change-over at a lower number of daily fractions than more slowly proliferating tissues. (It was estimated in the present work that the cell cycle time of the critical cells of the weevil could be as short as 10.3 h.)

12.7. Revision of the estimated repopulation rate for skin

The weevil results show that the rate of repopulation (expressed in rad/day) decreases exponentially with time between fractions (fig. 12). It is suggested that a similar effect occurs in the skin of mammals although the exponential decrease may occur rather more slowly with a "half-life" of between 2.5 and 5.0 days (section 11.3). It follows that after the mitotic delay, the average rate of repopulation for skin reactions during short intervals could be twice as great as the 7-day average rate of 25 rad/day estimated by Liversage [1969].

12.8. A revised interpretation of the increase in iso-effect dose when fraction number only is varied

Because the rate of repopulation increases as the fraction interval decreases, the increase in iso-effect dose observed for rapidly proliferating normal tissue on changing from, say, 5 fractions in 28 days to 21 fractions in 28 days, must now be considered as being partly due to an increase in the amount of repopulation occurring and not entirely due to intracellular recovery, as is generally accepted (section 11.4).

12.9. The design of future radiobiological experiments

The weevil results suggest that it is far more important to separate the effects due to the recovery of sub-lethal damage from those due to repopulation than to design experiments to determine the values of the supposed constants A and B in eq. (12). A and B are not constants, but are parameters which vary greatly with both fraction number and overall time. The design of such experiment is discussed in section 11.5.

12.10. The possible exploitation of repopulation effects

The above conclusions suggest that in rapidly proliferating tissues, cellular repopulation may be playing a much larger part in determining iso-effect relationships than has been generally accepted during the last decade, especially when the number of fractions is greater than, say, 10. Further research into the relationship between the rate of repopulation and the damage present in mammalian tissues may provide the information necessary to exploit the homeostatic mechanism believed to be present in normal tissues, but not in most tumours, to obtain a therapeutically useful differential effect.

Acknowledgments

We are most grateful to Dr. J.F. Fowler for his extremely helpful advice and encouragement during all stages of this project.

Invaluable advice on rearing and housing the insects was given by Mr. C.W. Coombs, Principal Scientific Officer at the A.R.C. Infestation Laboratory at Slough to whom we are also indebted for supplying the original 50 weevils who in turn supplied the 68,200 weevils used in these experiments.

The majority of the counting and scoring was carried out by Mrs. M. Squires whom we must also thank for so cheerfully undertaking many unenviable tasks associated with this work. Before she joined the project we received similar assistance from Mr. N. Salpadoru who, during his vacation from the University, worked here as a locum technician.

We are also most grateful to Mr. E.T. Hutchings and Mr. R. Coote who built the irradiation cabinet and partitioned off the constant temperature room so effectively.

We are indebted to the Departments of Medical Illustration at Charing Cross Hospital and St. Bartholomew's Hospital for their assistance in reproducing the diagrams.

We also wish to thank the staff of the Medical Physics Department under Dr. N.W. Ramsey and that of the Radiotherapy Department under Dr. D. O'Connell for their helpful co-operation and particularly for allowing us to irradiate weevils at times which were not always convenient to them.

A special word of thanks is due to Mrs. D. Houghton who made our original manuscript legible by typing it.

The financial assistance provided by the Charing Cross Hospital Clinical Research Sub-Committee who provided funds for the pilot experiments and Tenovus who paid the salary of Mrs. Squires and provided additional equipment, is most gratefully acknowledged.

We are grateful to the following for permission to reproduce figures: Drs Cornwell and Morris and the Atomic Energy Research Establishment for fig. 1; the Editors of the British Journal of Radiology for figs. 23 and 27; Dr. Douglas and colleagues, the Institute of Physics and John Wiley and Sons Ltd., for fig. 28; Drs. Hopewell and Wiernik and I.A.E.A. for fig. 30.

References

Abbott, W.S. 1925. J. Econ. Entomol. 18, 265.
Bedford, J.S. and Hall, E.J. 1963. Int. J. Radiat. Biol. 7, 377.
Berry, R.J., Wiernik, G. and Patterson, T.J.S. 1972. Brit. J. Radiol. 45, 793.
Cohen, L. 1960. Physical and Biological Parameters affecting due Reactions of Human Tissues and Tumours to ionizing Radiation. Ph.D. Thesis, University of Witwatersrand.
Cohen, L. 1968. Brit. J. Radiol. 41, 522.
Coombs, C.W. and Woodroffe, G.E. 1964. Entomol. Mon. Mag. 99, 145.
Cornwell, P.B. and Morris, J.A. 1959. UKAEA, Research Group Report No. AERE-R3065.
Denekamp, J. 1973. Brit. J. Radiol. 46, 381.
Douglas, B.G., Fowler, J.F., Denekamp, J., Harris, S.R., Ayres, S.E., Fairman, S., Hill, S.A., Sheldon, P.W. and Stewart, F.A. 1975. In: Cell Survival after Low Doses of

Radiation. (The Institute of Physics and John Wiley and Sons Ltd., England).
Dutreix, J., Wambersie, A. and Bounik, C. 1973. Europ. J. Cancer, 9, 159.
Elkind, M.M., Han, A. and Volz, K.W. 1963. J. Natl. Cancer Inst. 30, 705.
Elkind, M.M. and Sutton, H. 1959. Nature, 184, 1293.
Ellis, F. 1942. Brit. J. Radiol. 15, 348.
Ellis, F. 1967. In: Modern Trends in Radiotherapy, Series 1 (Butterworths, London) p. 34.
Ellis, F. 1970. Personal communication.
Ellis, F. 1968. In: Current Topics in Radiation Research, Vol. 4, (North-Holland, Amsterdam) pp. 357–397.
Ellis, F., Winston, B.M., Fowler, J.F. and DeGinder, W.L. 1969. Brit. J. Radiol. 42, 715.
Finney, D.J. 1952. Probit Analysis, 2nd Ed. (Cambridge University Press, London).
Fowler, J.F. 1965. Brit. J. Radiol. 38, 365.
Fowler, J.F. 1971. Brit. J. Radiol. 44, 81.
Fowler, J.F., Morgan, R.L., Silvester, J.A., Bewley, D.K. and Turner, B.A. 1963. Brit. J. Radiol. 36, 188.
Fowler, J.F. and Stern, B.E. 1963. Brit. J. Radiol. 36, 163.
Hall, E.J. 1972. Brit. J. Radiol. 45, 81.
Hall, E.J. 1968. Personal communication.
Hall, E.J. and Bedford, J.S. 1964. Radiat. Res. 22, 305.
Hegazy, M.A.H. 1969. Ph.D. Thesis, London University.
Holsti, L. and Taskinen, P.J. 1964. Acta Radiol. Scand. 2, 366.
Hopewell, J.W. and Wiernik, G. 1977. In: Radiobiological Research and Radiotherapy (IAEA-SM-212/44) IAEA, Vienna.
Jefferies, D.J. and Banham, E.J. 1961. UKAEA Research Group Report, No. AERE-R3503.
Jolles, B. and Mitchell, R.G. 1947. Brit. J. Radiol. 20, 405.
Kirk, J., Gray, W.M. and Watson, E.R. 1971. Clin. Radiol. 22, 145.
Lajtha, L.G. and Oliver, R. 1961. Brit. J. Radiol. 34, 252.
Lange, C.S. and Gilbert, C.W. 1968. Int. J. Radiat. Biol. 14, 373.
Liversage, W.E. 1969. Brit. J. Radiol. 42, 432.
Liversage, W.E. 1969a, Cell Tissue Kinet. 2, 269.
Liversage, W.E. 1971. Brit. J. Radiol. 44, 91.
Liversage, W.E. 1973. Ph.D. Thesis, University of London.
Martin, V.J., Burson, D.M., Bull, J.O. and Cornwell, P.B. 1962. UKAEA Research Group Report No. AERE-R3893.
Moulder, J.E. and Fischer, J.J. 1976. Cancer 37, 2762.
O'Connell, D. 1973. Proc. Roy. Soc. Med. 66, 938.
O'Connell, D., Joslin, C.A., Howard, N. Ramsey, N.W. and Liversage, W.E. 1967. Brit J. Radiol. 40, 882.
Oliver, R. 1964. Int. J. Radiat. Biol. 8, 475.
Orton, C.G. and Ellis, F. 1973. Brit. J. Radiol. 46, 529.
Paterson, E., Gilbert, C.W. and Matthews, J. 1952. Brit. J. Radiol. 25, 427.
Paterson, R. 1948. The Treatment of Malignant Disease by Radium and X-rays (Edward Arnold, London) p. 38.
Pendlebury, J.B., Banham, E.J., Cooper, B.E. and Bland, C.M. 1962. UKAEA Research Group Report No. AERE-R3641.
Riemann, J.G. and Flint, H.M. 1967. Ann. Entomol. Soc. Am. 60, 298.
Sambrook, D.K. 1962. Clin. Radiol. 13, 1.
Sambrook, D.K. 1963. Brit. J. Radiol. 36, 174.

Scanlon, P.W. 1960. Am. J. Roentgenol. 84, 632.
Sinclair, W.K. 1967. Radiation Research 1966. (North-Holland, Amsterdam) p. 607.

THE LOCALIZED DOSIMETRY OF INTERNALLY DEPOSITED ALPHA-EMITTERS

Erich POLIG

Kernforschungszentrum Karlsruhe, Institut für Genetik und für Toxikologie von Spaltstoffen, 7500 Karlsruhe, Fed. Rep. Germany

The stopping power is of central importance for the physical fundamentals of the dosimetry of internal α-emitters. Above 2 MeV particle energy the Bethe-Bloch theory is applicable and can be extended to lower energies if a correction for charge fluctuations is introduced. Stopping power values of protons or α-particles in elements are used for tissue as stopping medium by means of the Bragg additivity rule. Alternatively, tissue equivalent gases or solid materials may be used. The concept of linear energy transfer is related to the way energy is imparted around the particle trajectory and is closely connected to the theory of δ-ray generation. The Bragg-Kleeman rule provides a means of determining range-energy relationships in any material from a master curve. Over a certain range of energies α-particle range measurements can be fitted satisfactorily by a power function expression. Other analytical approximations useful for dose calculations are discussed. In deriving formulas for the dose(-rate) distribution emphasis is laid on situations were two separate media with a non-uniformly dispersed radioactive label are present. Results are discussed for spherical, cylindrical and plane separating interfaces and different assumptions for the geometry of the radioactive source.

Impartment of radiation energy to tissue volumes of limited size is a stochastic process. The meaning of the quantities "event size Y" and "local energy density z" together with its associated probability distributions is discussed. The probability distributions of the track segment length in a target, the LET-distribution in pathlength and the energy straggling distribution combine to determine the dose independent single event distribution of the local energy density $f_1(z)$. Another distribution function $f(z, D)$ characterizing the local energy density delivered to a sensitive volume, if the absorbed dose is D, may be calculated from $f_1(z)$ by a Monte Carlo procedure or as the result of a compound Poisson process. A measurement technique of Y and z spectra is explained and some applications of the concepts to particulate α-sources in tissue and to the irradiation of cells from contaminated bone are presented.

Autoradiography provides a means for the spatial localization of α-radiation and sources. Photographic emulsions and organic nuclear track detectors are the autoradiographic materials most frequently used. Their properties with respect to detection efficiency, resolution and track identification are discussed. Neutron induced autoradiography is a particularly sensitive method to detect very minute amounts of fissionable nuclides. Several procedures of making dose evaluations from autoradiographic detectors are described including visual counting, densitometric techniques and image analysis. A method of high track density measurements is explained.

Most of the experiments concerning the dosimetry of internally deposited α-emitters are dealing chiefly with the deposition of ^{226}Ra, ^{239}Pu and ^{241}Am in the skeleton. ^{226}Ra is present in bone volume as a rather uniform diffuse label and as hotspots at sites of bone growth. The results of autoradiographic and dosimetric investigations in man, dogs and mice are reviewed. Dose calculations are generally complicated by the varying fractional ^{222}Rn retention and the non-equilibrium between RaF and ^{222}Rn. Moreover, the retention kinetics of ^{226}Ra and its daughters has to be taken into account. A theoretical model for the burial of hotspots provides a simple estimate of the relative dose contribution from the diffuse and the hotspot component. ^{239}Pu is a bone surface-seeking nuclide with higher toxicity than the volume-seeker ^{226}Ra. Its deposition on surfaces is non-uniform. Results of dose determinations in soft tissue layers adjacent to bone for different times after uptake in man, dogs, rabbits and rats are reviewed. Part of the activity is deposited in marrow cells and may be harmful to the sensitive osteogenic cells too. The non-uniformity factors of the dose distribution are discussed. The burial processes which are treated from an experimental and theoretical viewpoint reduce the dose rates to the bone surfaces. A theoretical model demonstrates that there might be species specific differences in the RBE (Pu/Ra), suggesting an about three times higher RBE for man than for dogs. The results of experiments concerning the localized dosimetry of ^{241}Am in the rat skeleton are described.

Summarizing the conclusion is drawn that further investigations should be directed towards improvement of methods, a more detailed description of the changes in time, the burial phenomenon and interspecies comparisons. An interesting aspect of future work will be the linking of current mathematical tumour theories to refined dosimetric models.

CONTENTS

1. INTRODUCTION 192

2. ENERGY DISSIPATION AND ABSORBED DOSE 194
   2.1. Stopping power 194
   2.2. Linear energy transfer and δ-rays 200
   2.3. Alpha-particle range 203
   2.4. Dose distributions for idealized geometrical conditions 206

3. STOCHASTIC MICRODOSIMETRY 223
   3.1. Quantities and distributions 223
   3.2. Applications to internal alpha-emitters 244

4. EXPERIMENTAL PROCEDURES 249
   4.1. Emulsion autoradiography 249
   4.2. Dielectric track detectors 258
   4.3. Neutron induced autoradiography 263
   4.4. Track viewing and counting procedures 268

5. LOCALIZED DOSIMETRY OF RADIONUCLIDES IN THE SKELETON . . . 279
   5.1. Radium-226 279
   5.2. Plutonium-239 292
   5.3. Americium-241 313

6. CONCLUSIONS AND FUTURE OUTLOOK 317

ACKNOWLEDGEMENTS 318

REFERENCES 319

1. Introduction

At the outset of this article a comment concerning the use of the term "localized dosimetry" seems to be in order. The present situation of the so-called "microdosimetry" is characterized by two ways of thinking based on essentially different attitudes and notions. Classical microdosimetry is something like a transference of the concepts having evolved in radiation protection to a microscopic scale and is centered around the quantity "absorbed dose", being "the quotient of $d\bar{E}$ by dm, where $d\bar{E}$ is the mean energy imparted by ionizing radiation to the matter in a volume element and dm is the mass of the matter in that volume element" [ICRU, 1971]. In this form it reflects the interest in the quantitative description of the radiation field existing in unstructured matter and the distribution of the sources in space producing it. Such an approach, now widely applied in practical situations dealing with internal emitters, is also often referred to as "conventional" or "minidosimetry". An alternative approach is to consider radiation effects to be evoked from the impartment of energy to specific targets and takes into account the statistical fluctuations of energy transfer to tissue structures. This is microdosimetry in a strict sense, the concepts of which are now introduced hesitatingly into the set of methodical procedures instrumental in solving problems concerning the dosimetry of internal emitters. To catalyze this process a section on stochastic microdosimetry is contained in this article and the term localized dosimetry was chosen to unite the two ways of thinking under a common heading.

Alpha-emitting radionuclides constitute a radiation hazard to man if incorporated into his body and deposited in certain organs for a sufficient time to cause deleterious effects on cells or tissue structures. Because of the low penetrating power of α-particles with energies below 10 MeV (table 1) irradiation from external radionuclide sources normally will be absorbed in the outer layer of the skin before reaching living cells. Until now it is only from theoretical reasoning that in particular instances exterior α-irradiation might be considered as potential hazard [Harvey, 1971; Madhvanath et al., 1974; Al-Bedri and Harris, 1975].

The step from the average organ dose, a quantity still in use for toxicity comparisons, to the detailed dose distribution from a non-uniformly deposited radionuclide introduces immense complexity and greatly enhances the amount of labour necessary to arrive at useful results. This complexity arises from the intricate interplay of all factors involved, namely the changing pattern of distribution in an organ and the underlying physiological processes, the relationship between radiation sources and sensitive structures, the primary events of energy depositions, parameters characterizing cellular behaviour, etc. At present we are far from having achieved in any single instance a

Table 1
Physical characteristics of some alpha-emitters

Isotope	Energy [a] (MeV)	Half-life [a]	Range [b] in tissue (μm)
^{226}Ra	4.78 (95%) 4.60 (6%)	1602a	33.8
^{222}Rn (Em)	5.49 (100%)	3.82d	41.3
^{218}Po (RaA)	6.00 (100%)	3.05m	46.9
^{214}Po (RaC')	7.69 (100%)	1.64×10^{-4}s	66.9
^{210}Po (RaF)	5.31 (100%)	138.4d	39.4
^{232}Th	4.01 (76%) 3.95 (24%)	1.41×10^{10}a	26.2
^{228}Th (RdTh)	5.43 (71%) 5.34 (28%)	1.91a	40.4
^{224}Ra (ThX)	5.68 (94%) 5.45 (6%)	3.64d	43.2
^{239}Pu	5.16 (88%) 5.11 (11%)	2.44×10^4a	37.8
^{241}Am	5.49 (85%) 5.44 (13%)	458a	41.3
^{242}Cm	6.12 (74%) 6.07 (26%)	163d	48.1

[a] From Lederer et al. [1967] Table of Isotopes.
[b] Computed for mean particle energy from eq. (2.19) based on data of Walsh [1970].

nearly complete qualitative and quantitative description of the sequence of events leading from the deposition of a radionuclide in an organ to the manifestation of its harmful effects. Localized dosimetry could in principle play an essential role in such a description, but much remains still to be done.

Beyond its contribution to the elucidation of mechanisms, but intimately connected to it, localized dosimetry has its value in the scaling of radiation toxicities, thus setting up models which can be tested experimentally, or providing estimates and guidelines in cases where no experiments are possible. This aspect of employing the tools of localized dosimetry will be exemplified in section 5 dealing with the application of concepts to the deposition of radionuclides in the skeleton.

By the subsequent development of the matter treated in this article, starting from the physical principles of energy transfer to tissue and ending with the description of processes going on in real organs it was intended to illuminate the multidisciplinary character of this field of research. There is no doubt that just the amalgamation of ideas from several disciplines and the confrontation of different ways of thinking will prove beneficial for future progress, as it was in the past.

2. Energy dissipation and absorbed dose

2.1. Stopping power

The primary mechanism of physical energy dissipation of α-particles lies at the basis of any considerations about biological effects subsequently generated. Consequently, a quantitative description of energy transfer from the α-particle to tissue is an essential prerequisite for radiation dosimetry at the cellular level. The radiation dose D at a point P in tissue of density ρ may be calculated by the following fundamental expression [Roesch and Attix, 1968]:

$$D = \frac{1}{\rho} \int_0^{E_m} S(E)\, \Phi(E, P)\, dE , \qquad (2.1)$$

where S is the stopping power, i.e. the energy loss of the particle per unit pathlength, which generally depends on the particle energy E. $\Phi(E, P)$ is the spectral distribution of the particle fluence with respect to energy at point P [particles/(area × energy)] and E_m is the maximum energy of the particles considered. Eq. (2.1) elucidates the importance of an exact and quantitative knowledge of the slowing down of charged particles in tissue for any dose determination. The impartment of the energy of charged particles to biological matter is the consequence of several physical processes and the relative contribution of each process is determined mainly by the kinetic energy E of the particle. The combined effects of all these processes are summarized in the stopping power S. It is possible to specify energy ranges in which the assumption of a single mechanism of energy loss is appropriate to describe the slowing down of the particle. As all initial particle energies (E_0) to be considered here are below 10 MeV it is convenient to make the following division of energy ranges:

1. Low energy, below $\frac{1}{2} z^2$ (MeV) (z = particle charge) i.e. <2 MeV for α-particles. In this energy interval α-particles are subject to charge fluctuations, i.e. the positively charged α-particle may acquire orbital electrons from the stopping medium thereby reducing its effective charge. For all particle energies below $10^{-4}\, mc^2$ (m = particle mass, c = speed of light), i.e. below 0.4 MeV for α-particles, elastic collisions between the particle and whole atoms (nuclear collisions) make an important contribution to the stopping power.

2. Intermediate energy. This is the energy interval from 2 MeV upwards and extending beyond 4 GeV, covering, of course, the complete range of initial particle energies of all α-emitting nuclides of interest here (table 1). The essential mechanism of energy loss in this range is via Coulomb-interaction with atomic electrons resulting in excitations and ionizations of atoms or

molecules with roughly equal frequencies [Evans, 1962]. The secondary electrons (δ-rays) released after ionizations as caused by the primary α-particle produce further ionizations, excitations, radical formation, bond ruptures etc. [Katz et al., 1972).

It should be mentioned for completeness that nuclear interactions may occur in principle. However, due to the small probability of such events their contribution to the stopping power is negligible for the α-energies under consideration. The same is true for bremsstrahlung and polarization, the two other modes of energy transfer.

In the range of medium energy, where the interactions with the electrons of the medium are dominating, the stopping power S obeys the Bethe-Bloch formula [Bethe, 1930; Bloch, 1933],

$$S = \frac{dE}{dx} = -\frac{4\pi e^4 z^2 NZ}{m_e v^2} \left[\ln\left(\frac{2m_e v^2}{I(1-\beta^2)}\right) - \beta^2 - \Sigma_i\left(\frac{C_i}{Z}\right) \right], (\beta = V/C) . (2.2)$$

dE/dx is the energy loss per unit pathlength, v is the α-particle velocity, z is the particle charge, m_e and e is the mass and charge of the electron, respectively, N and Z are the number and charge of atoms in unit volume, respectively. I is an appropriate average over the excitation potentials of the different energy levels of the atom. The shell correction terms C_i play a role at low particle velocities and for light elements as stopping material. The effect of the shell correction amounts to approx. 10% of the total stopping power at low velocities [Bichsel, 1968]. Tables often give the quantities S/ρ (mass stopping power) or the stopping cross section $\epsilon = (1/n)(dE/dx)$ (n is the number of stopping atoms or molecules per unit volume) instead of S. The phenomenon of charge fluctuations below 2 MeV may be accounted for by introducing the effective charge z^* which then should be inserted instead of z into eq. (2.2). An expression for the effective charge was given by Barkas [1963] derived from emulsion measurements,

$$z^* = z[1 - \exp(-125 \beta z^{-2/3})] . \quad (2.3)$$

From eq. (2.3) it can be seen that for increasing values of β, z^* approaches z while for small β z^* decreases proportionally with β. Assuming a simple additivity of stopping effects eq. (2.2), though strictly valid for elements only, is applicable for compounds as well:

$$dE/dx = \Sigma_i (dE/dx)_i . \quad (2.4)$$

$(dE/dx)_i$ is the stopping power for the ith component. This rule, known as Bragg's additivity rule, implies that a mean excitation potential $\bar{I}$ may be defined:

$$\ln \bar{I} = \frac{\Sigma w_i Z_i \ln I_i}{\Sigma w_i Z_i} , \quad (2.5)$$

where w_i is the fraction of weight of the ith component. A similar expression should hold for the shell correction terms:

$$\bar{C}_i = \Sigma_j w_i C_{ij} . \qquad (2.6)$$

C_{ij} is the shell correction term for the ith energy level of the jth component. Table 2 shows the elemental composition of tissue and bone. It may be used for the determination of the mean excitation potential. The Bragg additivity rule is especially important for the determination of the stopping power for tissue. A direct measurement of the stopping power of α-particles would require the preparation of wet tissue samples of highly accurate thickness and handling the samples in a vacuum chamber with a monoenergetic α-particle beam. As this is not possible, the stopping power for tissue has to be inferred indirectly by one of the following alternative methods:

1. The stopping power of protons in the elements which constitute tissue may be used to calculate the corresponding values for tissue by means of the Bragg rule with a final transformation from protons to α-particles. Eq. (2.2) suggests that the last step may be accomplished by applying the relation:

$$S_\alpha = S_p \left[\frac{z_\alpha^*}{z_p^*} \right]^2 \qquad \text{for equal } \beta . \qquad (2.7)$$

z_α^*, z_p^* are the effective charges of the α-particle and the proton, respectively (eq. 2.3).

Table 2
The mean elementary composition of soft tissue and bone

Element	Atomic number	soft tissue [a]		cortical bone [b]		Excitation-potential I (eV) [c]
		weight (%)	g/cm^3	weight (%)	g/cm^3	
Hydrogen	1	10.4	0.106	3.4	0.068	18.7
Carbon	6	23.3	0.238	15.5	0.31	78
Nitrogen	7	2.5	0.026	4.0	0.08	85
Oxygen	8	63.1	0.644	44.1	0.882	88
Sodium	11	0.11	0.001	0.06	0.0012	143
Magnesium	12	0.01	0.0001	0.21	0.0042	154
Phosphorus	15	0.13	0.0013	10.2	0.204	172
Sulfur	16	0.20	0.002	0.31	0.0062	192
Calcium	20	0.02	0.0002	22.2	0.444	200
Total		99.77	1.02	99.98	2.0	

[a] Reference man ICRP Publication 23 [1975].
[b] Woodard [1962].
[c] Turner [1964].

2. A direct experimental determination of the stopping power of α-particles in elements avoids the particle transformation.

3. Experimental measurement of the stopping power in tissue equivalent gases avoids the application of the Bragg rule too. But there might still be some influences of chemical bonding and physical state.

4. Some plastics like Melinex or polycarbonate display a stopping behaviour similar to tissue. Therefore, they may be considered a substitute.

As pointed out by Reynolds et al. [1953], the applicability of the Bragg additivity rule may be questionable for the low energy range.

One of the early compilations of experimental data concerning the energy loss of charged particles, mainly protons, in matter below 10 MeV, was made by Whaling [1958], a work that is still referred to by many publications appearing more recently. Neufeld and Snyder [1960] determined the stopping cross section of tissue by application of the Bragg rule in the form:

$$\frac{dE}{dx} = -\Sigma N_i \overline{z^{*2}} \sigma_i , \qquad (2.8)$$

where σ_i is the specific energy loss of the ith component and was assumed to be a characteristic of the stopping material only (N_i is the number of atoms of type i per cm^3 of tissue). The computation of $\overline{z^{*2}}$, the average squared effective charge, was based on the experimental determination of $\overline{z^*}$ as a function of the particle velocity, and the known linear dependence of $\overline{z^{*2}}/z^2$ on $\overline{z^*}^2/z^2$. Rotondi [1968] calculated the stopping power of tissue by measuring the stopping power of the gases CH_4, O_2, N_2 and CO_2. His values for the stopping cross section are higher than those given by Neufeld and Snyder by as much as 50% in the low energy region (fig. 1), a discrepancy that may be attributed to the uncertainty in the determination of $\overline{z^{*2}}$ in the work of Neufeld and Snyder. Tissue was assumed to be composed of 10.1% H, 12.1% C, 4% N, 73.6% O (% by weight). There is a good agreement between Rotondi's values and the proton stopping power data of Whaling [1958], transformed to α-particles in tissue. Northcliffe and Schilling [1970] gave electronic stopping power and range data in the region between 0.01–12 MeV/amu, for several ions and materials, including the compounds polyethylene, Mylar and water. For the compounds the Bragg rule was applied. All data were derived from the two master stopping power curves of Al for solids, and Ar for gases. Walsh [1970] refined the Bragg additivity rule by taking into account charge exchange effects. His procedure was first to determine conversion factors for proton to alpha data of selected materials, for which data for both particles existed, and then averaging the conversion factors as a function of particle velocity. The proton data were taken from Reynolds et al. [1953], the alpha data from Kerr et al. [1966], and from experiments conducted by the author. The average ratio of alpha to proton stopping cross sections varies between 2

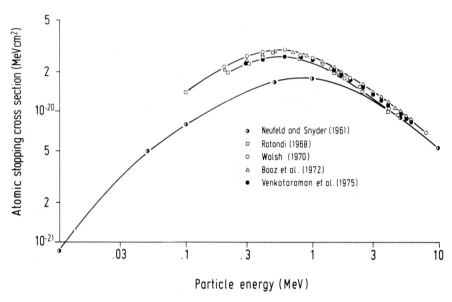

Fig. 1. Atomic stopping cross section $(1/n)$ (dE/dx) of α-particles in soft tissue.

and 4 in the proton energy range 25–500 keV. These conversion factors were then applied to proton data for the components of tissue. The application of the additivity rule yielded the stopping cross section for α-particles in tissue, as shown in fig. 1. The tissue composition assumed (63% H, 9.5% C, 1.36% N, 26% O, 9.46×10^{22} at./g tissue) differs drastically from that used in table 2 and by Neufeld and Snyder. The deviation between the stopping cross section of tissue and water amounts to 2% on an average. Walsh's stopping power values agree closely with Rotondi's in the range 0.1–10 MeV. Walsh and Pendergrass [1972] measured the residual energy of ^{241}Am α-particles, after the passage through varying thicknesses of absorbers of tissue equivalent plastic down to 1 MeV. A comparison with the results of the previously mentioned experiment revealed differences amounting up to 5–6% towards the end of the particle range, from which the authors concluded, that the physical state of the stopping medium and chemical bonding has an effect on the stopping power. Booz et al. [1972] used a tissue equivalent gas with composition as recommended by Rossi and Failla [1956] (10.19% H, 43.62% C, 3.52% N, 40.68% O) to determine the mass stopping power of α-particles in the energy range 0.2–6 MeV (fig. 1). Harley and Pasternak [1972] derived the stopping power in tissue as a function of the depth of penetration determined by alpha transmission measurements through polycarbonate foils. Polycarbonate is tissue equivalent in so far as it has a similar average excitation potential $\bar{I}$ (69 eV) as ICRP tissue (70 eV) [ICRP, 1959]. Therefore a correction for the electron density only has to be carried out. The energy transmission and stopping power curves were compared with theoretical curves from the Bethe-Bloch equation which was simplified for alphas of relatively low energy, according to the following expression:

$$\frac{dE}{dx} = \frac{0.1263}{E} \ln(7.99\,E) \qquad \text{(MeV/}\mu\text{m of tissue)}. \qquad (2.9)$$

For RaA (^{218}Po) α-particles, the results were in excellent agreement with the above equation which was additionally corrected for charge exchange in the region below 1.6 MeV. The measurements for RaC' (^{214}Po) α-particles showed some differences from the theoretically predicted values towards the end of the range which the authors attributed to a small error in the excitation potential. Al-Bedri and Harris [1975] compared the dependence of the residual particle energy in two tissue equivalent gases after passing tissue equivalent distances with that of the plastic Melinex. The effective atomic number of Melinex (6.24) agrees well with the value for tissue (6.54), and the electron densities are very similar too: 3.32×10^{23} g^{-1} for tissue and 3.14×10^{23} g^{-1} for Melinex. The composition of the gases was chosen as suggested by Rossi and Failla [1956] and Srdoč [1970]. While the agreement between the gases was found to be extremely good, there were some discrepancies compared to Melinex in the low energy domain. Armstrong and Chandler [1973] developed a computer program (called SPAR) for the computation of stopping powers and ranges for muons, charged pions, protons and heavy ions at energies from zero to several hundred GeV for any non-gaseous medium. The stopping power values published by Venkataraman et al. [1975] were based on data of Bourland et al. [1971] and Bourland and Powers [1971] for the gases H_2, N_2, O_2 and α-particles of 0.3–2 MeV, and on measurements of Wenger et al. [1973] with α energies of 2–5 MeV. The calculations were done for tissue with the composition formula $C_5H_{40}O_{18}N$ and water and show good agreement with the values of Walsh [1970] at least above 1 MeV particle energy. By means of the Bethe-Bloch formula (eq. 2.2) with the empirical charge correction and the theory of Lindhard and Scharff [1961], Kappos [1967a] computed the stopping power and residual energy of α-particles in tissue and bone. Fig. 2 shows the results as a function of the depth of particle penetration. A knowledge of the stopping power in bone is important, both for the calculation of radiation doses to cells residing in calcified tissue (osteocytes), and for dose determinations in soft tissue from the so-called "buried" activity, where the particle first penetrates a layer of bone to reach cells in the marrow. The tables of Barkas and Berger [1964] may be used to give, by appropriate transformation, values for the stopping power and ranges of α-particles above 4 MeV. Williamson et al. [1966] published comprehen-

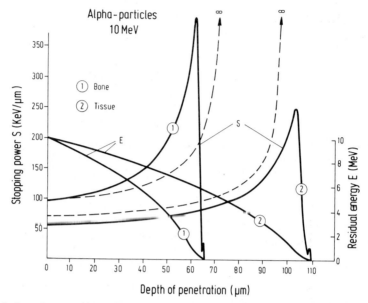

Fig. 2. Stopping power (S) and residual energy (E) as a function of depth of penetration in bone (1) and soft tissue (2). The dashed curves represent the stopping power derived from the range approximation $R = AE^m$ ($S_{\text{Tissue}} = 216E^{-0.5}$, $S_{\text{Bone}} = 159E^{-0.5}$) [Kappos, 1967a].

sive tables of the stopping power and ranges of ^{3}He particles and a broad variety of stopping media, which may be applied to α-particles by means of a simple change of the energy scale.

For practical use an appropriate averaging over the curves in fig. 1, excluding perhaps the measurements of Neufeld and Snyder, should yield stopping power values with an adequate accuracy, particularly if compared to other sources of error unavoidable in the dosimetry of internal α-emitters. As will be shown below (fig. 14), the deviations of dose-depth curves for very different and even crude approximations of the stopping power function are surprisingly small and not relevant over a considerable fraction of the range. A systematic investigation of the influence of the uncertainties in the stopping power on the absorbed dose and the stochastic quantities (section 3), especially towards the end of the particle range, does not yet exist, however.

2.2. Linear energy transfer and δ-rays

The Bethe-Bloch theory and its modifications quantitatively describe the slowing down of charged particles in matter, that is they are related to the state of the particle itself. From the viewpoint of the biological effects subsequently evoked, however, the mode of energy transfer to the surrounding medium is more relevant. To allow for this fact, the notion Linear Energy Transfer (LET, or simply L) was introduced. The concept of LET is quite distinct from stopping power, as the former takes into account if the energy is dissipated in the immediate neighbourhood of the track or if it is carried further away by means of δ-rays. Such a distinction may be essential, as the contribution of δ-rays may or may not be significant, depending on the size of the sensitive region and the LET [Elkind and Whitmore, 1967]. The International Commission on Radiation Units and Measurements (ICRU) defines the LET of charged particles as the quotient dE by dx, where dx is the distance traversed and dE is the mean energy loss due to collisions with energy transfers below some specified value Δ [ICRU, 1971],

$$L_\Delta = (dE/dx)_\Delta .\tag{2.10}$$

Thereby this definition specifies the amount of energy "locally" imparted. Here, "locally" means the environment of the particle track as determined by the cut-off energy Δ and the range energy relationship of the δ-electrons. L_∞ signifies the total LET which includes all energy losses and is therefore numerically equal to the stopping power, and L_{100}, for instance, is the linear energy transfer for a maximum energy of 100 eV [ICRU, 1971]. The subscript ∞ is used for convenience and should not be taken to mean, that an infinitely high energy transfer could occur. The ICRU [1970] report 16 gives a complete account of all aspects of linear energy transfer, including such questions as LET distributions and averages. Some of the δ-electrons produced by primary ionizations of the α-particle will be set in motion with sufficient kinetic energy to create their own track causing secondary excitations and ionizations. These δ-tracks are distinctly separate from that of the α-particle. According to Lea [1956] 5,207 primary ionizations per μm of tissue are caused by α-particles of 1 MeV, and 1,301 primary ionizations by 5 MeV particles. At the same particle energies 416 and 98 δ-rays of energy exceeding 100 eV are generated, respectively. Nearly half of the total number of ionizations are found in δ-rays of energy exceeding 100 eV. δ-electrons with lower energy have a range below 0.003 μm. Consequently, the ionizations caused by such secondary electrons are situated close to the α-particle track core and form clusters together with the primary ionizations. The Bethe-Bloch equation may be modified to give the following expression for L_Δ:

$$L_\Delta = \frac{2\pi e^4 z^{*2} NZ}{m_e v^2} \left[\ln \frac{2 m_e v^2 \Delta}{I^2(1-\beta^2)} - \frac{(1-\beta^2)\Delta}{2 m_e c^2} - \beta^2 - 2\Sigma_i \left(\frac{C_i}{Z}\right) \right] . \tag{2.11}$$

During its interaction with a heavy particle an electron may acquire a maximum amount of energy (w_{max}) in a head-on collision, and this upper limit may be calculated from kinematic considerations to be:

$$w_{max} = \frac{2m_e c^2 \beta^2}{1 - \beta^2} . \tag{2.12}$$

Eq. (2.12) indicates that the δ-rays become more confined to the track region, if the α-particle slows down and loses energy. Insertion of w_{max} for Δ in eq. (2.11) then yields the Bethe-Bloch equation. For 1 MeV and 5 MeV α-particles eq. (2.12) gives the values 0.54 keV and 2.73 keV for the maximum δ-ray energy, respectively. The whole spectrum of electrons set in motion as the result of the primary ionizations follows the well known δ-ray distribution law

$$dn = \frac{2\pi e^4 (z^*)^2 NZ}{m_e c^2} \frac{dw}{\beta^2 w^2} \qquad w \leq w_{max}, \tag{2.13}$$

$$dn = 0 \qquad w > w_{max}$$

where dn is the number of secondary electrons per unit length of path having energies between w and $w + dw$. The above expression is valid for energies less than 10 MeV/amu and is the simplified form of a more exact expression derived by Brad and Peters [1948]. An interesting conclusion may be drawn from eq. (2.12) and (2.13): the initial electron distribution is similar for all types of heavy ions with equal velocity. Only the absolute number of δ-rays varies with the square of the effective charge. The angle ϑ to the ion's path, at which the secondary electron is ejected, depends on the energy imparted and must be 0° for w_{max}. It can be shown that the relationship:

$$\cos^2 \vartheta = w/w_{max} \tag{2.14}$$

holds. An examination of the δ-ray distribution law (eq. 2.13) reveals that most of the δ-rays have energies much less than w_{max}. For this reason eq. (2.14) indicates that the majority of the δ-electrons leaves the ion's path perpendicularly. Based on the above considerations and assuming (a) that all electrons are ejected normally to the ion track, and (b) that the electron range is proportional to energy (with the constant of proportionality determined empirically), Butts and Katz [1967] derived the following expression for the δ-ray dose as a function of the distance r from the particle track:

$$D_\delta(r) = \frac{NZe^4}{m_e c^2} \cdot \frac{(z^*)^2}{r\beta^2} \left(\frac{1}{r} - \frac{1}{R_{max}} \right) \tag{2.15}$$

$$R_{max} = 6,230 \text{ g/cm}^2 \text{ erg} \cdot w_{max} .$$

Fig. 3 shows the δ-ray dose around the track in water for α-particles of several energies. The dose reaches extremely high values near the centre of the trajec-

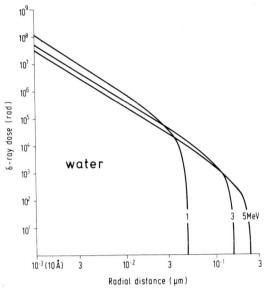

Fig. 3. δ-ray dose distribution in water around the trajectory of an α-particle with energy as specified. Calculations were made according to the formula of Butts and Katz [1967] (eq. (2.15) in the text).

tory, exceeding 100 Mrad around the Bragg-peak, and falls off very rapidly with distance. It is interesting to note that if $1/R_{max}$ can be neglected compared to $1/r$ the dose is independent of the precise form of the range energy relationship of electrons for small distances r.

The δ-ray theory of track structure is of value not only for dosimetric considerations but also for the quantitative description of various observations after irradiation with heavy ions. From a few basic assumptions, namely, that the dose distribution around an ion's path follows eq. (2.15), and that the response of the sensitive elements to δ-rays and their geometry is known, Katz and coworkers were able to describe the inactivation of dry enzymes and viruses by heavy ion irradiation [Butts and Katz, 1967] including a theory of RBE. The apparently universal applicability of this theory was further confirmed for the response of radiation detector materials in which atomic displacements play no important role [Katz and Butts, 1965; Katz and Kobetich, 1968, 1969] and for the inactivation of a variety of cell lines by heavy ion, fast neutron and pion bombardment [Katz et al., 1971; 1972].

2.3. Alpha-particle range

In spite of its apparent simplicity, the concept of particle range has its inherent difficulties, which led to the definition of several quantities, like pathlength, range, projected range, median projected range etc. A detailed discussion of the meaning of these quantities is out of place here because the differences are of minor importance for α-particles of low energy. Up to now there is no indication in the literature on the subject of mini- or microdosimetry of α-emitters that a distinction between the above mentioned parameters has turned out to be relevant. An α-particle may be absorbed by a nuclear reaction at any point on its normal path. The cross section for such an event is approximately equal to the geometrical nuclear cross section [Bichsel, 1968]. Consequently, there is no appreciable loss of α-particles by such events over distances below 100 μm. The other two factors, namely multiple scattering and the stochastic nature of the energy loss events are under practical circumstances negligible too. The range R of a particle of initial energy E_0 may be obtained by straightforward integration of the Bethe-Bloch equation,

$$R(E_0) = \int_{E'}^{E_0} \frac{dE}{S} + R(E'). \qquad (2.16)$$

The quantity R is called the range in the continuous slowing down approximation (csda), as the discrete events of energy transfer in random quantities are approximated by the smooth stopping power expression. The integration starts at some low energy E', corresponding to the low energy limit of validity of eq. (2.2) and the residual range $R(E')$ may be derived from empirical data.

A convenient and very useful method of computing range energy relationships in different materials from a master curve is the application of the famous Bragg-Kleeman rule. It states that for the ratio of ranges of particles with the same energy in two different media the following relation is valid:

$$\frac{R_1}{R_2} = \frac{\rho_2}{\rho_1} \left[\frac{\langle A \rangle_1}{\langle A \rangle_2} \right]^{1/2}. \qquad (2.17)$$

ρ_i is the density of the medium. The effective atomic number $\langle A \rangle$ for compounds may be calculated from

$$\langle A \rangle^{-1/2} = \Sigma f_i A_i^{-1/2}, \qquad (2.18)$$

where A_i is the atomic number of the ith component with weight fraction f_i. For water one obtains $\langle A \rangle = 9$, for soft tissue and compact bone with elemental composition as listed in table 2 the values are 5.62 and 8.37, respectively. Using the Bragg-Kleeman rule, Mays [1958] determined the range of 5.14 MeV α-particles in several materials including water (34.2 μm), collagen (31.1 μm) and compact bone (24.1 μm). It should be noted, however, that the approximation holds within 15% only. If, for instance, the range ratio in water and soft tissue is computed according to eq. (2.17) one obtains that the

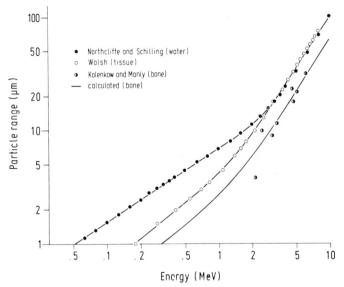

Fig. 4. α-particle range in water, tissue and bone. The range energy relationship in bone was determined by means of the Bragg-Kleeman rule and the data of Walsh [1970]. The computed curve is shown together with the measurements of Kolenkow and Manly [1967] in compact bone.

ranges in tissue should be shorter by 26% than in water, which certainly is too large a difference. Fig. 4 shows the range energy relationship of α-particles in water [Northcliffe and Shilling, 1970], soft tissue [Walsh, 1970] and compact bone. The latter curve was constructed by means of the Bragg-Kleeman rule and based on the data of Walsh. The range energy curve of Northcliffe and Schilling seems to overestimate the particle ranges at low energies. Brendle et al. [1975], when measuring the energy loss of α-particles in several compounds, obtained consistently lower range values than these authors. For the sake of convenience in analytical computations of dose distributions, a simple expression for the range energy relationship is often desired. Fig. 4 indicates that a fairly accurate approximation for energies above 1.5 MeV is represented by the analytical expression

$$R = AE^m. \qquad (2.19)$$

Fitting this expression to the experimental data of Walsh (fig. 4) yields $A = 3.62 \pm 0.22$ and $m = 1.43 \pm 0.04$ if R is in μm and E in MeV. The microdosimetric calculations of Charlton and Cormack [1962b] and Kappos [1967a] were carried out with $A = 3.08$ and $m = 1.5$. As will be seen below, however, some derivations of dose distribution formulas are based on the constant LET approximation, where $m = 1$ [Kononenko, 1957; Mays, 1958, 1960; Charlton and Cormack, 1962a, 1962b]. By appropriate differentiation of eq. (2.19) an analytical approximation for the stopping power may be obtained too. The error towards the end of the particle range becomes significant, however, as the analytical expression goes to infinity there (fig. 2). Zaidins [1974] used a more realistic two parameter expression as stopping power approximation, and a corresponding expression for the range which fit the experimental values more closely and nevertheless are still tractable computationally. He gave values of the adjustable parameters for protons, deuterons, tritons and a variety of target materials, but the determination of these parameters for tissue remains still to be done. Another approximate form of the range-energy relationship has been proposed by Kolenkow and Manly [1967]:

$$R = \frac{K_1}{2} E^2 + K_2 E . \tag{2.20}$$

In establishing this equation the authors were guided by the observation that the reciprocal of the stopping power as a function of the α-particle energy follows a straight line over a large range of energies. Based on experimental data in compact bone (fig. 4) they reported the values $K_1 = 1.04 \,\mu\text{m}/\text{MeV}^2$, $K_2 = 1.78 \,\mu\text{m}/\text{MeV}$. In dosimetric evaluations the values $K_1 = 1.46 \,\mu\text{m}/\text{MeV}^2$ and $K_2 = 4.58 \,\mu\text{m}/\text{MeV}$ were chosen for soft tissue [Kolenkow, 1967]. The above range-energy relationship implies that the often assumed constant range and stopping power ratio between soft tissue and bone is not strictly valid. With the aforementioned values of K_i the range ratio indeed varies between 2.3 and 1.88 for particle energies between 1 and 5 MeV.

2.4. Dose distributions for idealized geometrical conditions

Experimental dose determinations of α-emitters in tissue are nearly impossible. The reason is that many α-emitting nuclides are deposited non-uniformly in the relevant organs, even on a microscopic scale, and, consequently, the dose should be measured in volumes whose dimensions are small compared to the particle ranges. This situation would require a direct physical dosimetry in volumes with characteristic dimensions of 1 μm. Most information about dose distributions therefore was obtained from mathematical calculations assuming that:

1. there are at least two definite compartments separated by interfaces of known geometrical form (plane, spherical, cylindrical) and that the number of particles emitted per unit volume N is constant throughout such a compartment;

2. the emission of the α-particles is isotropic;

3. the particles are moving on straight paths, and

4. the energy dissipation along the particle's trajectory as dependent on LET may be described as a function of particle energy or distance traversed, suitable for analytical or numerical calculations. While the assumptions 2 and 3 seem to be acceptable a priori and very close to reality, assumption 1 and 4 may need some justification in particular instances, as errors introduced by these hypotheses may be considerable.

Kononenko [1957] made extensive calculations of mean doses and dose distributions for two media, separated by plane, spherical or cylindrical interfaces and labeled homogeneously, so that the number of decays per unit volume N_i in medium i ($i = 1, 2$) is constant. He further made the very important assumption that the LET along the particle's trajectory is constant. By considering the energy transported through an infinitesimal element $d\sigma$ of the separating interface from medium 1 to 2 and in the reverse direction he was able to derive expressions for the mean dose deposited. The mean values calculated are referring to boundary layers determined by the interface and the ranges R_i of the particles in either medium. The results for the different geometrical conditions may be summarized in the following general expression:

$$\bar{D}_1 = (1 - F) D_{\infty 1} + \frac{\rho_2 R_2}{\rho_1 R_1} F D_{\infty 2} . \qquad (2.21)$$

The mean dose $\bar{D}_1$ in the boundary layer of medium 1 can be expressed as the contribution of the doses of either medium far from the separating interface (equilibrium dose):

$$D_{\infty i} = \frac{E_0 N_i}{\rho_i} \qquad (E_0 = \text{particle energy}) \qquad (2.22)$$

with weighting factors as given in eq. (2.21). The factor in the second term of eq. (2.21) is partially composed of the ratio $\rho_2 R_2 / \rho_1 R_1$ which allows for the difference in stopping behaviour in the two media. F is a purely geometric factor. For a sphere with diameter d and medium 1 inside the sphere (fig. 5),

$$F(d, R_1) = \begin{cases} 1 - 3/8 \dfrac{d}{R_1} & \text{if } 0 \leqslant d \leqslant R_1 \\[2mm] \dfrac{3R_1}{4d} \left(1 - \dfrac{R_1^2}{6d^2}\right) & \text{if } R_1 \leqslant d \leqslant 2R_1 \; . \\[2mm] \dfrac{3}{4} \dfrac{d^2 R_1}{d^3 - (d - 2R_1)^3} \left(1 - \dfrac{R_1^2}{6d^2}\right) & \text{if } 2R_1 \leqslant d \end{cases} \qquad (2.23)$$

For an infinitely long cylinder with diameter d and medium 1 inside the

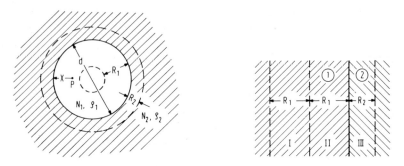

Fig. 5. Spherical cavity of diameter d and mass density ρ_1 embedded in a medium with density ρ_2. The corresponding particle ranges in the two media are R_1 and R_2, respectively.

Fig. 6. Two media with α-particle ranges R_1 and R_2 separated by a plane boundary. The average dose in region II is determined by contributions from regions I and III and region II itself.

cylinder:

$$F(d, R_1) = 1 - 0.7 \frac{d}{R_1} + 0.1625 \frac{d^2}{R_1^2} . \qquad (2.24)$$

Eq. (2.23) shows, that within a very small cavity ($d \ll R_1$) embedded in medium 2, F approaches unity and the mean dose inside the cavity is determined by the surrounding medium:

$$\bar{D}_1 = \frac{\rho_2 R_2}{\rho_1 R_1} D_{\infty 2} \qquad (\text{if } d \ll R_1) . \qquad (2.25)$$

The geometrical factor F for a plane interface may be obtained as the limit of eq. (2.23) for $\lim d = \infty$; then $F = 1/8$ and

$$\bar{D}_1 = \tfrac{7}{8} D_{\infty 1} + \frac{1}{8} \frac{\rho_2 R_2}{\rho_1 R_1} D_{\infty 2} . \qquad (2.26)$$

If, as shown in fig. 6, $\bar{D}_1$ is the mean dose in region II, then eq. (2.26) states, that the contribution of region I to the mean dose in II is $\tfrac{1}{8} D_{\infty 1}$ and the contribution from region III is made up of the same factor 1/8 and an additional factor related to the density ratio and the different ranges in 1 and 2. The second term in eq. (2.26) becomes dominating if

$$N_2 > 7 N_1 (R_1/R_2) , \qquad (2.27)$$

which for a bone-tissue interface with medium 2 as bone means, that the concentration of the isotope should be 11 times higher in bone than in soft tissue.

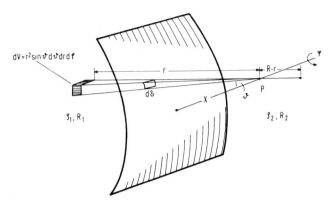

Fig. 7. Geometry for calculating the energy dissipation at P near a boundary separating two media of densities ρ_1 and ρ_2 and α-particle ranges R_1 and R_2, respectively. $x =$ shortest distance to the interface.

To calculate the depth-dose distribution across an interface separating two media with density ρ_1 and ρ_2 and particle ranges R_1 and R_2, respectively, one considers the energy dissipated at point P, situated at a distance x normal to the interface, from particles emitted in the infinitesimal volume $dV = r^2 \sin\vartheta \, d\vartheta dr d\varphi$ (fig. 7). The total pathlength R traversed by the particles lies between R_1 and R_2 depending on the relative positions of the particle origin, separating interface and point P. The dose at P then is obtained by summing up the contributions of the infinitesimal volume elements to yield the general expression

$$D = \frac{1}{4\pi\rho} \iiint_V N(r, \vartheta, \varphi) \left(\frac{dE}{dr}\right)_P \sin\vartheta \, d\vartheta \, dr \, d\varphi . \qquad (2.28)$$

The limits of integration depend on the geometrical form and dimensions of the radioactive deposit. D is the dose or dose rate, if N is the number of particles emitted per unit volume or per unit volume and unit time, respectively. The easiest way to solve eq. (2.28) for a particular irradiation geometry is to assume a linear range energy relationship (constant LET approximation), i.e. $dE/dr = E_0/R =$ const. Kononenko [1957] calculated the dose distribution in the constant LET approximation for plane and spherical boundaries. It is evident a priori that the dose distribution is a function of relative coordinates, i.e. that the magnification or reduction of all dimensions and ranges by a constant factor does not change the functional dependence. For a spherical boundary with medium 1 inside the sphere (fig. 5) the dose distribution in medium i ($i = 1, 2$) has the general form

$$D\left(\frac{x}{R_i}\right) = (1 - S_i) D_{\infty i} + \frac{\rho_j R_j}{\rho_i R_i} S_i D_{\infty j} \qquad \begin{array}{c} i, j = 1, 2 \\ i \neq j \end{array}, \qquad (2.29)$$

which is very similar to eq. (2.22) for the mean values. However, the geometry factor S_i is now a function of the relative coordinates:

$$S_i = S_i\left(\frac{x}{R_i}, \frac{d}{R_i}\right) \qquad i = 1, 2. \tag{2.30}$$

As the irradiation geometry is asymmetrical, $S_1 \neq S_2$. The calculations of Kononenko yielded the following results for the medium inside the sphere:

$$S_1(u, v) = \begin{cases} \frac{1}{2}\left[(1-u)\left(1 + \frac{1+u}{2(v-2u)}\right) + \left(u + \frac{u^2}{v+2u}\right)\ln u\right] & \text{for } v > 1 \\ \frac{1}{2}\left[2 - \frac{v}{2} - \left(u + \frac{u^2}{v+2u}\right)\ln\left(\frac{v}{u} - 1\right)\right] & \text{for } v < 1 \\ 0 & \text{for } u > 1 \end{cases}$$

$$u = \frac{x}{R_1}, \qquad v = \frac{d}{R_1} \tag{2.31}$$

and similar expressions for S_2. Fig. 8 gives the curves S_i for both sides of the spherical boundary. The geometry factor $P(u)$ for a plane boundary is obtained as the limit of eq. (2.31) for lim $v = \infty$.

$$P(u) = \begin{cases} \frac{1}{2}[1 + u(\ln u - 1)] & \text{for } u \leq 1 \\ 0 & \text{for } u > 1, \end{cases} \tag{2.32}$$

and the dose distribution may be determined by eq. (2.29), with S substituted by P. If $N_1 = 0$ eq. (2.29) yields,

$$D\left(\frac{x}{R_1}\right) = \frac{\rho_2 R_2}{\rho_1 R_1} P\left(\frac{x}{R_1}\right) D_{\infty 2}, \tag{2.33}$$

an expression which is identical to the formula derived by Spiers [1949] and Munson [1950] for the specific ionization from secondary electrons released by X-ray irradiation. The same factor $P(u)$ describes the dose distribution in medium 2 if x is taken to be the (positive) distance from the boundary and the relative distance u equals x/R_2. As can be seen from eqs. (2.29) and (2.32) there is a discontinuity of the dose distribution at the boundary surface, even if $D_{\infty 1} = D_{\infty 2} = D_\infty$. In this case,

$$D_1(0) - D_2(0) = \frac{D_\infty}{2}\left(\frac{\rho_2 R_2}{\rho_1 R_1} - \frac{\rho_1 R_1}{\rho_2 R_2}\right). \tag{2.34}$$

By means of the Bragg-Kleeman rule (eq. 2.17) the conclusion may be drawn

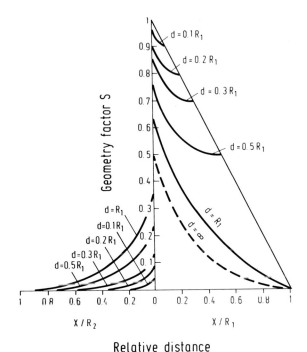

Fig. 8. Geometry factors S_1 (right side) and S_2 (left side) for a spherical interface of diameter d [Kononenko, 1957].

that the above difference is positive if the relation $\langle A \rangle_2 > \langle A \rangle_1$ holds for the effective atomic numbers of the two media, otherwise it is negative or zero. For $\rho_1 = \rho_2$, $R_1 = R_2$ and $N_2 = 0$, the contribution of a layer of activity of infinitesimal thickness dx on the boundary surface can be computed by differentiation of eq. (2.33) with respect to x taking into consideration the condition that $N_2 dx = n_A$, where n_A is the number of particles generated per unit area. This leads to an expression derived by Mays [1958],

$$D\left(\frac{x}{R}\right) = -\overline{D} \ln\left(\frac{x}{R}\right), \qquad (2.35)$$

where $\overline{D}$ is the average dose in the layer with thickness R.

$$\overline{D} = \frac{n_A E_0}{2R\rho}. \qquad (2.36)$$

Hindmarsh et al. [1958] considered the situation where charged particles (electrons) are generated uniformly throughout bone slabs of thickness Δ smaller than the particle range and calculated the dose in the adjacent soft

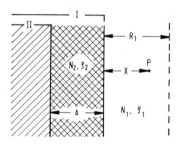

Fig. 9. Geometry for calculating the dose (or dose rate) at P in medium 1 due to a contaminated layer of thickness Δ in medium 2. The calculation is reduced to the difference I–II of the contributions from two infinite media. N_1 and N_2 are the number of decays per unit volume (and per unit time) in 1 or 2, respectively.

tissue with the constant LET approximation. The irradiation geometry is as depicted in fig. 9. The expression of Hindmarsh et al. can be derived from the formula of Kononenko for the plane boundary. Fig. 9 illustrates that the dose at point P may be written as the difference of the contributions of volume I and volume II. By introducing the new geometry factor $L[(x/R_1), (\Delta/R_2)]$ one has in medium 1:

$$D\left(\frac{x}{R_1}, \frac{\Delta}{R_2}\right) = \frac{\rho_2 R_2}{\rho_1 R_1} L\left(\frac{x}{R_1}, \frac{\Delta}{R_2}\right) D_{\infty 2}, \qquad (2.37)$$

where $L[(x/R_1), (\Delta/R_2)]$ may be expressed in terms of the geometry factor for the plane boundary,

$$L\left(\frac{x}{R_1}, \frac{\Delta}{R_2}\right) = P\left(\frac{x}{R_1}\right) - P\left(\frac{x}{R_1} + \frac{\Delta}{R_2}\right). \qquad (2.38)$$

It should be noted that not Δ/R_1 but the appropriate relative coordinate Δ/R_2 should appear in eq. (2.38). This is immediately evident if one sets $x = 0$ and $\Delta = R$, as in this case L must be $1/2$. By a similar procedure the factor L for medium 2 can be deduced:

$$L\left(\frac{x}{R_2}, \frac{\Delta}{R_2}\right) = \begin{cases} 1 - P\left(\frac{x}{R_2}\right) - P\left(\frac{\Delta}{R_2} - \frac{x}{R_2}\right) & \text{if } x \leq \Delta \\ \\ P\left(\frac{x}{R_2} - \frac{\Delta}{R_2}\right) - P\left(\frac{x}{R_2}\right) & \text{if } x > \Delta \end{cases}. \qquad (2.39)$$

Increasing the thickness Δ of the slab beyond R_2 has no further effect on the dose at any point in medium 1 if the specific activity in the slab is constant.

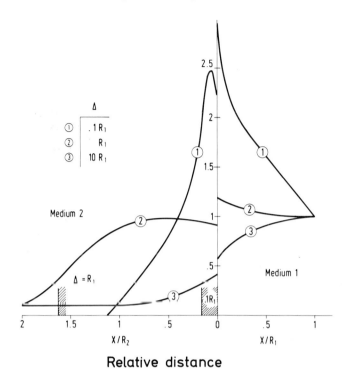

Fig. 10. Dose distribution across a plane interface (relative scale) separating medium 1 with equilibrium dose $D_{\infty 1} = 1$ due to a uniform label, from medium 2 in which a contaminated layer of thickness Δ is located adjacent to the interface. The amount of activity in the layer was kept constant ($N_2 \Delta$ = const.) for the three values of Δ chosen.

The maximum dose within the slab however increases until it attains the value $D_{\infty 2}$ for $\Delta \geqslant 2R_2$. Similarly, the dose at the surfaces of the slab, which can be calculated from eqs. (2.37) and (2.39) setting $x = \Delta$:

$$L\left(\frac{\Delta}{R_2}\right) = \frac{1}{2} - P\left(\frac{\Delta}{R_2}\right), \tag{2.40}$$

reaches its maximum value $D_{\infty 2}/2$ for $\Delta \geqslant R_2$. Again there is a discontinuity of the dose distribution across the boundary. The difference of the dose values on either side may be determined by eq. (2.29) for $x = 0$ and substituting S by L (with $D_{\infty 1} = 0$).

$$D_1(0) - D_2(0) = \left(\frac{\mu_2 R_2}{\rho_1 R_1} - 1\right)\left(\frac{1}{2} - P\left(\frac{\Delta}{R_2}\right)\right) D_{\infty 2}. \tag{2.41}$$

The step in the dose distribution approaches zero if Δ becomes small, and attains its maximum for $\Delta \geqslant R_2$. Fig. 10 represents the dose distribution across the boundary for a contaminated surface layer of thickness $0.1R_1, R_1$ and $10R_1$, respectively, in medium 2, and a homogeneously labeled medium 1. The values of $D_{\infty 2}$ for the three thicknesses were adjusted so that the amount of activity remains constant. Thus, the effect of increasing non-uniformity of the activity concentration on the dose values near the boundary is illustrated. The geometry factors $P(u)$ which were calculated under the assumption of an exponential range-energy relationship (eq. 2.19) (to be discussed below) were taken from Howarth [1965b] (table 4).

Spiers [1968] compared the average dose $\bar{D}_s$ in a layer of thickness aR_1 in medium 1 for a surface deposition at the boundary, to the average dose $\bar{D}_v$ from the same amount of activity distributed uniformly throughout a layer of thickness R_2 in medium 2. This dose ratio $\bar{D}_s/\bar{D}_v$ has a special importance with regard to the question of the relative hazards in the skeleton from incorporated surface- and volume-seekers. In this particular case medium 2 is bone, the flat interfaces are the endosteal surfaces and medium 1 signifies soft tissue. For the ratio $\bar{D}_s/\bar{D}_v$ he obtained the equation:

$$\left(\frac{\bar{D}_s}{\bar{D}_v}\right)_a = \frac{\Delta}{R_2} \frac{1 - \ln a}{1 - a(0.75 - 0.5 \ln a)} . \tag{2.42}$$

The original expression of Spiers was slightly modified in eq. (2.42) to hold for volume distributions of depth Δ larger than R_2. The numerical values of $\bar{D}_s/\bar{D}_v$ for several values of a are given in table 3. The ratio is approximately 4 over a large fraction of the irradiated volume in medium 1 ($\Delta = R_2$). Assuming that the activity is distributed uniformly over trabecular bone and approximating the trabeculae by parallel slabs of thickness $T = 288$ μm the values in the last column of table 3 result. Δ was set to $T/2$, as the deposition is on both surfaces of the trabecular slab. The trabecular thickness chosen is a mean value for the femur of man [Spiers et al., 1971]. Table 3 shows, that the ratio is then considerably larger amounting to 18.4 on the average. In their attempt to approximate trabecular bone by a simple geometrical configuration of hard and soft tissue, Hindmarsh et al. [1958] made similar calculations for cylinders of infinite length but finite thickness of the cylinder walls. The interior of the cylinder was considered to represent soft tissue while the cylinder itself was assumed to consist of bone. The geometry factors calculated, however, are valid only for the special case of points along the axis of the cylinder. Mays [1960] treated the special problem of a contaminated layer of finite thickness Δ, or infinitely thin, located at a distance d from the boundary plane in medium 2 (bone). This is the situation one encounters in the case of surface-seeking α-emitters where the initial endosteal surface depositions become "buried" consecutively by the apposition of new bone. The geometry factor of Mays, derived by direct integration, may be obtained by considering the dose from a slab of thickness $\Delta + d$ and subtracting the

Table 3
Ratio of average doses to depth aR_1 in medium 1 due to a surface deposition (D_s) or a volume deposition (D_v) extending to depth Δ in medium 2 [a]

Range fraction a	D_s/D_v	
	$\Delta = R_2$	$\Delta = 114\ \mu m$
0.1	4.08	19.4
0.2	3.78	18.0
0.3	3.70	17.6
0.4	3.71	17.6
0.5	3.75	17.8
0.6	3.81	18.1
0.7	3.87	18.4
0.8	3.94	18.7
0.9	3.98	18.9
1.0	4.00	19.0

[a] Adapted from Spiers [1968].

contribution from the slab extending from the boundary surface to depth d (fig. 11). Then one has for the dose in medium 1 (tissue) from a contaminated layer in bone

$$D_1\left(\frac{x}{R_1}, \frac{d}{R_2}, \frac{\Delta}{R_2}\right) = \frac{\rho_2 R_2}{\rho_1 R_1}\left[L\left(\frac{x}{R_1}, \frac{d+\Delta}{R_2}\right) - I\left(\frac{x}{R_1}, \frac{d}{R_2}\right)\right] n_{\infty 2}, \quad (2.43)$$

which, by insertion of eq. (2.38) for L and eq. (2.32) for P becomes identical to the expression of Mays:

$$D_1\left(\frac{x}{R_1}, \frac{d}{R_2}, \frac{\Delta}{R_2}\right)$$

$$= \frac{\rho_2 R_2}{\rho_1 R_1}\frac{1}{2}\left[\frac{\Delta}{R_2} + \left(\frac{x}{R_1} + \frac{d}{R_2}\right)\ln\left(\frac{x}{R_1} + \frac{d}{R_2}\right) - \left(\frac{x}{R_1} + \frac{\Delta+d}{R_2}\right)\ln\left(\frac{x}{R_1} + \frac{\Delta+d}{R_2}\right)\right].$$

(2.44)

By convention the second and third terms in parenthesis on the right side of eq. (2.44) have to be set to -1 if the arguments of the logarithms exceed 1. A close inspection of eqs. (2.44) and (2.38) for the slab of thickness Δ reveals, that they are of similar type. In fact, moving the slab in medium 2 a distance d away from the interface has the same effect on the dose at a point P with distance x in medium 1, as if moving this point a distance $R_1/(R_2 d)$ away from the interface. The limiting case of an infinitely thin layer may be

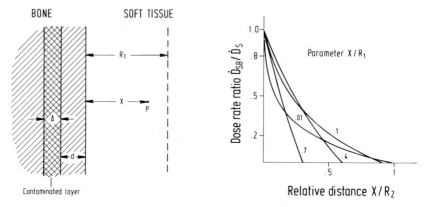

Fig. 11. Geometry for calculation of the dose at point P in medium 1 (soft tissue) due to a contaminated layer of thickness Δ located in medium 2 (bone) at a distance d from the interface separating the two media.

Fig. 12. Decline of dose rates at points with relative coordinates x/R_1 (= 0.01, 0.1, 0.4, 0.7) with increasing depth of burial x/R_2. On the ordinate the ratio of dose rates $\dot{D}_{SB}$ for a buried plane to the initial dose rate $\dot{D}_S(x/R_2 = 0)$ is plotted.

obtained from eq. (2.44) if Δ goes to zero while the product $N_1\Delta$ remains constant and equal to n_A, the specific area deposition. Then

$$\dot{D}_1\left(\frac{x}{R_1}, \frac{d}{R_2}\right) = -\frac{n_A E_0}{2R_1\rho_1} \ln\left(\frac{x}{R_1} + \frac{d}{R_2}\right). \tag{2.45}$$

The logarithm has to be taken as zero if its argument exceeds unity. It is interesting to see, how the dose rate $\dot{D}_1$ changes at a specific point in medium 1, if the contaminated interface becomes "buried" in medium 2. Eq. (2.45) gives the dose rate $\dot{D}_1$ instead of D_1, if n_A represents the number of particles produced per unit area and unit time. In fig. 12 the ratio of the dose rate of a radioactive plane at relative distance d/R_2 to the "initial" dose rate from a plane at the boundary between the two media $\dot{D}_{SB}/\dot{D}_S$ is plotted for three relative coordinates x/R_1 as parameters. The decline of the dose rate becomes steeper as the distance of the point lies beyond $x/R_1 = 0.3$. For points very close to the separating interface the dose rate declines rapidly for small values of d/R_2, but then decreases slowly (fig. 12). Spiers [1953] evaluated the dose variation in soft tissue enclosed between two slabs of bone. The expression for this geometrical arrangement may be derived readily by appropriate application of eqs. (2.33) or (2.37).

Charlton and Cormack [1962a, b] modified one of the basic assumptions of the theory discussed above, namely the assumption of constant LET. Bearing in mind that the LET of a 5 MeV α-particle varies by a factor of more than 2.5 from the starting point to the Bragg-peak (fig. 2) this assumption seems to be an oversimplification. The modified theory was applied to both secondary electrons produced by X-rays [Charlton and Cormack, 1962a] and α-rays [Charlton and Cormack, 1962b]. Expression (2.19) was used as range-energy relationship with coefficient m and factor A adapted appropriately to the case of α- or electron irradiation. Differentiation of eq. (2.19) yields an empirical LET-dependence of the form

$$\frac{dE}{dx} = -\frac{E_0}{mR^{1/m}} \cdot \frac{1}{(R-x)^{1-1/m}}, \qquad (2.46)$$

where R is the particle range and x the distance from the origin of the particle. By proper selection of m, normally 1.5 for α-particles and 1.75 for electrons, this analytic expression fits the Bragg-curve rather closely over a considerable part of the range. However, the discrepancy towards the end of the particle range is enormous as expression (2.46) goes to infinity there (fig. 2). Inserting eq. (2.46) in (2.28) yields geometry factors for the different irradiation conditions. For a plane interface, for instance, one obtains the integral

$$P\left(\frac{x}{R_1}\right) = \frac{1}{2} \int_0^{\arccos(x/R_1)} \left(1 - \frac{x}{R_1 \cos \vartheta}\right)^{1/m} \sin \vartheta \, d\vartheta, \qquad (2.47)$$

instead of eq. (2.32), and similar integrals for spherical and cylindrical cavities [Charlton and Cormack, 1962a, b; Aspin and Johns, 1963; Howarth, 1965a, b]. As these integrals obviously cannot be solved in closed form, Charlton and Cormack calculated them numerically. As an example the geometrical factors for radium and various cylindrical cavities were derived and the results compared to Spiers' [1953] parallel plate calculations and to the values of Hindmarsh et al. [1958] for the centres of cavities (fig. 13). If radium is in equilibrium with its daughters, the mean energy of the α-particles is 5.75 MeV with a corresponding range in tissue of 44 μm (fig. 4). The figures of Spiers are generally lower than those of Charlton and Cormack, as has to be expected from mere qualitative considerations (fig. 13). Additionally the variable LET calculations with monotonically increasing LET causes an enhancement of the factors especially near the axis. Consequently the factors of Hindmarsh et al. assume intermediate values.

Howarth [1965a, b] derived rapidly convergent series expansions of the corresponding integrals and published extensive tables of the numerical values of the geometry factors for plane, spherical and cylindrical boundaries. Pro-

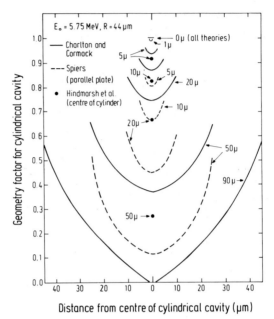

Fig. 13. Geometry factor for soft tissue cavity of varying width. The cavity is approximated by an infinite cylinder [Hindmarsh et al., 1958; Charlton and Cormack, 1962] or by an infinite slab [Spiers, 1953] and is surrounded by bone uniformly labeled with ^{226}Ra in equilibrium with its daughters. Dimensions in the figure indicate the diameter of the cylinder or the thickness of the slab. – – – and ● using the constant LET, ——— the variable LET approximation (eq. (2.46) in the text).

ceeding from the formula obtained for the plane interface

$$P\left(\frac{x}{R_1}\right) = \frac{1}{2}\left(1 - \frac{x}{R_1}\right)^{1/m}$$

$$\times \left[1 - \frac{x}{R_1} - \frac{x}{mR_1}\sum_{n=1}^{\infty}\left[\left(1 - \frac{x}{R_1}\right)^n \middle/ \left(\frac{1}{m} + n\right)\right]\right], \qquad (2.48)$$

he was able to express the corresponding factor for the spherical interface in terms of $P(x/R_1)$. As the expression is very intricate, we renounce to present it here. The tables of Howarth are reproduced in abbreviated form in tables 4 through 6. An alternative approach was adopted by Harvey [1971], who approximated the Bragg-curve by a parabola with two parameters adjusted appropriately to obtain a close fit. By application of this empirical approximation, equations for the dose rate from a point source and from a plane source were derived. If N is the source strength expressed in particles emitted per unit time, the expression for the dose rate $\dot{D}$ at a distance x from a point

Table 4
Geometry factors for a plane interface

x/R	Howarth [a]	Kappos [b]		
		4 MeV	5.3 MeV	7 MeV
0.0	0.5000	0.504	0.501	0.510
0.1	0.3658	0.351	0.353	0.361
0.2	0.2790	0.252	0.277	0.282
0.3	0.2109	0.179	0.215	0.202
0.4	0.1555	0.119	0.144	0.149
0.5	0.1101	0.073	0.092	0.106
0.6	0.0732	0.039	0.054	0.068
0.7	0.0439	0.017	0.026	0.032
0.8	0.0217	0.0036	0.007	0.012
0.9	0.0066	0.0003	0.0004	0.0014
1.0	0.0000	0.0000	0.0000	0.0000

[a] From Howarth [1965b], Br. J. Radiol. 38, 51.
[b] From Kappos [1967], Biophysik 4, 137.
x = distance normal to interface, R = particle range.

source is

$$\dot{D} = \frac{NE_0}{4\pi\rho R}\left[\frac{a}{x^2} + \frac{3b}{R^2}\right], \qquad (2.49)$$

where a and b are the empirical parameters determined by the fit to the Bragg-curve, with the condition $a + b = 1$. For the case $a = 1$ eq. (2.49) changes to the equation for the constant LET approximation. This correlates well with the meaning of a and b, as these parameters were selected so, that a signifies that part of the total energy absorbed at constant LET, while b signifies the fraction of energy absorbed at variable LET. Similarly one gets an expression for the plane source corresponding to that of Mays [1958] if $b = 0$, and with an additional correction term for $b > 0$.

Experimental dose rate measurements with thermoluminescent LiF were undertaken by Harvey and Townsend [1971] to ensure that the theoretical predictions were not widely different from the measured dose rates in a tissue equivalent substance (Melinex), adjacent to a plane source of α-particles. The experimental results were in reasonable agreement with theory but suggested a somewhat lower dose towards the end of the range and a higher dose near the surface. Another equation describing the depth-dose curves for α-particles was published by Walsh and McRee [1971]. These authors also compared the two cases of a plane source with uniformly distributed activity, and the same amount of activity contained in discrete aggregates situated in the plane, and

Table 5
Geometry factors for a spherical interface [a]

d/R	0	0.1	0.2	0.3	0.4	0.5	0.6	0.7	0.8	0.9	1.0
						$x/(d/2)$					
0.1	0.983	0.978	0.975	0.973	0.971	0.969	0.968	0.967	0.967	0.967	0.966
0.3	0.948	0.932	0.923	0.916	0.911	0.906	0.903	0.900	0.899	0.898	0.897
0.5	0.911	0.885	0.869	0.857	0.848	0.841	0.835	0.831	0.828	0.826	0.825
0.7	0.871	0.834	0.812	0.795	0.782	0.772	0.764	0.758	0.754	0.751	0.750
0.9	0.826	0.779	0.750	0.729	0.712	0.699	0.689	0.681	0.676	0.672	0.671
1.1	0.773	0.714	0.681	0.656	0.636	0.621	0.608	0.599	0.592	0.589	0.587
1.3	0.731	0.655	0.608	0.573	0.548	0.532	0.520	0.509	0.502	0.498	0.497
1.5	0.700	0.609	0.550	0.503	0.465	0.435	0.415	0.408	0.402	0.398	0.397
1.7	0.676	0.571	0.502	0.445	0.397	0.355	0.320	0.293	0.280	0.282	0.282
1.9	0.658	0.540	0.461	0.396	0.340	0.289	0.243	0.201	0.162	0.132	0.136
						x/R					
3.0	0.600	0.469	0.380	0.306	0.242	0.185	0.134	0.088	0.049	0.017	0
4.0	0.575	0.442	0.352	0.278	0.215	0.160	0.112	0.071	0.037	0.012	0
5.0	0.560	0.426	0.336	0.263	0.201	0.148	0.102	0.064	0.033	0.011	0
8.0	0.537	0.403	0.314	0.242	0.182	0.132	0.089	0.055	0.028	0.009	0
20.0	0.515	0.380	0.292	0.223	0.165	0.118	0.079	0.048	0.024	0.007	0
40.0	0.507	0.373	0.286	0.217	0.160	0.114	0.076	0.046	0.023	0.007	0
∞	0.500	0.366	0.279	0.211	0.156	0.110	0.073	0.044	0.022	0.007	0

[a] From Howarth [1965b], Br. J. Radiol. 38, 51.
x = distance within sphere normal to interface, d = diameter of sphere, R = particle range.

Table 6
Geometry factors for a cylindrical interface [a]

d/R	\multicolumn{11}{c}{x/(d/2)}
	0	0.1	0.2	0.3	0.4	0.5	0.6	0.7	0.8	0.9	1.0
0.1	0.966	0.960	0.957	0.954	0.952	0.950	0.949	0.948	0.947	0.947	0.947
0.3	0.897	0.879	0.869	0.860	0.854	0.849	0.845	0.842	0.840	0.839	0.838
0.5	0.828	0.798	0.780	0.767	0.756	0.747	0.740	0.735	0.732	0.730	0.729
0.7	0.751	0.719	0.693	0.674	0.658	0.646	0.636	0.629	0.624	0.621	0.620
0.9	0.697	0.643	0.609	0.584	0.563	0.547	0.534	0.524	0.518	0.514	0.513
1.1	0.648	0.576	0.531	0.497	0.471	0.451	0.434	0.422	0.414	0.409	0.407
1.3	0.622	0.534	0.475	0.428	0.389	0.360	0.339	0.324	0.313	0.307	0.305
1.5	0.604	0.502	0.434	0.378	0.330	0.289	0.255	0.231	0.217	0.209	0.206
1.7	0.591	0.477	0.401	0.338	0.284	0.237	0.195	0.160	0.131	0.118	0.144
1.9	0.581	0.455	0.372	0.303	0.245	0.194	0.148	0.109	0.074	0.044	0.032
	\multicolumn{11}{c}{x/R}
3.0	0.550	0.417	0.328	0.256	0.196	0.144	0.100	0.063	0.033	0.011	0
4.0	0.538	0.403	0.315	0.243	0.184	0.133	0.091	0.056	0.029	0.009	0
5.0	0.530	0.396	0.307	0.236	0.177	0.128	0.087	0.053	0.027	0.008	0
8.0	0.519	0.384	0.296	0.226	0.168	0.120	0.081	0.049	0.024	0.008	0
20.0	0.508	0.373	0.286	0.217	0.160	0.114	0.076	0.046	0.023	0.007	0
40.0	0.504	0.369	0.282	0.214	0.158	0.112	0.075	0.045	0.022	0.007	0
∞	0.500	0.366	0.279	0.211	0.156	0.110	0.073	0.044	0.022	0.007	0

[a] From Howarth [1965b], Br. J. Radiol. 83, 51.
x = distance within cylinder normal to interface, d = diameter of cylinder, R = particle range.

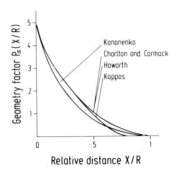

Fig. 14. Comparison of geometry factors $P(x/R)$ for a plane interface derived from different approximations of the LET-dependence. Kononenko [1957]: Constant LET. Charlton and Cormack [1962b], Howarth [1965b]: Power function approximation (eq. (2.47) in the text). Kappos [1967a]: Bethe-Bloch theory, charge correction and elastic collision theory.

gave a numerical example for the relative dose rates in the two cases.

Probably, the most accurate calculations of the geometry factor P for a plane boundary were carried out by Kappos [1967a]. For the LET-dependence the Bethe-Bloch equation (eq. 2.2) was assumed to be valid for energies above 2 MeV. In the range 10 keV to 2 MeV the stopping power was determined by interpolation and extrapolation of experimental data of protons using the conversion factors of Whaling [1958]. For very low energies the stopping power was computed according to the theory of Bohr [1948] and Lindhard and Scharff [1961]. The values of P are compiled in table 4 and show that P is dependent on the initial particle energy too. P increases with increasing energy and for a constant x/R this dependence is most pronounced towards the end of the range. A comparison between the factors P obtained under the different assumptions and taking Kappos' data as exact reveals (fig. 14), that Kononenko's theory underestimates P for small and intermediate distances, and slightly overestimates for x/R above 0.6. The theory of Charlton, Cormack and Howarth leads to values exceeding the curve of Kappos above $x/R = 0.4$. Consequently, it is not immediately obvious if the exponential LET-dependence is really an improvement over the simpler constant LET theory with respect to this particular geometrical situation. It should be stressed, however, that such differences must not be overemphasized, as for all practical applications the uncertainties introduced by the incomplete information about the biological factors are preponderant and exceeding by far the errors of the physical dose calculations.

3. Stochastic microdosimetry

3.1. Quantities and distributions

The way of calculating doses on a microscopic scale, i.e. in regions comparable to or lower than the α-particle range, as described in the foregoing section, has some inherent limitations. From the viewpoint of this type of microdosimetry (which may be termed classical or non-stochastic microdosimetry) only such situations are of interest, in which an inhomogeneous distribution of the particle sources exists, yielding non-constant dose or dose rate values. This approach is preferred in many practical applications because of its simplicity and it was successful for the correlation of some observed radiation effects to the amount of energy deposited. However, it neglects two essential aspects: (i) the fundamental nature of the process of energy deposition itself, which basically is a sequence of stochastic events confined to a small region around the track core, and (ii) the fact that biological substrates are composed of functional entities such as cells, organelles, molecules, thereby necessitating the introduction of the concept of the biological "target". This concept rests on the notion that the energy causing a specific biological effect originates in a volume, the "sensitive region", that is in most if not in all instances no larger but probably a good deal smaller than the cell. The term "sensitive volume" or "target" should be interpreted in a general sense not only as simply identifiable with the volume of some known biological entity, likely to respond to lesions produced therein, but should be conceived of as a formal target. This "formal target" then may define the volume where the probability of action of the primary events of energy deposition to produce a biological effect is non-zero [Zimmer, 1961], an equivalent volume corresponding to a target with indefinite boundary [Lea, 1956] or a diffusion distance etc. The energy deposited in individual sensitive sites is a stochastic quantity and averaging the energy density over a large number of targets is required to yield the dose values computed with the aid of the expressions in the preceding section. Kellerer [1975] established a quantitative criterion under what conditions the local dose variations in a site must be taken into account and when the characterization of radiation exposure by the quantity "absorbed dose" is inadequate. The shaded regions in fig. 15 specify for what combinations of site diameter and absorbed dose D the relative deviations of the local dose in the site from the mean value D exceed 20%. This surprisingly stringent criterion reveals that there is practically no justification for applying the absorbed dose concept with good conscience to most of the experiments dealing with internal α-emitters.

While the concept of LET turned out to be adequate and sufficient for the classical treatment of microdosimetry, this is no longer the case for a com-

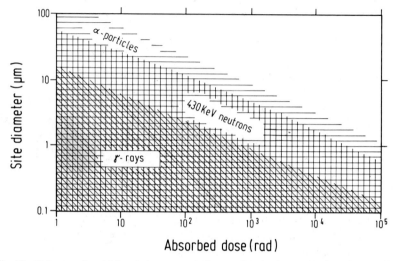

Fig. 15. Criterion for taking into account the stochastic dose variations in spherical targets. The shaded regions for α-particles, 430 keV neutrons and γ-rays indicate for which combinations of site diameter and absorbed dose (D) the mean deviations of the local dose from D exceed 20% [Kellerer, 1975].

plete description of the stochastic process of energy deposition in small volumes. The LET represents a mean energy loss per unit length interval. The real energy loss for a certain track segment is a stochastic quantity which may, especially in the case of low LET and small sensitive volumes, deviate considerably from its expectation value. This phenomenon will be discussed below. Another important complication are δ-rays which, because of their spread-out from the track core, lead to a diffuse track structure. Even for α-particles where the δ-ray range is always small, it may be long compared with the characteristic dimensions of a sensitive structure and consequently the energy transferred by a primary event within the target may be absorbed outside. In special cases, e.g. for electrons towards the end of their range, track curvature becomes a pronounced effect and if the radius of curvature is comparable to the mean distance between the loci of the primary events, the LET concept completely loses its meaning. Moreover, depending on the size of the target, the LET may vary considerably within the target itself, in particular where the variation of LET is large, i.e. in the region of the Bragg-peak. Kellerer and Chmelevsky [1975d] made a careful analysis of the applicability of LET, based on calculations of the contribution of the individual stochastic factors, such as track segment variation in the target volume, LET-distribution, straggling, and δ-rays, to the total relative variance of energy deposition. The factor with the largest contribution to the relative variance

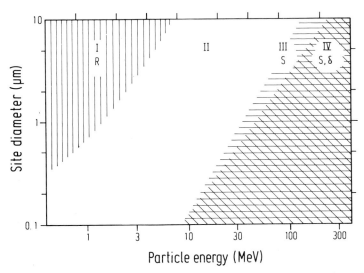

Fig. 16. Diagram showing the parameter space in which the LET-concept is applicable (region II). Parameters are the α-particle energy and the diameter of a spherical site. In regions I the finite particle range and in region III and IV energy straggling and δ-ray escape have to be taken into account [Kellerer and Chmelevsky, 1975d].

was regarded as decisive for the description of the energy dissipation process. Straggling and energy loss by δ-rays becomes important for particle energies above 8 MeV (fig. 16). For energies below this limit the variation of LET, and for larger size diameters the incomplete particle traversals due to the finite particle range are the dominating factors. Because of inadequacies of the LET-concept summarized above, Rossi and coworkers introduced a completely new set of quantities, the exact definitions of which were laid down later on in report 19 of the International Commission on Radiation Units and Measurements [ICRU, 1971]. Up to now several comprehensive articles were issued presenting a general outline of stochastic microdosimetry [Rossi, 1961, 1966, 1967a, 1967b, 1968; Kellerer and Chmelevsky, 1975a, b, c; Hug and Kellerer, 1966; Spiers, 1970; Numakunai, 1974]. In the following the aspects of stochastic microdosimetry, related to the energy deposition by α-particles, will be emphasized.

The microdosimetric quantity providing a link to the LET-concept is the event size, Y, where by event the passage of a charged particle through or near a test volume is meant. The energy E imparted within the volume under consideration may result from the direct interaction of the particle with the material in the volume or from δ-electrons originating in the vicinity from particles with glancing incidence, or, depending on the shape of the test

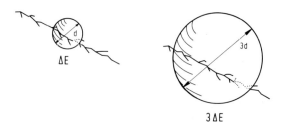

Fig. 17. Schematic drawing of events demonstrating that the event size Y is nearly independent of the size of the spherical target for tracks having similar eccentricity [Rossi, 1968].

volume, from both processes. Usually for the sake of simplicity and because of its isotropic properties a spherical shape is chosen. Then the ratio

$$Y = E/d \qquad (d = \text{sphere diameter}) \tag{3.1}$$

is approximately independent of the size of the sphere in a certain range of d. This can be demonstrated by fig. 17 in which the trajectory of a particle is drawn, crossing two spheres with different diameters but with the same relative distance to the centre. The track segment within the volume is then proportional to the sphere diameter and consequently Y is a constant. For charged particles of LET which is practically constant over the size of the sphere and passing on parallel straight lines the distribution of Y then corresponds directly to the distribution of the track segments. The latter may be computed from simple geometrical considerations to be

$$p(x) = \begin{cases} 2x/d^2 & \text{for } 0 \leqslant x \leqslant d \\ 0 & \text{otherwise} \end{cases} \tag{3.2}$$

(x is the track segment length), i.e. a triangular distribution as shown in fig. 18 [Rossi, 1960, 1968]. The problem of the pathlength distribution from random intersections of straight lines was solved for more general conditions, namely for prolate spheroids [Biavati, 1966a], for spheres and divergent particle rays [Biavati, 1966b] and for cylinders [Wilson and Emery, 1967; Birkhoff et al., 1970; Kellerer, 1971]. Kellerer [1971] derived an important theorem relating the sum distribution $F(x)$ of the lengths of isotropic chords in a convex body to the track segment length produced by intersections in the body in an isotropic uniform field of straight tracks, whose track lengths R have a sum distribution $G(R)$. Then the sum distribution of the track segment length s is

$$P_s(x) = K[F(s) \int_s^\infty G(R) \, dR + G(s) \int_s^\infty F(x) \, dx]. \tag{3.3}$$

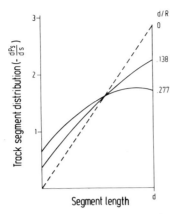

Fig. 18. Track segment distributions for three different values of d/R (d = site diameter, R = particle range). The dashed line shows the triangular distribution for parallel chords intersecting the sphere. The parameters d/R = 0.138 and 0.277 were chosen so as to correspond to a 5 μm and 10 μm diameter site in soft tissue crossed by ^{239}Pu α-particles.

For the special case of a monoenergetic field of α-tracks with a single fixed range R eq. (3.3) reduces to:

$$P_s(s) = K[(R-s)F(s) + \int_s^\infty F(x)\,dx] \qquad (s \leqslant R). \qquad (3.4)$$

$$K = \bar{R} + \bar{l}, \qquad (3.5)$$

where $\bar{R}$ is the mean particle range ($R = \bar{R}$ for eq. 3.4) and $\bar{l}$ is the mean chord length, for which Czuber [1884] showed (by expanding Cauchy's theorem) that for an arbitrary convex body of volume V and surface S the relationship

$$\bar{l} = 4V/S \qquad (3.6)$$

holds. The sum distributions $P_s(s)$, $F(s)$, $G(R)$ designate the probability of the corresponding random variables to exceed the values s and R, respectively, contrary to the usual habit to specify the probability for an occurrence below these values. Eqs. (3.3) and (3.4) may be rewritten to show the relative contributions of the four different kinds of track segments, namely those from particles starting and ending in the volume (insiders), particles crossing (crossers), particles starting outside and stopping in the volume (stoppers) and particles starting in the volume and stopping outside (starters). Fig. 18 shows the track segment distributions for a sphere with diameter d and particle tracks with length R and the two different ratios d/R = 0.138 and 0.277, respectively. With increasing ratio d/R the relative contribution of small track segments increases due to starters and stoppers and the distribution differs more and more from the triangular form. Insiders are only possible for $d/R > 1$. It is important to note that the definition of the event size Y may be extended to hold for arbitrary convex bodies if one defines

$$Y = \tfrac{2}{3} E/\bar{l} , \tag{3.7}$$

where $\bar{l}$ is determined by eq. (3.6). This definition differs by the factor 2/3 from that given by Rossi [1967a]. It makes eq. (3.7) to reduce to eq. (3.1) for a sphere. From the sum distributions (eqs. 3.3 or 3.4, respectively) the relative frequency of track segments per unit interval of s may be derived by differentiation.

It is evident, merely from an inspection of track segment distributions like those shown in fig. 18, that even under the assumption of constant LET a broad variety of local energy increments from single events Y may be observed. Consequently, if a biological effect depends on the amount of energy deposited in a specific target volume or becomes manifest only if a critical threshold of deposited energy is exceeded, a complex dependence of the effect frequency on LET must be expected, whereas describing the effects in terms of Y-spectra and local energy density is more adequate for the nature of the processes involved.

For the LET two distributions are defined, namely the distribution of track length in LET, $t(L)$, giving the fraction of track lengths at a LET equal to L per unit LET interval, and the distribution of dose in LET, $D(L)$, designating the fraction of dose deposited at a LET equal to L per unit LET interval. The first moments of these distributions, called the track average LET and the dose average LET,

$$\bar{L}_T = \int_{L_{min}}^{L_{max}} Lt(L)\, dL \qquad \text{track average LET}$$

$$\bar{L}_D = \int_{L_{min}}^{L_{max}} LD(L)\, dL \qquad \text{dose average LET} , \tag{3.8}$$

[Rossi, 1966, 1967a] normally differ widely [Elkind and Whitmore, 1967]. Similar functions exist describing the distribution of dose in Y, $D(Y)$, and the distribution of event frequency in Y, $p(Y)$, specifying the relative number of events at Y, per unit interval of Y. The derivation of the probability average $\bar{Y}_p$ and the dose average $\bar{Y}_D$ can be carried out in close analogy to eq. (3.8).

$$\bar{Y}_p = \int_0^{Y_{max}} Yp(Y)\, dY \qquad \text{probability average}$$

$$\bar{Y}_D = \int_0^{Y_{max}} YD(Y)\, dY \qquad \text{dose average .} \tag{3.9}$$

In general $p(0) > 0$, i.e. there are events depositing no energy in the test volume. This may be due to the passage of sparsely ionizing particles through small volumes or in larger volumes if the particle incidence is very close to the edge. As with the corresponding LET averages, the two quantities $\overline{Y}_p$ and $\overline{Y}_D$ are normally different. $D(0)$ is always zero. The simple relationship

$$D(Y) = p(Y) Y / \overline{Y}_p \tag{3.10}$$

allows to change from one distribution to the other [Rossi, 1966, 1967a]. The distributions discussed above are normalized and therefore by definition dose independent. The absolute number of events Y per unit interval, however, must be a function of dose. It is this quantity that can be directly assessed by experimental measurement and from which $p(Y)$ and $D(Y)$ must be derived. If in an experiment a total number $m(Y)$ of events are recorded at event size Y per unit interval of Y, in a large number N_0 of spheres at a dose D, then

$$n(Y) = m(Y)/N_0 \cdot D \tag{3.11}$$

is the event spectrum referring to a single sphere and normalized to unit dose. By means of $n(Y)$ the event frequency integral $\Phi(Y)$ may be computed. It specifies the occurrence of events having a size in excess of Y in a single sphere, per unit of absorbed dose.

$$\Phi(Y) = \int_Y^{Y_{max}} n(Y) \, dY. \tag{3.12}$$

$\Phi(Y)$ is a monotonically decreasing function of Y and $\Phi(0)$ is the total number of events of any size occurring per unit of absorbed dose. Y_{max} equals $L_{max} \cdot d$ if the variation of LET across the sensitive volume or straggling effects are neglected. The event probability spectrum $p(Y)$ is obtained by simply normalizing $n(Y)$.

$$p(Y) = \frac{n(Y)}{\int_0^{Y_{max}} n(Y) \, dY}. \tag{3.13}$$

From the definition of Y (eq. 3.7) and $\overline{Y}_p$ (eq. 3.9) it is evident that there must be a direct relationship between $\overline{Y}_p$ and the dose D. If on the average a total number N_0 of events are registered in a sensitive volume V then

$$D = 6 \overline{Y}_p \cdot N_0 / S \cdot \rho \tag{3.14}$$

where S is the surface of the convex volume and ρ its mass density. Then it can be shown easily that

$$\Phi(Y) = \frac{S \cdot \rho}{6\overline{Y}_p} \int_Y^{Y_{max}} p(Y)\,dY, \qquad (3.15)$$

and, as $p(Y)$ is normalized

$$\Phi(0) = S \cdot \rho/6\overline{Y}_p. \qquad (3.16)$$

The usefulness of the unnormalized distribution function $\Phi(Y)$ lies in the fact that it contains the information about the relative effectiveness of different radiation in producing certain biological changes, especially in instances where the induction of the effect under consideration may be described under the hypothesis that an event threshold Y_{th} exists and a single event is responsible for the manifestation of the effect. $\Phi(Y)$ was therefore defined by integrating over the upper tail of $p(Y)$, as the relationships become then especially simple. It may be readily verified that for the simple model specified above the fraction of targets not showing the effect follows the single exponential relationship known from classical hit theory. Hence

$$D_{37} = 1/\Phi(Y_{th}). \qquad (3.17)$$

Abandoning the hypothesis of an event threshold on the other hand and assuming that any event is effective, eq. (3.17) represents an estimate of the target size. For a spherical target for instance the minimum diameter is determined by

$$\Phi(0, d_{min}) = 1/D_{37}. \qquad (3.18)$$

The probability of occurrence of large events increases with LET and consequently one may conclude, on the basis of the one-event model, that the effectiveness of high LET radiation to produce effects with a threshold energy is larger than those of low LET radiation [Rossi, 1967a].

While the stochastic quantity Y was introduced to replace its analogous quantity of the classical theory (the LET – both quantities have equal dimensions) the stochastic quantity corresponding to the macroscopic dose is the local energy density z. If m is the mass of the target volume and E is the energy imparted after a series of events, then

$$z = E/m. \qquad (3.19)$$

In report 19 [ICRU, 1971] the energy imparted is defined as the sum of the energies (excluding rest energies) of all those directly and indirectly ionizing particles which have entered the volume, minus the sum of the energies (excluding rest energies) of all those directly and indirectly ionizing particles which have left the volume, plus the net energy produced or expended in any transformations of nuclei and elementary particles which have occurred within the volume. Each single event Y contributes a certain energy increment ΔE or equivalently an increment Δz to the local energy density, equal to

$$\Delta z = 4Y/(S \cdot \rho). \qquad (3.20)$$

z is by definition dependent on the dose D, the size of the test volume and the characteristic properties of the radiation. Considering a single target out of a large ensemble of targets, exposed to a radiation field, the quantity z may deviate widely from the dose D. The average z taken over a large number of targets, however, should approach D very closely. For low LET radiation, where even for relatively small doses a large number of events may have occurred in the sensitive volume, the probability of finding z in a single target close to D is relatively high. For high LET radiation and low doses, but few targets may have experienced a passage at all, but in those hit z may exceed D by orders of magnitude. In fact, the distribution of z, in general being dose dependent, may become nearly independent of dose for very low doses and if the distribution refers only to targets hit. For very high doses or increasing target size the local energy density z again approaches the macroscopic dose. The distribution of the increments of local energy density Δz, often termed single event spectrum $f_1(z)$, is independent of dose and, as is obvious from eq. (3.20), is closely related to the event spectrum $p(Y)$. Indeed one has

$$f_1(z) = \frac{S \cdot \rho}{4} p\left(\frac{S \cdot \rho}{4} z\right). \qquad (3.21)$$

$f_1(z)$ plays a central role in microdosimetry, as a knowledge of this spectrum allows to derive the dose dependent z-distribution $f(z, D)$ for any target, i.e. the relative frequency per unit interval of z with which a local energy density z is produced by one, two, or more events. In accordance with what has been stated above the following relationships hold for the averages of Δz and z, respectively:

$$\bar{z} = \int_0^{z_{\max}} f(z, D) z \, dz = D$$

$$\bar{z}_F = \int_0^{\Delta z_{\max}} f_1(z) z \, dz = \frac{4\bar{Y}\rho}{S \cdot \rho}. \qquad (3.22)$$

$\bar{z}_F$ is the frequency mean of the single event spectrum. If a macroscopic dose D was applied, $\bar{z}_F$ determines the mean number $D/\bar{z}_F$ of absorption events in the sensitive volume. $zf_1(z)dz$ is then a measure of the fraction of dose delivered at z per unit interval of z and correspondingly

$$\bar{z}_D = \frac{1}{\bar{z}_F} \int z^2 f_1(z)\,dz \tag{3.23}$$

may be interpreted as the energy mean of z.

$\bar{z}_D$ turns out to be a decisive quantity in the theory of dual radiation action [Keller and Rossi, 1971, 1972, 1973; Rossi and Kellerer, 1973]. These authors showed that if the number n_ϵ of primary lesions in a target is proportional to z^2, then one obtains for the number of primary lesions as a function of the macroscopic dose D the expression

$$n_\epsilon(D) = K(\bar{z}_D D + D^2). \tag{3.24}$$

This equation displays the different effectiveness of different kinds of radiation via $\bar{z}_D$, which is very small in the case of low LET radiation and large for high LET radiation. Equating the effects according to eq. (3.24) for two radiations yields the iso-effect ratio between the two types of radiation and thus a functional dependence of RBE on dose.

Up to this point all quantities, namely z, $f(z, D)$, d, D necessary for a complete description of local energy deposition on a microscopic scale have been introduced. However, the question arises, how to obtain information about the distribution of the basic quantities event size Y, and local energy density z, which constitute the starting point for practical applications. A direct experimental measurement of Y or z distributions would require dosimeters of a size equal to the size of the biological target, i.e. smaller or at most equal to the dimensions of a cell. To avoid the almost insuperable technical difficulties one would encounter in manufacturing such small dosimeters, Rossi and Rosenzweig [1955] developed a method utilizing two important principles advanced by Failla. One of these is, that cellular or subcellular volumes may be magnified to convenient dimensions (several centimeters) if this enlargement is compensated for by a corresponding reduction in density, which consequently necessitates the use of a gas. The other is, making the essential parts of such dosimeters, both wall and gas, tissue-equivalent. By an appropriate reduction of gas pressure the linear magnification factor obtainable with such a measurement system may range up to 100,000. A spherical proportional counter constructed along the lines mentioned above could simulate spherical sensitive volumes of unit density in tissue down to 0.1 μm. Spherical proportional counters of the Rossi-type were used to measure the z-spectra of ^{60}Co-γ-rays and neutrons of different energies [Rossi, 1961]. But there are shortcomings associated with the performance of proportional gas counters. Only such energy losses are detectable, leading to ionizations in the gas, which means that the minimum amount of energy loss is that required for the production of an ion pair (~30 eV). Besides the various stochastic factors instrumental to the true shape of the z-spectrum there are additional factors of technical nature. The number of ions produced along a particle track segment is fluctuating (Fano fluctuations) even if the total energy loss along the track segment is considered to be constant. The electron avalanche formation is of stochastic nature as well and introduces additional fluctuations. Both these factors contribute to the pulse heigh spectrum of the counter, which is actually measured, but their influence can be made small [Kellerer, 1968]. Another factor, namely the possible difference in the energy transfer process (density effect) for condensed matter and gaseous media is of minor importance [Rossi, 1967a]. For densely ionizing particles of low energy, e.g. α-particles from internal emitters, only wall-less proportional counters are applicable, because of the low penetration of the radiation and for other reasons [Gross et al., 1969a]. The main types of spherical counters developed are the simple wall-less counter with two rod-like or spherical electrodes [Glass and Braby, 1969], the ring counter [Gross et al., 1969a, b], the grid walled counter [Gross et al., 1969a, 1970]. Wilson [1969] designed a cylindrical wall-less counter. The ring counter consists of a helical cathode and a concentric anode wire, both defining a cylindrical region of amplification and a concentric ring which together with two field tubes determines the collecting volume. The grid counter has the ring replaced by a tissue equivalent grid of spherical shape, to overcome the uncertainties and poor definition of the shape of the collecting volume associated with the other counter types [Gross et al., 1969a]. The principle of performance may be demonstrated by fig. 19, displaying schematically the electric field and equipotential lines of the wall-less counter constructed by Glass and Braby [1969]. Electrons set free by the passage of an ionizing particle are accelerated in the electric field towards the anode, following the electric field lines if their drift velocity is much greater than their thermal diffusion rate. The latter condition may be fulfilled by an appropriate choice of the electrode potentials for a desired gas pressure. The collecting volume is determined by a guard electrode which intercepts all field lines outside the counter region marked by the dashed line. Electrons generated outside this volume are blocked by the guard electrode and prevented from reaching the anode and being registrated. Gross et al. [1969a, 1970] used both the ring counter and the grid walled counter to determine the event spectra for α-rays from an ^{241}Am source. Though the particle beam was uncollimated, the track segment spectra were negligibly different from those of a parallel beam. Fig. 20 shows the results obtained for the three simulated sphere diameters 0.5 μm, 1 μm, 2 μm. The data are displayed using as abscissa Y the quantity $E/(2/3\ d)$. The ordinate gives the relative number of events adjusted to have approximately equal peak heights for the three curves. Normalization to unit area was not possible because of uncertainties with regard to the

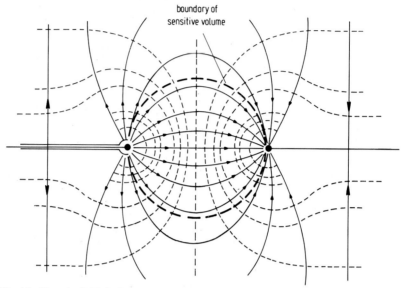

Fig. 19. Electric field (solid) and equipotential (dashed) lines of the wall-less proportional counter designed by Glass and Braby [1969] to measure energy deposition spectra of α-particles. The goblet-shaped guard electrode on the left defines the collecting volume (dashed heavy line).

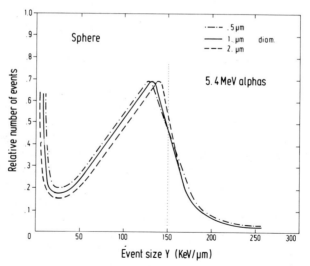

Fig. 20. Unnormalized event spectra for ^{241}Am α-particles determined by means of a wall-less proportional counter. The curves correspond to simulated tissue spheres of 0.5 μm, 1 μm and 2 μm diameter. The dotted line indicates the geometrical cut-off value corresponding to a triangular distribution (see fig. 18) [Gross et al., 1970].

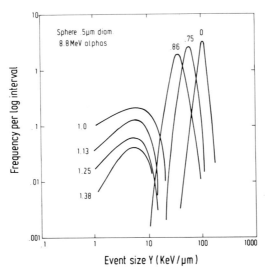

Fig. 21. Event size distribution in a 0.5 μm diameter sphere for a collimated beam of α-particles (grid-walled counter). The ordinate represents the product of frequency and event size. The parameters indicated at the curves specify the relative distance of the beam from the center of the sphere in terms of the sphere radius (>1 particles passing outside, <1 particles crossing the sphere) [Gross et al., 1970].

low energy tails of the curves. Obviously the measured spectra deviate considerably from the triangular shape corresponding to the chord lengths distribution (fig. 18). The large contribution of low energy events is undoubtedly due to α-rays protruding into the sampling volume from particles passing nearby the counter region. This explanation was corroborated by investigating the dependence of the low energy part on the incident particle energy, which showed that the relative contribution of small events increases with particle energy [Biavati et al., 1968]. In fig. 21 the event spectra of a collimated particle beam are drawn on a double logarithmic scale for several distances of the beam from the center of the sphere [Gross et al., 1970]. Curves 1.0–1.38 depicting the events caused by beams located outside the sphere represent the contribution of δ-rays to the event spectrum and are of very similar shape with stationary peak values. While the spectrum 1.0 shows the simultaneous effects of several δ-rays, there is only a contribution from single δ-rays at the larger distances (1.13–1.38). The peaks of the curves 0 to 0.86 from beams crossing the sphere shift to smaller values because of the decrease in mean chord lengths for increasing distance from the center of the sensitive volume. The spread of curve 0 from particles traversing through the center may be attributed to energy straggling and shows up in fig. 20 as a rounding of the sharp cut-off edge expected theoretically. The width of curves 0.75 and 0.86 is additionally due to chord length variations because of the finite thickness of the particle beam. Curves like those in fig. 21 may be combined by multiplication with appropriate weighting factors [Gross et al., 1969a] to reproduce the spectrum of a broad parallel beam. These weighting factors have to allow for the fact that the relative contributions of different beam sections vary as the square of the distance from the centre. Glass and Braby [1969] obtained an event spectrum similar to that in fig. 20 with the so-called Martini-counter and were able to explain the high energy tail by calculating a theoretical distribution under the assumption that the true energy loss along a track segment fluctuates with a variance of 0.57 times the expectation value.

Besides an experimental measurement, the event spectrum or its equivalent the single event distribution $f_1(z)$ (eq. 3.21) may, at least in some instances, be determined on a theoretical basis. For α-particles which are moving along straight paths the two essential, and in many instances the only relevant, stochastic variables are the track segment length and the variable LET. The problem of calculating the single event spectrum $\hat{f}_1(z)$ then reduces to evaluating the combined influence of these factors which are characterized by the LET-distribution in track length $t(L)$ and the sum distribution of the chord length $P_s(s)$, respectively. By convoluting the two distributions one obtains [Kellerer, 1968, 1969]:

$$\hat{f}_1(z) = -m \int_{L_{min}}^{L_{max}} t(L) \frac{dP_s}{ds}\left(\frac{mz}{L}\right) \frac{dL}{L}. \tag{3.25}$$

This expression is strictly valid only, if one may equate a specific track segment s to mz/L, which is true for volumes small enough to make the LET practically constant across the diameter. In case where the sensitive volume is so large that the variation of LET must not be neglected $\hat{f}_1(z)$ may be derived according to an analysis given by Caswell [1966] who calculated the distribution of the energy deposition of heavy recoils in a spherical cavity irradiated with neutrons. His analysis dealing with the even more complicated situation of a continuous distribution of initial energies may be adapted to an isotropic field of monoenergetic α-particles. From fig. 20 it is obvious that in the case of small target volumes the fluctuations of energy loss along the track segment have considerable influence on the shape of the single event spectrum at large values of z and one might suspect from the figure that there is a finite probability of observing events exceeding the cut-off value of Y by a factor of two. Looking at the energy deposition process in detail as characterized by the expectation value of the energy loss per unit pathlength, the LET, one has to recognize that there are two additional stochastic variables combining to cause the fluctuations of energy loss observed around its mean value $\overline{E} = Ls$. These are the number of primary collisions, i.e. the number of encounters between the incident α-particle and the electrons in the stopping material along the track segment, and the energy ϵ imparted to an individual electron. The number of primary collisions is distributed according to the Poisson distribution, as the primary collisions are statistically independent events [Kellerer, 1968, 1969]. If $\bar{\epsilon}$ is the mean energy deposited in single collisions then the probability $p(\mu)$ of observing exactly μ collisions is,

$$p(\mu) = \exp(-\overline{E}/\bar{\epsilon}) \frac{(\overline{E}/\bar{\epsilon})^{\mu}}{\mu!} . \qquad (3.26)$$

The classical distribution $w(\epsilon)$ of energy transferred in single primary events was already given in eq. (2.13) in a form suitable for the determination of the δ-ray dose. The actual distribution deviates at low energies ϵ from the $1/\epsilon^2$ dependence because of resonances of the collision cross section at the excitation energy levels [Bichsel, 1969]. To insure agreement with the measured stopping power values, a low energy cut-off value ϵ_{min} has to be assumed too. Then the first moment of $w(\epsilon)$ may be calculated from

$$\bar{\epsilon} = \int_{\epsilon_{min}}^{\epsilon_{max}} \epsilon w(\epsilon) \, d\epsilon . \qquad (3.27)$$

$\bar{\epsilon}$ depends critically on the choice of the spectrum $w(\epsilon)$. Experimental results of Rauth and Simpson [1964] indicate values for $\bar{\epsilon}$ of typically 60 eV. Kellerer [1967] used a semi-empirical distribution $w(\epsilon)$, combining the classical distribution for high energies and the results of Rauth and Simpson [1964] for low energies, to determine the straggling distribution of 5.75 MeV α-particles in water. $\bar{\epsilon}$ has the value 62.5 eV for this distribution while according to the classical theory a value of 9.2 eV would result. This example strongly suggests the inadequacy of the classical distribution. The energy mean $\overline{\epsilon^2}$, on the other hand, is rather insensitive to the shape of the distribution at low energies,

$$\overline{\epsilon^2} = \int_{\epsilon_{min}}^{\epsilon_{max}} \epsilon^2 w(\epsilon) \, d\epsilon = \frac{\epsilon_{max}}{2 \ln(\epsilon_{max}/I)} , \qquad (3.28)$$

and therefore the expression (3.28) computed by inserting the classical distribution is rather close to observed values. It must be noted that ϵ_{max} should be adjusted to lower values for very small targets, to allow for δ-ray escape [Kellerer, 1967, 1969]. The straggling distribution $s(E, \overline{E})$, specifying the probability of observing an energy loss E per unit energy interval if the mean energy loss is $\bar{E}$ is the solution of a compound Poisson process:

$$s(E, \bar{E}) = \Sigma p(\mu) w_\mu(E), \qquad (3.29)$$

where $p(\mu)$ is determined by eq. (3.26). Eq. (3.29) may be interpreted as the sum of the probability of single exclusive events, namely the passage of a particle which deposits in exactly μ primary collisions the energy E. A summation over μ then yields the probability of energy loss E for any number of primary collisions. The distributions $w_\mu(E)$ for μ primary collisions can be constructed by means of the single collision distribution. This is accomplished mathematically by subsequently "folding" the elementary distribution $w(\epsilon)$ to arrive at the higher order distributions. The procedure implies the use of the following recursion formula:

$$w_\mu(E) = \int_{\epsilon_{min}}^{E} w_{\mu-1}(E - \epsilon) w(\epsilon) \, d\epsilon. \qquad (3.30)$$

The equation above is easily understood. The convolution integral on the right side sums the probability of the individual events, where by $\mu - 1$ foregoing collisions the energy $E - \epsilon$ was transferred, and in the "last" collision an energy ϵ to add to a total energy loss of E. Consequently, by integration the probability of an energy loss E in μ collisions for any apportionment of E between the "first" $\mu - 1$ and the "last" collision results. The mathematical procedure is especially easy if convolution is carried out by using the Fourier or Laplace transforms of the functions, as then convolution of the original functions is equivalent to a simple multiplication of the transformed functions [Kellerer, 1967, 1968, 1969]. The variance of $s(E, \bar{E})$ is mainly determined by $w(\epsilon)$, the primary collision spectrum. Consequently, it would be inadequate to describe straggling by a fluctuating number of collision events with a fixed energy loss, i.e. as a pure Poisson process. Indeed one can show that the relative variance, being a measure of the spread in the distribution, contributed by the fluctuations in the number of collisions is $\bar{\epsilon}/\bar{E}$, while the relative variance of $s(E, \bar{E})$ is $\overline{\epsilon^2}/(\bar{\epsilon} \cdot \bar{E})$. As $\overline{\epsilon^2}/\bar{\epsilon}$ is 429 eV for 5.75 MeV α-particles [Kellerer, 1967] it follows that the total relative variance is about 7 times larger than that of the pure Poisson process. The relation above also shows that the relative variance decreases with increasing mean energy loss $\bar{E}$ and this is the reason why for densely ionizing radiation and large target volumes energy straggling may be neglected. A rough calculation shows that for a 5.75 MeV α-particle crossing a 5 μm cell diameter the square root of the relative variance is 0.03, indicating a relative narrow distribution of E around $\bar{E}$, while for a 1 μm target diameter this quantity has a value of 0.18, indicating considerable fluctuations. A precise specification of the primary event spectrum $w(\epsilon)$ is difficult for very small targets, as in this case the cut-off value for the δ-ray energies has to be adjusted to the distance between track and boundary of the sensitive site, and thus becomes a function of the position along the track depending on the size and the shape of the site. Generally, one may assume that the primary event spectrum depends in some way on a coordinate x describing uniquely the relative position of the locus on the track where the primary event takes place and the wall of the site. The event spectrum $w(\epsilon, x)$ then specifies the energy ϵ deposited within the site and has no bearing on the energy lost by the primary particle. Roesch and Glass [1971] were able to prove that, even in the more general case, the energy deposition along a track segment may be described by a compound Poisson process and that there exists an effective distribution $w(\epsilon)$ independent of the position along the track which may be calculated from the distributions $w(\epsilon, x)$.

Another important point is the fact, that even for two passages with the same amount of energy deposited in the site, the number of ionizations is different. Taking into account the fluctuations in the ion yield means, going one step deeper in the hierarchy of the stochastic variables and may be of relevance for the quantitative analysis of radiation effects which are known to be due to ionic interactions.

If the straggling distribution $s(E, \bar{E})$ was determined by insertion of eq. (3.26) and eq. (3.30) into eq. (3.29), the single event distribution of the local energy density can be calculated as the sum of conditional probability terms each representing the probability of the occurrence of an energy loss z if $\bar{z}$ is the mean energy density expected from the track segment length and the LET·

$$f_1(z) = m \int s(mz, m\bar{z}) \hat{f}_1(\bar{z}) \, d\bar{z}. \tag{3.31}$$

$\hat{f}_1(z)$ has to be taken from eq. (3.25). Baily and Steigerwalt [1975] pointed out that in many cases of interest for a constant LET, only some pathlengths out of the complete distribution function should be selected to compute the resulting distribution $f_1(z)$ with sufficient accuracy.

In theoretical and experimental investigations of energy straggling results are often compared with the Vavilov theory of straggling. Vavilov [1957] obtained with a rigorous analytical treatment a solution for "thin" absorbers, i.e. absorbers in which the energy loss is much smaller than the energy of the incident particle. His calculations were based on the classical primary event spectrum but yielded very complex expressions. The Vavilov distribution is characterized by a skewed shape with the mean energy loss $\bar{E}$ larger that the most probable energy loss. Experimental determinations of the straggling distribution have been carried out by Booz et al. [1971] for α-particles of 0.27 to 5.3 MeV passing through layers of tissue equivalent gas with thicknesses varying between 0.1 μm and 1.0 μm, and by Hogeweg and Barendsen [1971]

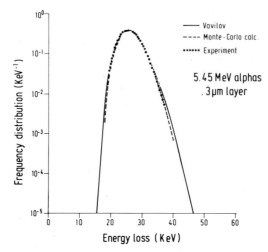

Fig. 22. Spectral distribution of energy transfer to a tissue layer of 0.3 μm thickness. Comparison of experimental results, Monte Carlo calculations, and Vavilov distribution [Booz et al., 1971].

with a tissue equivalent cylindrical counter for several α-energies. In addition Booz et al. [1971] made Monte Carlo calculations implementing the detailed succession of steps in the energy transfer, including the stochastic of excitation and ionization, δ-ray ejection angle, pathlength to the wall of the sensitive site and electron transmission to the walls. Their experimental results show that the distribution of the energy deposition for α-particles with energy around 5 MeV is within the experimental errors identical with the Monte Carlo calculations and the Vavilov distribution for absorbing layers down to 0.3 μm (fig. 22). For 0.1 μm thickness there is still good agreement between Monte Carlo calculations and experiment but the Vavilov distribution displays a pronounced high energy tail. The deviations between the Monte Carlo calculations and the Vavilov theory become more and more pronounced at decreasing thickness of the layer, owing to electron transport outside the sensitive region, as in this case the maximum electron range exceeds the dimensions of the layer (fig. 23). Hogeweg and Barendsen [1971] observed a discrepancy in the reverse order, i.e. the measured values were exceeding the Vavilov distribution at the high energy portion of the curve. For completeness it should be noted, that in the case of many collisions along the track segment $s(E, \overline{E})$ is a Gaussian distribution with mean $\overline{E}$ and variance

$$\sigma^2 = 4\pi e^4 z^{*2} NZx \qquad (x = \text{track segment length}). \qquad (3.32)$$

The variance is independent of particle energy but depends on the charge

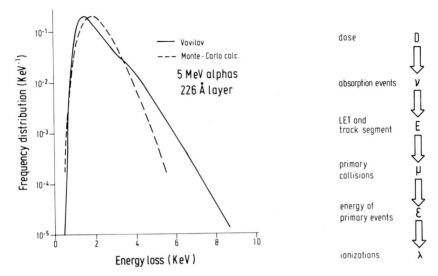

Fig. 23. Spectral distribution of energy transfer to a tissue layer of 226 Å thickness. Comparison of Vavilov distribution and Monte Carlo calculations. Measurements were not possible [Booz et al., 1971].

Fig. 24. Sequence of stochastic events and quantities originating from the non-stochastic quantity absorbed dose (D). The primary events with energy ϵ may lead to a fluctuating number of ionizations (λ). The number of primary collisons (μ) along a particle track segment is a random variable depending on the LET and the length of the segment. Finally the absorbed dose which is the mean of the specific energy imparted to the target is determined by the number ν of absorption events, i.e. passages of α-particles through or near the target.

state of the particle. Fig. 24 summarizes the interrelationship of the different stochastic variables starting with the macroscopic dose D. The decision to what level in the "hierarchy" of the stochastic variables one has to go down, must be based on the nature of a specific experimental or theoretical problem at hand. If, for instance, the distribution of dose in cells situated in a radiation field of internal α-emitters is required it will be appropriate to take into consideration the variation in LET and the track length distribution. Straggling may be neglected in this case or allowed for by assuming the Gaussian limit of the straggling distribution (eq. 3.32).

If the single event spectrum $f_1(z)$ is known from calculations using eq (3.31) or from experiment, it is possible to calculate the dose dependent distribution $f(z, D)$ or its associated sum distribution $F(z, D)$. Some general properties of $f(z, D)$ however may be stated without a detailed knowledge of $f(z, D)$ [Hug and Kellerer, 1966]. For small doses D, i.e. $D \ll \bar{z}_F$, $f(z, D)$ has

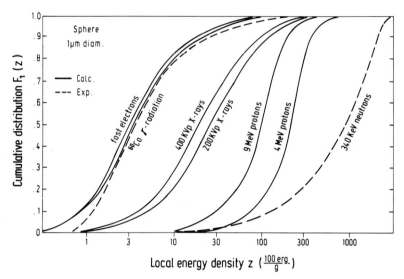

Fig. 25. Cumulative distribution of the local energy density z for different radiations and a tissue sphere of 1 μm diameter (single event distribution) [Hug and Kellerer, 1966].

essentially the form $f_1(z)$, as multiple passages are very improbable. But as only a fraction $\bar{z}_F/D$ of a total ensemble of targets receives a hit, it follows that

$$f(z, D) = \left(1 - \frac{\bar{z}_F}{D}\right) \delta(z) + \frac{\bar{z}_F}{D} f_1(z). \tag{3.33}$$

$\delta(z)$ is the δ-function. For large doses ($D > 20\, \bar{z}_D$) the distribution of the local energy density approaches a Gaussian distribution with mean D and variance $D \cdot \bar{z}_D$. Hence, an exact calculation of $f(z, D)$ is only necessary for the intermediate dose range. In fig. 25 a series of computed and measured sum distributions $F_1(z)$ of single events are plotted for different radiations. With increasing LET the curves are shifting to the right, indicating an enhancement in the local energy density per passage. Such curves or their derivative $f_1(z)$ form the basis for the calculation of $f(z, D)$ described in the following. There are essentially two computation procedures worked out by Kellerer [1966, 1968; Hug and Kellerer, 1966]. One of them is a Monte Carlo procedure based on random path calculations in a $D - z$ plane. To cover several orders of magnitude both in D and z the axes are divided in a double logarithmic scale. Both the increments in local energy density, and in time (the latter is equivalent to a dose increment) are chosen at random according to the single event distribution $f_1(z)$ or the exponential distribution, respectively. Fig. 26 illustrates the result after the calculation of one

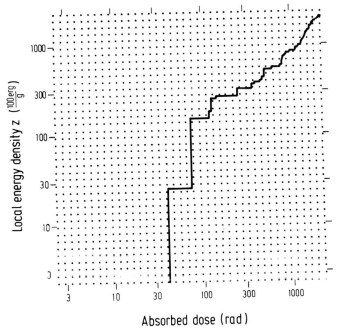

Fig. 26. Random path illustrating the Monte Carlo procedure of calculating the distribution function of the energy density $f(z, D)$. The curve shown represents an example for 200 kV$_p$ X-rays and a spherical tissue volume of 1 μm diameter. The cumulative distribution $F(z, D)$ is determined on each of the plotted points in the double-logarithmic raster and after generating a large number of paths [Hug and Kellerer, 1966].

path corresponding to the energy deposition in a spherical volume of 1 μm diameter by 200 kV$_p$ X-rays. For very high doses the path approaches a straight line with an angle of 45 degrees signifying that for a large number of events z deviates only slightly from D. If such path calculations are gone through repeatedly, adding after each run one unit to every raster point lying above the path, one finally obtains after appropriate normalization a set of sum distribution functions. An alternative approach is to exploit the fact that $f(z, D)$ may be considered to be the result of a compound Poisson process

$$f(z, D) = \sum_{\nu=0}^{\infty} \exp(-D/\bar{z}_F) \frac{(D/\bar{z}_F)^{\nu}}{\nu!} f_{\nu}(z). \qquad (3.34)$$

$f_{\nu}(z)$ is the distribution of the local energy density under the condition that ν events have occurred. Then eq. (3.34) is easily interpreted as the sum of terms each representing the probability per unit interval of z of occurrence of exactly ν events, if the mean number of events is $D/\bar{z}_F$ and the local energy density of all ν events is z. The functions $f_\nu(z)$ are generated by subsequent folding of the single event spectrum

$$f_\nu(z) = \int_0^z f_{\nu-1}(z-x) f_1(z) \, dx . \qquad (3.35)$$

The interpretation of this integral is completely analogous to eq. (3.30). Eq. (3.34) becomes especially simple for the Fourier transforms $f^*(t, D)$ and $f_\nu^*(t)$ (the asterisk denotes the transformed functions). Convolution is then equivalent to simple multiplication

$$f_\nu^*(t) = f_1^*(t)^\nu \qquad (3.36)$$

and, as can easily be verified

$$f^*(t, D) = \exp(D[f_1^*(t) - 1]/\bar{z}_F) . \qquad (3.37)$$

The computational algorithms of the compound Poisson process are the body of a versatile FORTRAN computer program developed by Kellerer [1968] which is applicable both to straggling and microdosimetric problems. A summary of the quantities and functions employed in microdosimetry is given in a paper by Kellerer and Rossi [1969] which the reader should consult if a rapid understanding of some essential points, as discussed above, is required.

At present relatively little information regarding the questions and problems of internal α-emitters in conjunction with the concepts of stochastic microdosimetry is obtainable. In particular, to provide a link to minidosimetry, there is some need to develop rapid computational procedures for the determination of energy deposition spectra adequate to fast scanning techniques. Such work would bring up questions related to (i) the precise measurement of single event spectra for specific geometries of α-irradiation and its analytical representation, (ii) simplifications of the compound Poisson procedure to allow explicit analytical calculations, and (iii) an analysis of the errors introduced thereby and by the imprecise knowledge of the dose.

3.2. Applications to internal alpha-emitters

Microdosimetric concepts were applied with the aim to elucidate the possible mechanisms of radiation injury. This was done particularly for the interpretation of survival curves and the determination of a lower limit for the sensitive target volume [Kellerer, 1966; Hug and Kellerer, 1966; Rossi, 1967a] and then developed further to the theory of dual radiation action [Rossi, 1971; Kellerer and Rossi, 1972], including a theory of RBE. While these applications are based on the large background of experimental data concerning the inactivation of cultured cells, there are only a few publications at present dealing with the specific problems of internal emitters in terms of stochastic microdosimetry. Moreover the complexities introduced by considering effects on cells in an intact organism impose severe limitations on any mechanistic interpretation of dose-effect relationships, even if effects and doses may be quantified, so that one normally is restricted to a formal characterization of the radiation field and a hypothetical target by means of the local energy density z.

Kappos [1967b] carried out a computer analysis of energy deposition in spherical volumes of different diameters. For the relatively large targets considered ($d \geqslant 1$ μm) he disregarded energy straggling and assumed that the α-paths are straight lines. To test the influence of the lateral energy deposition by δ-rays he recalculated some event spectra assuming that the α-ray energy is spread over a cylindrical volume around the particle paths with constant energy density, or with an energy density decreasing with distance from the track core. The radionuclide was assumed to be distributed uniformly around the target volume or concentrated in a second phase which is bone-equivalent with respect to α-particle stopping. The Y-distributions from the homogeneous deposition of the radionuclide show pronounced maxima for 10 MeV α-particles which shift to higher Y for decreasing particle energies, as may be expected from the increase of the track average LET, and becomes broader. A comparison between the results for 10 μm and 1 μm sphere diameters reveals the influence of finite target size, the maximum Y being around 250 keV/μm for 1 μm spheres and 170 keV/μm for 10 μm spheres. Fig. 27 shows the Y-spectra resulting from a contamination of a surface layer of bone of 1 μm thickness in a sensitive volume of 2 μm diameter located in the adjacent soft tissue at varying distances from the layer. This is a situation characteristic for the surface-seeking isotopes like ^{239}Pu and ^{241}Am. In the immediate vicinity of the surface, the most probable event is close to the value resulting from the mean LET and a passage through a sphere diameter (136 keV/μm). The peaked distribution flattens out for larger distances and shows approximately constant probability in a broad range of events for 20–30 μm distance. At large distances near the particle range, a peak appears at small event sizes Y, due to incomplete passages (stoppers). Kappos, however, points out that there is some uncertainty about the contribution near the end of the particle trajectory, as the stopping power is not precisely known there and because the assumption of straight paths is no longer valid. The simple cylinder model of δ-ray energy deposition has qualitatively the same effect on all Y-spectra in fig. 27, namely a distinct enhancement of small event probabilities for cylinder diameters of 0.1 μm and larger. Kappos also evaluated the influence of multiple passages on the energy deposition spectra and

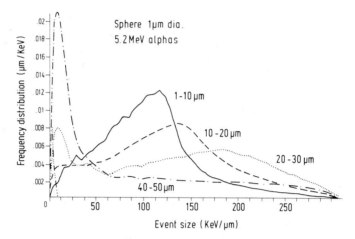

Fig. 27. Frequency distribution of event size Y in a 1 μm diameter tissue sphere. The sites are assumed to be located in marrow at certain distances from a plane bone surface. Bone is labeled in a surface layer of 1 μm depth. The frequency function is determined for class intervals ΔY of 5 keV/μm [Kappos, 1967b].

demonstrated a typical example of 10 μCi incorporated in the human skeleton and a sphere diameter of 10 μm. In this case 20% two-hit events have to be expected within 20 h, but the energy deposition spectrum is only slightly affected, exhibiting a shift to larger values of Y. In situations of very inhomogeneous deposition of the radionuclide, however, multiple passages may play a decisive role at specific sites in tissue and must not be disregarded.

The model devised by Tisljar-Lentulis et al. [1975, 1976] takes into account the inhomogeneous distribution of activity in particulate sources. Tissue was simulated by a three-dimensional regular matrix of cells and concentric nuclei with 10 μm and 8 μm diameter, respectively. The sources were located at the centres of the nuclei having an activity of 27 pCi, equivalent to one disintegration per second, and emitting isotropically 5.14 MeV α-particles of range 40 μm. The cells were in close-packed arrangement and the specific activity of tissue was varied by changing the number of sources. Differential energy loss values of α-particles in water were used for the cell nuclei, the cells and the intercellular space; the δ-ray effect and straggling were neglected. For α-particles of this range and with the tissue structure assumed each point-like source irradiates 379 cells and for the computations a problem time of 5.236 min was selected to insure isotropic particle emission. The z-spectra in fig. 28 for different specific activities are similar in shape shifted by a factor corresponding to the fraction of irradiated cells which increases for increasing specific activity. The discontinuous appearance

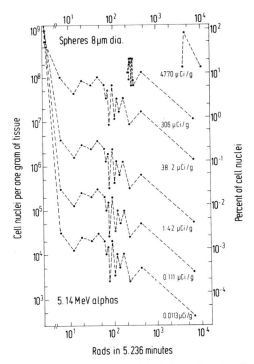

Fig. 28. Specific energy delivered to cell nuclei of 8 μm diameter for point sources of ^{239}Pu α-particles, each with an activity of 27 pCi. Parameters signify the specific activities. The broken lines connect groups of nuclei receiving a certain amount of energy. The ordinates on the left and right specify the number of nuclei per gram of tissue or the percentage of nuclei, respectively, in which the local dose corresponding to the points is given off [Tisljar-Lentulis et al., 1975].

of the curves stems from the fact that, instead of treating the track segment lengths as a stochastic variable, a mean energy deposition was calculated from averaging over the length distribution. Under these circumstances the curves represent the effect of LET variation and of multiple passages. Because of the regular tissue structure only discrete values of z occur which are connected with broken lines in fig. 28. The mean value of z is 72.5 rad in 5.236 min for the cell nuclei. Above 53 μCi/g the irradiation zones around the point sources begin to overlap and consequently the z-spectrum changes (fig. 28).

The model of Roesch [1975] differs from that of Tisljar-Lentulis et al. with respect to the structure of tissue and the assumption about the source locations. The particulate α-sources of ^{239}Pu were assumed by Roesch to be placed at random in the organ in which the sensitive sites (cells) are randomly distributed as well. This leads one to introduce a double-stochastic process characterized on the one hand by the distribution of the local energy density for a definite relative position of site and sources and, on the other hand, by the stochastic nature of site localization. The latter factor gains importance the more non-uniform the distribution of activity is. If v_{dV} is the probability of choosing a site in the volume element dV and $f_{dV}(z)$ the distribution of specific energy in dV, then the probability density $g(z)$ for all sites in the volume V is

$$g(z) = \int_V v_{dV} f_{dV}(z) \, dV, \tag{3.38}$$

where v_{dV} has to be normalized appropriately. The above relation reduces to $g(z) = f(z)$ if $f(z)$ is not dependent on the choice of the volume element dV, as is the case in charged particle equilibrium. Explicit calculation of eq. (3.38) requires the knowledge of the single event densities $f_1(dV', z)$ specifying the energy density distribution, if it is known that one decay occurred in a particulate located in dV'. For a pointlike source $f_1(dV', z)$ is a function of the distance between source and site and was determined from the track length distribution in spherical sites of 4.6 μm diameter by convolution with a Gaussian distribution with known variance [Glass and Braby, 1969] (see above). The contribution of secondaries from α-particles passing nearby was accounted for by adding a low specific energy component to the spectrum. In fig. 29 the distributions $g(z)$ are shown for particulates that emitted 1, 10, 10^2, 10^3 and 10^4 α-particles per particulate. The mean dose was kept constant at 75 rad and thus the density of particulates is inversely proportional to the number of α-particles per particulate ranging from $9.1 \times 10^8 \, g^{-1}$ to $9.1 \times 10^4 \, g^{-1}$. As the abscissa in fig. 29 is plotted on a logarithmic scale, the ordinate has to be transformed too in order to achieve a normalization of the areas under the curves. The decreasing area under the curves for increasing activity of the sources reflects the fact that, due to concentration in fewer sources, a growing number of sites receives no hit at all. On the other hand, the shift of the curves to large values of z indicates, that most of the sites within the range of the sources are subject to multiple passages, whereas the similarity in shape of the three curves for 1, 10 and 100 decays per particulate indicates that energy deposition in single events is prevailing. The percentages of sites in which energy deposition occurred were 48, 44, 35, 16, and 2.2% and the corresponding mean specific energies for these sites were 156, 171, 214, 470, 3,490 rad, in the order of increasing number of decays per particulate. With the average dose chosen, the shapes and positions of the curves in fig. 29 are determined mainly by the α-particles from a single particulate, i.e. there is no site localization for which the probability of receiving hits from two or more different particulates is distinctly above zero. The effect of "cross-fire" from two or more particulates would be seen at higher

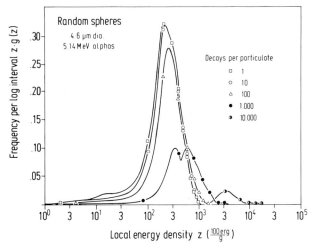

Fig. 29. Specific energy density spectra in 4.6 μm diameter tissue spheres located at random in tissue, where randomly positioned particulate sources of ^{239}Pu α-particles are deposited. The density of the sources is varied inversely proportional to the number of decays per particulate, so as to keep the mean dose constant at 75 rad. For the sake of normalization the ordinate is transformed according to the logarithmic scale of the abscissa. The decreasing area under the curves indicates a reduction in the number of spheres hit by α-particles [Roesch, 1975].

absorbed doses. It follows from the model of Roesch that, if the radiation sensitivity K is defined as $1/D_{37}$ and refers to the number of sites not hit, then K decreases with increasing number of decays per particulate and with decreasing diameter of the sites. This definition of sensitivity assumes that inactivation of the site is caused by a single passage of an α-particle.

4. Experimental procedures

4.1. Emulsion autoradiography

Evidently, the problems of dose determination on a microscopic scale are closely linked to the physical and chemical properties of radiation detectors available to meet two basic requirements:

(i) The ability to register events (decays, particle passages) efficiently and with a high degree of integration over time. In other words, the registration of events has to be summed up somehow to provide adequate statistical accuracy if the directly measured quantities have to be converted to rates (specific activites, particle fluxes, dose rates etc.). This is especially important for the small concentrations of radioactivity normally encountered in studies of internal α-emitters.

(ii) The detectors should provide the necessary spatial resolution to achieve a precise localization of events and source distributions, the latter being of particular importance in the case of very non-uniform deposition of radioactivity in tissue.

Photographic emulsions especially designed for the purpose of registering tracks of α-particles not only fulfill these two basic requirements but have additional advantages too. Many aspects of photographic autoradiography were treated in detail in several useful monographs [Gude, 1968; Rogers, 1969; Fischer and Werner, 1971; Gaham, 1972] particularly emphasizing β-autoradiography. Unfortunately, α-autoradiography is sometimes considered to be a trivial matter and, therefore, only little space is devoted to its problems in the literature. The following paragraphs should convince the reader that this attitude is by no means justified, at least as far as quantitative α-autoradiography is concerned.

The preparation of samples for autoradiographic studies begins with the fixation and embedding of those organs or parts of organs to be investigated. Several materials and embedding procedures have been developed and applied among which are paraffin and celloidin embedding, used for soft tissue samples, and embedding in methyl methacrylate, butyl methacrylate or resins, preferably used for bone samples. In principle any embedding medium useful in conventional histological techniques is suitable for autoradiography, provided there are no chemographic effects, i.e. a chemical influence on the autoradiographic image effecting an additional stimulus or an inhibition of the development of silver grains, if the embedding medium is in contact with the emulsion during exposure. Methyl methacrylate seems to be given preference in many laboratories now, owing to its superiority over several other materials [Von Seggen et al., 1973]. Having prepared tissue sections of appropriate thickness by means of a microtome, or "infinitely" thick specimen blocks with a flat polished surface, autoradiographic exposure may begin by bringing source and emulsion into contact. There are essentially three mounting techniques among which the proper one has to be chosen according to the type of emulsion available and the kind of evaluation the autoradiographic image is to be used for.

1. The contact technique is preferably used with emulsions supported by a glass plate or a flexible plastic backing. The autoradiographic image and the source are separated after exposure and any correspondence between the histological structure and the track distribution is lost, provided no special arrangements for a realignment of source and image have been made. This method has the advantage of quantitative evaluations of autoradiographs without the disturbing interference of the histological structures.

2. The stripping film technique is the most commonly used autoradiographic method. The Kodak AR 10 film is perhaps the best known representative of this class of emulsions. It consists of an emulsion of 4 μm thickness reinforced by a 10 μm thick gelatin layer which in turn is mounted on a glass plate support. Both the emulsion and its gelatin backing are stripped off from the glass slide and floated onto the tissue sample in water, where they shrink during drying and thus get into close contact with the tissue section.

3. The liquid emulsion technique has two variations to be distinguished by the manner in which the tissue sections are mounted. Coating the microscope slide with the emulsion and mounting the tissue samples thereafter is often referred to as "mounting procedure". With the "coating procedure" the emulsion is gently brushed onto the sections already mounted on the glass support, or the coating is produced by dipping into the liquid emulsion. Liquid emulsions are normally delivered as a solidified gel and must be melted before use.

Most of these methods and emulsion types were frequently used in dosimetric and distribution studies of α-emitting radionuclides such as ^{226}Ra [Arnold and Jee, 1959; Marshall and Finkel, 1959; Rowland and Marshall, 1959; Lloyd, 1961], ^{239}Pu [Arnold and Jee, 1957; Jee and Arnold, 1961; Twente and Jee, 1961; Jee et al., 1962; Bleaney, 1967; Vaughan et al., 1967; James and Kember, 1970; Schlenker and Marshall, 1975] and ^{241}Am [Williamson and Vaughan, 1963; Durbin et al., 1969; Seidel, 1975], just to mention a few. Table 7 lists some of the common emulsions especially suitable for α-autoradiography

The exposure time of autoradiographs depends, besides the irradiation geometry, on such factors like specific activity in the tissue, size of the fields

Table 7
Emulsions for alpha-autoradiography

Manufacturer	Type	Grain diameter (μm)	Method
Ilford	GO	0.27	Liquid
Ilford	KO	0.20	Liquid
Ilford	LO	0.15	Liquid
Ilford	C2		Contact
Kodak	NTA	0.22	Contact, liquid
Kodak	NTB	0.29	Contact, liquid
Kodak	AR 10	0.20	Stripping
Agfa-Gevaert	Scientia Nuc 3.07	0.07	Contact

in which track densities have to be determined, statistical errors that should be tolerated for the track density determination, procedure of measurement etc. A general rule cannot be given. If, for instance, the size of the measurement area and the specific activity are kept constant then the product $r\sqrt{t}$ is constant as well (r = relative statistical error, t = exposure time), i.e. for an improvement of the statistical accuracy by a factor of 2 the exposure time has to be extended four times. For very prolonged exposure times, i.e. in the case of samples with very low specific activities, latent image fading has to be taken into consideration as the ability of emulsions to integrate events over time is limited. Usual measures to keep latent image fading at a minimum are careful drying of films, exposure at low temperatures (4°C) and at low humidity. Additionally, exposure in an inert gas atmosphere may prove advantageous. Generally, the limitations imposed by fading are not so severe with α-autoradiographs where tracks are counted, as with the grain counting procedures of β-autoradiography. After development of the latent image according to the recommendations of the manufacturer, individual α-tracks may be viewed under a microscope or, if the local track densities are high enough, the autoradiographic image may even be visible with the naked eye (gross autoradiographs). An α-particle passing an emulsion renders every grain along its path developable. Depending on the inclination of the particle's trajectory to the surface of the emulsion and on the emulsion thickness one sees a more or less short projection of the track onto the emulsion surface. Only a few, if any, of the tracks exhibit the full particle range, which is for instance 22 μm for an 5.3 MeV α-particle [Herz, 1969]. Fig. 30 shows as an example a stripping film autoradiograph of a rat femur section. The tracks originating mainly from the endosteal surfaces, indicate clearly the nature of ^{241}Am(III) as a surface-seeking radionuclide. One advantage of track autoradiography with emulsions is the possibility to locate the radiation source very precisely. Though in principle beginning and end of a track are indistinguishable, the observer gains this information from the arrangement of all neighbouring tracks by intuition. Point sources of radioactivity are an especially obvious case, as they give rise to a star-shaped track pattern with a more or less circular core of high grain density, whose diameter depends on the source strength and the exposure time. One of the disadvantages of emulsion autoradiographs, especially with regard to automatic track counting, is the disrupted appearance of the tracks (fig. 30). Because of the random distribution of the silver halide grains in the matrix, the developed grains along the particle's path form "blobs" separated by gaps. This may result in erroneous track counting, if the instrument used for evaluation regards every isolated track segment as a separate track. The number of grains developed along the full range of a particle depends, apart from the particle energy, on emulsion parameters like mean grain diameter, grain density, sensitivity of the grains

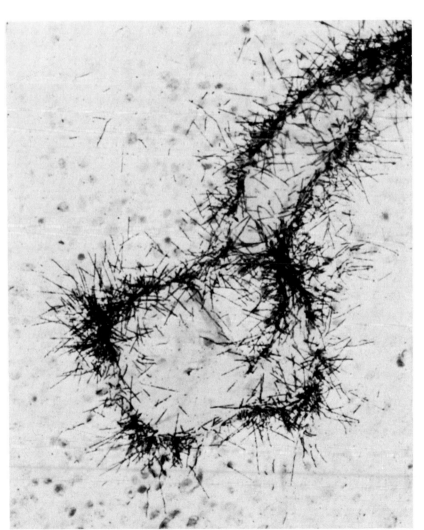

Fig. 30. Stripping film autoradiograph of a rat bone section. The image shows a piece of trabecular bone (femur) with α-particle tracks originating from ^{241}Am deposited on bone surfaces. 30 μCi/kg ^{241}Am(III), 22 days exposure time, ×250. [Courtesy of Dr. A. Seidel.]

etc. By means of the theory of emulsion statistics [Barkas, 1963] one obtains, for instance, for Ilford G5 emulsion with a mean grain diameter of 0.28 μm that 55 grains are traversed on the average by an particle of 5.3 MeV. This value is an upper limit of the number of developed grains. Of the full particle range (22 μm) only about 45% is lying within the silver halide crystals, which in turn gives a mean energy of 43.8 keV to render a single grain developable. This is a considerable waste of energy, as only something around 30 eV is necessary to trigger the development of a grain. In principle it is possible to use the primary particle energy more effectively, thereby reducing the exposure times by orders of magnitude. This can be accomplished by converting the particle energy to photons by means of stopping in a phosphorescent medium like silver activated zinc sulfide [Hsieh et al., 1965; Matsuoka et al., 1967]. However, it must be doubted whether the present methods of producing gross autoradiographs of low level sources can be adapted so as to yield the high geometrical resolution required for mapping the microdistribution of a radionuclide.

In fig. 31 the schematic representation of the irradiation geometry is shown. For the sake of generality an absorbing layer is assumed to be sandwiched between source (tissue section) and emulsion. Let r be the ratio of the particle range in the absorber to the range in the source and taken as constant for all particle energies. Then it can be shown by geometrical reasoning that the track density σ_T in the emulsion (tracks per unit area) and the activity concentration ρ_A (decays per unit volume) are related by the expression

$$\sigma_T = \rho_A \frac{T}{2} \left[1 - \frac{q/r}{R} - \frac{T}{2R} \right] \qquad T \leqslant R - q/r, \tag{4.1}$$

(R is the particle range in the source, T, q is the section and absorber thickness, respectively). The second and third term in brackets constitute the relative self-absorption of the absorber and the source, respectively. The self-absorption of the absorbing layer exceeds that of the source if $q > rT/2$. For air as absorber this means, that a gap thickness as large as 0.5 mm only gives rise to 1% absorption, or in other words, from the viewpoint of registration efficiency there is no need for a very close contact between emulsion and source. The main purpose of placing absorbers between tissue section and emulsion is to improve the spatial resolution of the autoradiograph, especially if a precise quantitative localization of an inhomogeneous deposit of activity by a measurement procedure not aided by the intuitive capabilities of the human observer is required. Quite clearly, an improvement of the spatial resolution can only be achieved at the expense of a loss in registration efficiency. Fig. 32 displays the relative decrease of the track density as a function of the relative

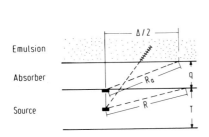
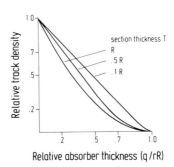

Fig. 31. Diagram of the irradiation geometry for a source-absorber-emulsion detector arrangement. R and R_a are the α-particle ranges in the source and the absorber, respectively. Δ is the maximum observable spread of a point source.

Fig. 32. Relative track density as a function of the relative absorber thickness q/rR (q = absorber thickness; R, R_a = particle range in source and absorber, respectively, $r = R_a/R$).

absorber thickness q/rR for several section thicknesses. The registration efficiency falls off more rapidly for "thick" sections than for thin ones, which is easy to see, as for thick sections an increasing fraction of the sections does not contribute to the track density on the detector surface, if the absorber thickness is enhanced. While air gaps between emulsion and source have only a small influence on the detection efficiency, the effect on the spatial resolution may be drastic. In the notation of fig. 31 the maximum spread Δ of a point source located anywhere within the tissue section is

$$\Delta = 2q\sqrt{\left(\frac{R}{q/r}\right)^2 - 1} \qquad (q/r < R). \qquad (4.2)$$

Assuming an absorber thickness of $q = 0.2\,R$, a range ratio of $r = 0.5$ and a "thick" source the spread diminishes from $\Delta = 2\,R$ without absorber to $0.91\,R$ with absorber, and the track density decreases to 36% of the original value. If, however, the absorber is assumed to be an airgap of the same width but with a range ratio of $r = 1,500$, a maximum spread of $\Delta = 3,000\,R$ has to be expected and consequently the autoradiographic image will have a rather diffuse appearance. The problem of resolution in α-track autoradiographs has not been thoroughly accounted for in the literature. This may be partially due to the fact that visual or semi-automatic evaluation, controlled by a human observer prevails. The question of spatial resolution becomes important, however, if automatic measurement techniques are to substitute human work. It should be possible for this case to adapt the theories of Doniach and Pelc [1950] and of Lamerton and Harris [1954] to take into account the special features of automatic track counting. These authors set up criteria for the lower limit of resolution based on the distribution of the grain density, which are especially suited for β-autoradiography.

For an "infinitely" thick source, i.e. a source thicker than the particle range R, eq. (4.1) reduces to

$$\sigma_T = \frac{\rho_A}{4} \frac{(R - q/r)^2}{R}, \qquad (4.3)$$

which by setting $q = 0$ is equivalent to the expression derived by Mays [1958] and earlier by Schaefer [1948]. Eq. (4.1) may be readily specialized to hold for a contaminated surface oriented perpendicularly to the section plane, if ρ_A is substituted by the specific area deposition of activity (decays per unit area) and σ_T by the lineal density of tracks seen on both sides of the surface. If the stopping power is different on both sides of the surface this has to be taken into consideration by a straightforward correction of the resulting formula [Bleaney, 1967]. The source strength of pointlike aggregates of activity in a tissue section may be calculated directly by counting the number of tracks in a star, if the sections are thin compared to the particle range. For thick sections and aggregates with equal source strength Q spread uniformly throughout the thickness, the mean number $\bar{N}$ of α-particles reaching the film per unit time is [Mays, 1958],

$$\dot{\bar{N}} = \frac{Q}{2} \left(1 - \frac{T}{2R}\right) \quad (Q = \text{number of decays per unit time}), \qquad (4.4)$$

which can be verified from eq. (4.1) too. A test of the hypothesis to have only sources of equal activity may be provided by measuring the track distribution of a sufficient number of sources, which must be Poissonian if the hypothesis is true [Vaughan et al., 1967]. An alternative and perhaps easier way would be to calculate the depth of localization of individual stars from their diameter which together with the number of tracks gives Q.

Lindenbaum et al. [1967] and Lindenbaum and Russel [1972] demonstrated that quantitative autoradiography with emulsions may replace gross radiochemical assay in particular instances where not high accuracy but additional information about the localization of a radionuclide is desired. These workers evaluated the track density in liver sections from animals injected with ^{239}Pu(IV) (30% ultrafiltrable). The sections were coated with Ilford L4 liquid emulsion. As a lower limit of the track length 8.3 μm was chosen and under this condition the registration efficiency was 2% for an emulsion of 1 μm thickness. For short intervals after injection no track aggregation was observed and the figures for the specific activity varied from 53% to 88% of those determined by a standard radiochemical procedure [Schubert et al., 1961]. At later times after injection there is a tendency towards the formation of aggregates appearing as stars. The autoradiographically determined specific activity then overestimated the true values by 146% to 311% [Lindenbaum et al., 1967]. By variable exposure to ^{241}Am and ^{239}Pu on sections of mouse liver Lindenbaum and Russel [1972] again showed that only a relatively small disparity exists between the results of quantitative autoradiography and liquid scintillation assay [Lindenbaum and Lund, 1969]. They also demonstrated a fairly linear relationship between track density and exposure time. The differences in the registration efficiency compared to the more exact methods were attributed by the authors to the unevenness of the emulsion thickness and tissue surface and subjective errors in track counting. The unavoidable fluctuations of emulsion thickness which are characteristic for the liquid emulsion technique indeed make this method appear inappropriate for quantitative work. The problem of subjective errors in visual track counting is a serious one, not only because it necessitates straining and tiresome work, but also because of the limited capability of track recognition. Particles entering the detector surface nearly perpendicularly leave tracks hardly distinguishable from background grains if viewed by a microscope. So the human observer develops an inclination to register only those tracks the projected length of which exceeds a certain subjective limit. This explains why sometimes the registration efficiency of photographic emulsions was found to be lower than 100% [Bleaney, 1967]

One of the most impressive demonstrations of the capabilities and power of the method of quantitative autoradiography with emulsions is the work of Rotblat and Ward [1956]. In order to estimated the tissue dosage from α-particles of ^{232}Th and ^{226}Ra not in equilibrium with their daughter products, these authors measured the relative activities of each of the elements of both decay series by α-particle track analysis in Ilford C2 emulsions. From a geometrical analysis it can be shown that the distribution of track length l in a thick emulsion layer of a uniformly labeled tissue section is peaked at $l = R - c$ (ϑ, T, m) (R is the particle range in emulsion) with c depending on the maximum angle of the tracks evaluated (ϑ), the section thickness (T) and the ratio of the stopping power in emulsion and section (m). For groups of α-particles of sufficiently different energies the contributions of each single particle source may be unfolded from the superposition of all track length distributions yielding the relative activities of the elements present in tissue. Rotblat and Ward measured tracks with angles up to 30° to the emulsion surface. The track length analysis provided a measure of the parent ^{232}Th relative to the group of decay products originating from ^{228}Th and considered to be in equilibrium among each other. For the radium series limits for the relative activities could be given.

4.2. Dielectric track detectors

The observation that permanent latent damage is produced along the paths of charged particles in certain classes of organic polymers, glasses and crystals has led to a rapidly expanding field of research and a host of practical applications. In particular the possibility to "develop" the latent damage regions to a microscopic or even macroscopic size by a simple chemical procedure promoted the introduction of this kind of non-photographic track registration. Today solid state nuclear track detectors (SSNTD) may be considered as an alternative to photographic emulsions and in particular cases such as presence of a high background of unwanted low LET radiation SSNTDs are the method of choice. The principles and widespread applications of SSNTDs have been reviewed in several articles [e.g. Stolz and Dörschel, 1968; Fleischer et al., 1965a, 1969, 1972]. The recent book of Fleischer et al. [1975] gives a complete account of all aspects of the physics and technology. The mechanism of damage formation is not yet well understood but it looks as if the ion explosion spike model of Fleischer et al. [1965b] is based soundly on current experimental evidence. According to this model, the traversal of a charged particle creates a narrow region depleted of electrons by primary ionizations. Subsequently the net positive charge produces displacements of the atoms via Coulomb repulsion, thereby generating interstitial defects and vacancies along the track core. In organic polymers it is believed that, in addition to these effects, bond breaking of the chain molecules plays a major role [Fleischer et al., 1965b]. Several criteria have been proposed to describe quantitatively under what conditions — charge and speed of particles, properties of the material — developable tracks may be expected [Fleischer et al., 1967; Katz and Kobetich, 1968; Benton and Nix, 1969]. All of these criteria have some limitations and inconsistencies with experimental results. The damage around the particle's path is confined to a maximum diameter of about 100 Å and "latent" insofar, as only visible by electron microscopic techniques. This zone, however, if treated by an appropriate chemical like alkali hydroxide solutions (organic polymer detectors) or hydrofluoric acid (glass detectors), is preferentially dissolved with an etch rate (v_T) along the track core, larger than the bulk etch rate (v_B) of the unirradiated material. Within the scope of this simple model the track should evolve as a conical etch pit with cone angle ϑ, where $\sin\vartheta = v_B/v_T$ (fig. 33). The intersection of the cone with the etched bulk surface is an ellipse with semiminor and semimajor axes determined by the angle of incidence ϕ, the cone angle ϑ and the etched distance in the bulk material $v_B t$ [Henke and Benton, 1971]. In reality v_T is not a constant but generally increases with ionization rate. The variation of v_T has to be accounted for in more exact models of track evolution kinetics, as described by Somogyi and Szalay [1973] and Paretzke et al. [1973].

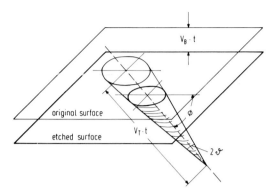

Fig. 33. Schematic view of track etching in solid state nuclear track detectors. Because of the different etch rates in the unirradiated bulk material (v_B) and along the particle trajectory (v_T) an etch cone evolves with cone angle ϑ ($\sin\vartheta = v_B/v_T$). ϕ is the angle of inclination of the particle trajectory to the detector surface.

The track etching kinetics entails two important aspects which distinguish the SSNTDs from photographic emulsions. Firstly, the track size can be controlled within a wide range by the kind of etchant used, concentration, temperature, and etching time. Normally, it is found that the track diameter varies roughly linearly with etching time as shown by Somogyi et al. [1968] and Polig [1975a] for α-tracks in cellulose nitrate, and by Blanford et al. [1970] for fission fragment tracks in cellulose acetate butyrate. The track diameter kinetics of fission tracks in glasses seems to follow the $t^{1/2}$ dependence (t = etching time) for highly concentrated etching solutions [Somogyi, 1966]. Such curves of the etch kinetics provide a means to control the track size. If, for instance, high track densities have to be evaluated one normally prefers to produce small etch pits to avoid overlapping or saturation of the registration capability. Unlike emulsion detectors, there exists a lower limit of the angle of particle incidence ϕ, below which no particles are registered. This angle is the cone angle ϑ and it is immediately clear from fig. 33 why for smaller angles of incidence no etch pits can be developed. This so-called critical etch angle varies with etching temperature, as does v_T/v_B. The existence of a critical angle explains, irrespective of other possible causes, why the registration efficiency of SSNTDs must always be smaller than that of emulsions for otherwise identical irradiation conditions. The obvious advantages of SSNTDs over photographic emulsions are:

1. Practically no latent image fading at room temperature, even for plastic detectors [Benton, 1968; Piesch and Sayed, 1974].

2. No background enhancement during prolonged exposures.

3. Mechanical and chemical stability.

4. Simple track geometry.

It is particularly the last point that makes SSNTDs especially suitable to automatic track counting procedures. At present the most sensitive detector material known is cellulose nitrate, in which even tracks of 55 keV protons were revealed [Jones and Neidigh, 1967; Lück, 1974]. A special cellulose nitrate detector is the LR 115 produced by Kodak-Pathé, consisting of a 10 μm red dyed cellulose nitrate layer on a plastic base whose properties were extensively tested by Costa-Ribeiro and Lobão [1975]. This foil has the unique feature that the dye is removed by etching through the track holes. The etch pits then appear as bright spots and have a high contrast against background if viewed with blue light. It must be doubted, however, whether this detector is suitable for the quantitation of high track densities as, in order to perforate the layer of dye, relatively large holes have to be etched out. Other organic polymers like cellulose acetate, cellulose acetate butyrate and polycarbonate are also used. The sensitivity of the latter, however, is too low to record α-tracks.

Up to now relatively few studies dealing with α-autoradiography by means of SSNTDs were made. Cole et al. [1970] prepared autoradiograms from mouse bone to map the microdistribution of monomeric ^{239}Pu. The sections were coated with a nitrocellulose solution produced according to a recipe of Benton [1968]. By appropriate dilution of the stock solution with ethyl acetate the dipping procedure could be controlled so as to generate a 3 μm thick detector coating over the sections. As a result well defined track distributions were obtained with an etch pit diameter of approx. 1 μm. The authors reported even repeated etching (6.5 N NaOH, 24°C) to be possible if the exposure time turned out to be too short. The resolution of the cellulose nitrate film was found to be at least as good as that of AR 10 stripping film, but with the lower detection efficiency to be expected. The dipping procedure of Cole et al. seems to be attractive because of its simplicity compared to the liquid emulsion technique and the possibility to adjust the exposure time by repeated etching. The effect of the cellulose nitrate solution on prestained sections has not been investigated so far. In contrary to photographic emulsions, staining after exposure seems to be impossible. For quantitative applications the detector coating must be produced with even and sufficiently reproducible thickness. The thickness is a critical parameter with the coating technique, as the detector itself acts as absorber and the detection efficiency varies with the thickness. Simmons and Fitzgerald [1970] made an intercomparison between Kodak Type A plates, AR 10 stripping film and a nitrocellulose detector coating of 15 μm and 30 μm, with respect to the spatial resolution. The detectors were tested by means of ^{239}Pu line sources and measuring the width of the blackened region of silver grains or etched tracks. The cellulose nitrate films showed a marked improvement in resolu-

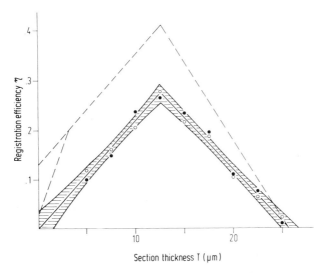

Fig. 34. Registration efficiency η as a function of section thickness T with trabecular bone sections and an additional 6 μm Mylar foil as absorber. The dashed lines show the theoretically expected values of η which differ for the two etching temperatures 20°C (○) and 30°C (●) for $T < 3.1$ μm. The steeper line refers to 30°C. The confidence region for probability 0.95 is hatched [Polig, 1975a].

tion over the photographic emulsion and, as was to be expected, the resolution could be enhanced by casting thicker films. As uneven detector thickness and underlying tissue structures are two main obstacles in quantitative autoradiography with automatic counting systems, Polig [1975a] developed special detector samples by coating microscope slides with a cellulose nitrate solution, prepared with slight modifications according to Benton [1968]. After drying, the detector samples were exposed in contact with radioactive bone sections from rat femora and etched in 6.5 N NaOH at 20°C and 30°C. The advantage of the contact technique is, that variations in detector thickness (20–70 μm in Polig's experiment) have no effect on the detection efficiency. To quantify the extent of self-absorption in embedded sections of trabecular bone, the detection efficiency was determined by means of placing bone sections of varying thickness and an additional Mylar foil (6 μm thick) between the detectors and superficial radiation sources of ^{239}Pu electrodeposited on stainless steel planchets. Fig. 34 shows that the detection efficiency referring to 2π geometry attains a maximum at a section thickness of 12.5 μm and falls off to zero for a source located directly on the surface of the Mylar foil ($T = 0$). This is quite in contrast to the behaviour of photographic emulsions and can be explained by the existence of a critical value of

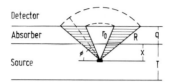

Fig. 35. Irradiation geometry for exposure of solid state nuclear track detectors. R = etchable range, r_0 = unrevealed range. The hatched sectors determine the solid angle for detectable particle trajectories [Polig, 1975a].

LET below which no etchable damage is formed along the particle's trajectory. This LET limit corresponds to a geometrical distance r_0 that can be inferred from a Bragg-curve (fig. 2), specifying the length of the trajectory extending from the starting point of the particle that cannot be revealed. Fig. 35 illustrates the situation by means of a simple scheme of the irradiation geometry [Thomas et al., 1973; Polig, 1975a]. Both the etchable particle range R and r_0 define the solid angle in which particle trajectories may be revealed by etching. From geometrical considerations one gets for the detection efficiency η:

$$\eta = \begin{cases} \dfrac{q/r}{r_0} - \sin\vartheta + \dfrac{x}{r_0} & \text{for } 0 \leqslant x \leqslant R\sin\vartheta - q/r \\[1em] \left(\dfrac{1}{r_0} - \dfrac{1}{R}\right)(x + q/r) & \text{for } R\sin\vartheta - q/r \leqslant x \leqslant r_0 - q/r . \\[1em] 1 - \dfrac{q/r}{R} - \dfrac{x}{R} & \text{for } r_0 - q/r \leqslant x \leqslant R - q/r \end{cases} \quad (4.5)$$

ϑ is the critical angle of etching, r is the ratio of particle range in absorber to range in source, q is the absorber thickness. According to eq. (4.5) the function describing the depencence of η on the thickness x of the absorbing layer is made up of three linear portions. Determining the parameters r_0 and R from the abscissa of the maximum and the point where the curve falls off to zero, and computing the curves for η shows that the theoretical values generally exceed the measured ones (fig. 34). This discrepancy was also observed by Thomas et al. [1973] who made similar measurements with cellulose triacetate detectors and aluminium absorbers, and had to introduce a reduction factor 0.5, not explainable on the basis of the simple model, to obtain a good fit to the experimental data. It must be emphasized that allowing for the small effect of bulk etching, which removes a thin layer from the detector surface, and admitting values of r other than unity gives no consistency either. May be that a refined model taking into consideration some optical criteria of track visibility and including the more exact features of the etch kinetics like variation of v_T with LET will result in an agreement between experiment and theory. Nevertheless, even without a theoretical interpretation, the empirical curve of fig. 34 provides factors K_D converting the measured track density at a specific site to dose rates for any section thickness T. Under the assumption of a "locally" homogeneous distribution of activity around the field of measurement one has for the dose rate $\dot{D}$

$$\dot{D} = K_D \frac{\sigma_T}{\rho_M \cdot t} \qquad \begin{array}{l} t = \text{exposure time} \\ \rho_M = \text{density of tissue} \end{array} \qquad (4.6)$$

with the dose conversion factor [Polig, 1975a],

$$K_D = E_0 \bigg/ \int_0^T \eta(x)\,dx \qquad E_0 = \text{particle energy} . \qquad (4.7)$$

The integral in eq. (4.7) has to be calculated from fig. 34. K_D is specific for the particular absorber material and thickness used. At present there exists no analysis of the errors introduced by the hypothesis of a "locally" homogeneous distribution, as such an estimate has to await an exact theoretical description of the registration efficiency. One may argue, however, that the error should be tolerable as one can imagine several locally inhomogeneous distribution patterns yielding in effect nearly the same dose rates as the hypothesis of homogeneity. One particular example is the pattern of point sources of radioactivity distributed with uniform density throughout the tissue section.

From an inspection of fig. 34 it is clear that for bone sections thinner than 25.5 μm it is advantageous with respect to the optimum registration efficiency, to place additional absorbers between detector and source. Thereby the section is brought to a distance of maximum η. It is possible to show that the optimum absorber thickness q decreases linearly with increasing section thickness [Polig, 1975a].

4.3. Neutron induced autoradiography

The fact that some heavy α-emitters undergo fission in a flux of thermal neutrons was utilized to develop a very rapid and sensitive autoradiographic technique. This method was called neutron induced autoradiography (NIAR) because, as a result of exposing the tissue sections in the thermal column of a reactor, for every fissioned atom two fission fragments are produced which are detectable by SSNTDs. As the fission fragments have a high LET, a much broader variety of detector materials is available than for the direct α-autoradiography, including the less sensitive organic polymers, glasses and mica. A

Table 8
Fission cross sections σ_f for thermal neutrons [a]

Nuclide	σ_f(barn) [b]	$2.89 \cdot \sigma_f \cdot t_{1/2}$ (cm² sec)
^{232}U	75.2	4.91×10^{-13}
^{233}U	531.1	7.69×10^{-9}
^{235}U	582.2	3.74×10^{-5}
^{237}Pu	2,400	2.73×10^{-14}
^{238}Pu	16.5	1.32×10^{-13}
^{239}Pu	742.5	1.65×10^{-9}
^{244}Cm	1.2	1.97×10^{-15}
^{249}Cf	1,660	5.30×10^{-11}
^{251}Cf	4,300	3.52×10^{-10}
^{252}Cf	32	7.64×10^{-15}

[a] From Seelmann–Eggebert et al. [1974] Nuklidkarte.
[b] 1 barn = 10^{-24} cm².
$t_{1/2}$ = half-life of the nuclide.

decisive parameter for choosing the NIAR technique is the fission cross section σ_f, specifying an imaginary target area which a thermal neutron has to hit in order to induce fission. The probability of fissioning a single atom is then equal to the product of neutron flux Φ and fission cross section. Table 8 lists the fission cross sections for some α-emitting nuclides of biological relevance. The usefulness of NIAR cannot be judged from the fission cross sections alone, however, but has to be inferred from the ratio of the generation rates of fission products to α-particles emitted. It is easily shown that this ratio, being a measure of the relative advantage of NIAR over α-autoradiography, is equal to $2.89 \cdot \sigma_f \cdot t_{1/2} \cdot \dot{\Phi}$, where $t_{1/2}$ is the half-life of the nuclide and $\dot{\Phi}$ the flux density (neutrons per unit area and unit time). If the values listed in the third column of table 8 are multiplied by the available flux density $\dot{\Phi}$, the ratio of the generation rates is obtained, or in other words, the factor by which the exposure time in the neutron flux has to be reduced to get nearly equal track densities on the detectors. Here it must be stressed that the registration characteristics of fission fragments differ from that of α-particles for two reasons: The charge of the fission fragments entering the detector surface is not a unique quantity but follows a distribution law. The same is true for the initial energies. As highly ionizing particles fission fragments have a shorter mean range than the α-particles of the same nuclide. This proves to be advantageous if a high resolution of the mapping of the nuclide distribution is demanded. As the last column of table 8 shows, there are some nuclides for which NIAR may be applied with particularly high benefit even if only a relatively low neutron flux density $\dot{\Phi}$ is available. These nuclides are ^{235}U and

^{239}Pu: exactly the elements being investigated by NIAR so far. In a flux density of 10^{10} neutrons cm^{-2} sec^{-1} autoradiographs of ^{239}Pu can be produced in an exposure time 16 times shorter than for an approximately equal track density of an α-autoradiograph. For flux densities usual in reactors, ranging from 10^{11} to 10^{13} neutrons cm^{-2} sec^{-1} exposure time or minimal concentrations detectable are several orders of magnitude smaller. But table 8 also shows that a profitable application of NIAR requires rather high neutron flux densities for some elements even though their fission cross section is high (^{237}Pu, ^{249}Cf). For high neutron fluxes the problem of heating up the detectors and track fading by annealing gains importance and necessitates the use of detector materials with high annealing temperature.

Hamilton [1968] seems to have been the first to realize the usefulness of NIAR as an alternative to conventional autoradiography in the life sciences. He applied this method to determine the distribution of uranium in samples of animal tissue, plants, and natural waters using a polycarbonate detector. For the measurement of the concentration of ^{239}Pu in bone and several soft tissue samples he reported a practical detection limit of less than 10^{-6} pCi per microscopic field of view. Becker and Johnson [1970] exposed bone samples of beagle dogs mounted on glass slides and calculated the average activity of ^{239}Pu per square centimeter (a) according to the equation

$$a = \frac{K \cdot \sigma_T \cdot A \cdot S}{N_0 \cdot \sigma_f \cdot \Phi}, \tag{4.8}$$

where σ_T is the track density, A is the atomic weight (239), S is the specific activity (0.0617×10^{12} pCi/g for ^{239}Pu), N_0 is Avogadro's number. K is a factor that characterizes self-absorption and fission fragment detection efficiency. The authors also pointed to a technical difficulty serious for the detection of very minute amounts of ^{239}Pu, namely the natural uranium content of microscope slides used either as support for the samples or as track detectors. Consequently a considerable background of etch pits from uranium fission fragments is present. Auxier et al. [1975] and Kienzler [1976] observed the same phenomenon. The former, however, reported that quartz glass exhibited no measurable fission fragment background.

If the dose rate due to α-particles of energy E_0 has to be calculated directly from the recorded fission track density σ_T under the assumption of a "locally" homogeneous distributon of the nuclide, expression (4.8) may be developed further to have the form of eq. (4.6), but now the dose conversion factor reads as,

$$K_D = \frac{0.693\, E_0}{t_{1/2} \cdot \dot{\Phi} \cdot \sigma_f \int_0^T \eta(x)\, dx}. \tag{4.9}$$

The registration efficiency $\eta(x)$ as a function of the depth of penetration behaves differently from that of α-particles (fig. 34) decreasing approximately linearly with absorber thickness [Kienzler, 1976]. This suggests that $r_0 = 0$ (eq. 4.5), i.e. the particle trajectories of fission fragments are etchable in full length in polymer and glass detectors.

The NIAR technique has been used by several investigators to study the microscopic localization of ^{239}Pu in the femur and vertebrae of rabbits [Bleaney, 1968, 1969], in the vertebrae and ulnae of dogs [Jee, 1970, 1972; Jee et al., 1972, 1975] and rats [Storr et al., 1975]. Schlenker and Oltman [1973, 1974] analyzed the ^{239}Pu and ^{235}U distribution in human bones. Most of these experiments were performed to measure the specific surface deposition of the radionuclide on trabecular surfaces, to quantify the "burial" phenomenon, and to calculate dose rates in the region of the endosteal cell populations. All these workers used commercial polycarbonate plastic foils as detectors (LEXAN, MAKROFOL), which were etched after exposure in KOH or NaOH at varying concentrations, temperatures and time intervals. To localize precisely the track etch pits with respect to their position relative to the bone and marrow phase, some of these authors took advantage of an amazing phenomenon, the generation of a printed image of the bone structure in the polycarbonate foil under suitable irradiation conditions. The so-called "detailed NIAR" method, developed and routinely applied by Jee [Jee, 1970; Jee et al., 1972], produces autoradiographs in which the mineralized tissue appears darker than the soft tissue phase. Fig. 36 exhibits such a detailed NIAR autoradiograph in which the fission fragment tracks are clearly seen as black dots or needles, delineating the endosteal surfaces of bone, represented by the shadowed region (B), or being buried by the apposition of new bone and thereby shifted away from the surface. The advantages of this procedure are obvious, as it eliminates the need of a precise realignment of tissue section and detector foil which both have to be separated during etching. The major cause of the optical contrast between the two phases is a variation in the optical density due to small, shallow etch pits. Kleeman and Lovering [1967] observed similar "prints" when they used LEXAN plastic foils to register fission tracks of uranium in rocks. The detector surface which had been in contact with the rock section exhibited a detailed print of the rock structure, mapping grain boundaries and cracks with remarkable accuracy. The origin of the plastic print is still somewhat miraculous. Jee [1972] could only obtain a bone image if prior to etching the detectors were stored for several weeks after exposure. This led him to conclude that (n, α) reactions are not responsible for the generation of the image. Schlenker and Oltman [1973], however, demonstrated that the prints are present immediately after exposure to the thermal neutron flux. Attempting to explain the mechanism of the formation of the prints, Kłeeman and Lovering [1970]

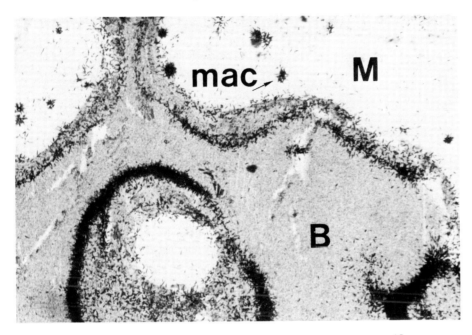

Fig. 36. Neutron induced autoradiograph produced by irradiation with 10^{18} neutrons/ cm^2. 6 μm thick section of the distal femur of a beagle injected with 0.015 μCi/kg ^{239}Pu and sacrificed 746 days after i.v. injection. The detector was subjected to the ARIE procedure (×150). B = bone, M = marrow, mac = macrophage [Fellows et al., 1975; courtesy of Prof. W.S.S. Jee].

correlated the observed etch pit density with the content of B and Li in some minerals. Their work yielded strong evidence that indeed a (n, α) reaction with ^{10}B, ^{6}Li, ^{17}O, ^{32}S or ^{33}S may be responsible for the generation of the image, the etch pits being caused by α-particles or light recoils. To overcome the difficulties of demonstrating very minute amounts of ^{239}Pu in tissue, Fellows et al. [1975] modified the detailed NIAR method to arrive at what they call now the ARIE procedure (ARIE = anneal-reirradiate-etch). Tissue samples of low Pu content require such large neutron fluxes to produce an adequate number of fission fragments, that the bone image becomes obscured by excessive background pitting of the polycarbonate foils. The optimum neutron flux for generating the bone image was found to be 10^{15}–10^{16} neutrons cm^{-2}. To insure both the production of the image and the track distribution with equal quality, Fellows et al. subjected the detector-source combination to an annealing procedure (149°C, 10–15 min) after exposure to 10^{17} neutrons cm^{-2} and again exposed the samples to a flux of 1.5 × 10^{15} neutrons cm^{-2}. This second exposure regenerates the bone image which together with all background pits was erased by the annealing procedure. Annealing, on the other hand, only slightly affects track registration as it results in a shortening of the fission tracks only. One embarrassing observation was noted by Fellows et al., namely the occasional appearance of a negative image. For an unexplained reason bone areas are lighter than marrow in some samples. This observation might perhaps shed new light on the current hypotheses of image formation.

The study of Auxier et al. [1975], dealing with a variety of detector materials, was directed towards an improvement of the spatial resolution of ^{239}Pu localization by means of the fission track method. Uranium oxide microspheres embedded in cellulose nitrate and with diameters ranging from 4 to 40 μm served as particulate sources of fission fragments. The spatial resolution of a radiation source achievable by fission tracks may be anticipated to be better than by α-tracks because of the shorter particle range of the former. Bleaney [1969] determined the mean range in embedded tissue as 14 μm, Fellows et al. [1975] reported values of 20–25 μm in polycarbonate. Improvement of the resolution was found to be possible by viewing the etch pits with a scanning electron microscope (SEM), as this instrument images only the surface, i.e. the entrance location of the tracks. The apparent track cluster diameters resulting from the fission fragments of the spheres was then reduced by a factor of two, compared to the usual microscopic observation. Placing absorbers between source and detectors turned out to be the most promising procedure. Layers of polymer absorbers up to a thickness of 18 μm are satisfactory for LEXAN as detector, while for mica the optimum thickness was found to be 13–15 μm. Owing to its large critical angle of 35° glass has a built in collimating effect. Consequently, glass detectors were recommended by the authors as the material of choice for high resolution autoradiography, if used without absorbers and after short etching intervals.

4.4. Track viewing and counting procedures

Most of the quantitative autoradiographic work with photographic emulsions was done by visual counting of single α-tracks under a microscope. To distinguish individual tracks in areas of high radiation exposure, large optical magnifications are usually employed, necessitating the use of immersion objectives. The reason that the bright field transmitted light method was nearly always used may be tradition or easy availability of the optical equipment, combined with some ignorance about the advantages offered by the incident light techniques [Gullberg, 1957; Dörmer, 1967]. The flexibility of the combined incident dark field and transmitted bright field illumination allowing to control the relative contrast between the emulsion track grains (appearing as bright spots) and the background, is of special importance in stripping or liquid emulsion autoradiography with stained tissue beneath the detector [Green et al., 1975]. For etched holes in SSNTDs the same microscopic techniques are generally feasible, but requiring lower magnification only as the tracks are of relatively large size. Two additional and less common microscopic viewing methods are by polarized light, as the etch pits effect depolarization, and by interference contrast, which results in a brilliant relief image of the etched holes on a coloured background [Piesch, 1969]. Another method of contrast enhancement consists of filling the etch pits with fluorescent dyes and observing with UV-light under dark field illumination [Fink, 1973].

Visual track counting demands utmost concentration and patience from the observer and, therefore, is considered a tedious and demoralizing task. For these reasons and in order to get rid of subjective counting errors several technical auxiliary means came into use relieving the investigator of the tiresome and monotonous parts of the work. The density of α-tracks at a specific site of the detector is a function of the blackening of the emulsion plate which in turn may be related to the light transmission measured with a microdensitometer. Such a densitometric method was devised by Twente et al. [1958; Twente and Jee, 1961] using a Leitz Panphot microscope and a Macbeth Ansco densitometer. A low power objective was chosen to insure a great depth of field, which is mandatory if all grains along the tracks in a thick emulsion (Ilford C2) have to be in focus. Measured with a 70 μm aperture, a precise linearity between the track density and the photometer reading could be ascertained [Twente and Jee, 1958]. This finding may be explained by the fact that the optical density in fields with up to one thousand tracks was still very low. A comparison of the results obtained with the densitometer to visual track counting showed up excellent agreement with maximum deviations of 11%. Another test of the reliability of the densitometric method involved comparison with the radiochemically determined content of ^{239}Pu in thoracic vertebrae [Twente and Atherton, 1959; Twente and Jee, 1961] and yielded a 76% recovery of the radiochemical data, which the authors considered to be good. A better agreement was not to be expected because of the variation of the ^{239}Pu content in adjacent vertebrae, uncertainties in the endosteal surface determination, requisite for calculating the total bone content from a limited number of densitometer measurements and ignoring a certain fraction of activity not deposited on endosteal surfaces. A similar densitometric technique was applied by Marshall and Finkel [1959] to measure the microscopic dose distribution from ^{45}Ca, ^{90}Sr and ^{226}Ra in several bones of mice. Later on, the microdensitometric procedure of track counting was refined and automatized by Lloyd et al. [1966] using a flying spot scanner (Chloe) linked to a computer. With this machine scanning of the specimens is performed by a moving light source represented by a 15 μm diameter spot of light from a cathode-ray tube. Again the amount of light attenuation sensed by a photomultiplier is correlated with the local track density, which on the other hand is a function of the dose rate at a specific site. The "Microanalyzer" developed by Marshall et al. [1973] is a three-microscope synchronous scanner and was employed for the microscopic analysis of bone sections burdened with radium and other radionuclides. It allows to correlate three images simultaneously (e.g. UV, X-ray and autoradiograph) by densitometric measurements in a wide density range. Alignment of these images is achieved by adjustment of the three stages relative to each other under subjective visual control by means of a superposition of the separate microscopic images with the aid of an optical bridge. The sophisticated electronic circuitry provides the display of counts, areas, densities, density integrals and -distributions and a built-in function generator makes possible to correct for non-linear response of autoradiographic films.

James [1969] used a microdensitometer as well to measure the specific surface deposition of ^{239}Pu from the blackening of K2 emulsion by α-tracks. The instrument was calibrated by means of a bone surface model consisting of ^{241}Am, electrodeposited on an aluminium disc with known specific area deposition, and embedded perpendicularly to the sample surface in araldite resin. The peak photometer reading obtained by scanning across the lining of the tracks with circular apertures of 8.5 μm diameter proved to vary exactly linearly with exposure time or, equivalently, the specific surface concentration. As James pointed out, however, this method critically depends on the assumptions about the source geometry which in practice may deviate considerably from that of the model source used to calibrate the instrument. Consequently, errors up to 30% in the determination of the ^{239}Pu surface activity have to be envisaged. To avoid the problems introduced by assumptions about the source geometry he devised a "target oriented" method of dosimetry, independent of any hypotheses about source locations and taking into account the stochastic nature of the problem [James, 1969; James and Kember, 1970]. The target contemplated is the cell nucleus of 5 μm diameter approximated by cylinders of 3 μm height and 5 μm diameter. The cylinder is defined by the thickness of the AR 10 stripping emulsion and by an eyepiece graticule with nine circles and a reference line (fig. 37). While the reference line is aligned with the bone surface, the position of the three rows of circles in distance of 5, 12.5 and 20 μm from the reference line define the mean locations of the osteoblast, pre-osteoblast and mesenchymal cell population, respectively. This experimental setup has a twofold purpose. Firstly, α-particle traversals through nuclear volumes are directly amenable to measurement without recurring to computations on the basis of idealizations. Secondly, dose values may be evaluated in the target themselves, no matter what the radiation field is. From eq. (2.1), which represents the general rela-

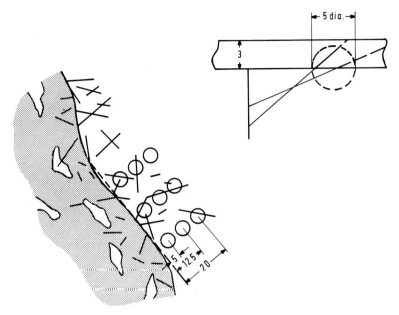

Fig. 37. Diagram showing the positioning of an eyepiece graticule in making flux measurements at a bone surface (lower left). The dashed marker line coincides with the surface. Cell nuclei are simulated by cylindrical volumes defined by the thickness of the emulsion (3 μm) and the 5 μm diameter circles in the eyepiece graticule (upper right). All distances in the figure are in μm [James, 1970].

tionship between dose D and the energy spectrum of the particle flux $\Phi' = d\Phi/dE$, an expression for the constant stopping power approximation is readily derived. Then with $S(E) = S =$ const. and $d\Phi/dE = \Phi/E_m$ eq. (2.1) yields

$$D = (\Phi \cdot S)/\rho . \qquad (4.10)$$

As not Φ, but the number of tracks ϕ crossing the target volume is observed ($\phi = \Phi \cdot S_s$, S_s is the projected area of the sphere), eq. (4.10) changes by appropriate substitution of Φ by ϕ and of S by $\overline{\Delta E/l}$ ($\bar{l}$ = mean chord length)

$$D = \phi \, \overline{\Delta E}/m \qquad (m = \text{mass of the target}), \qquad (4.11)$$

an expression which is intuitively obvious, because ϕ represents the number of target hits and $\overline{\Delta E}/m$ the mean dose delivered per hit. For the latter quantity James [1972] obtained a value of 109 rad on the basis of computer calculations for a uniformly distributed α-emitter of 5.14 MeV (^{239}Pu). For that matter, it may be noteworthy that the constant stopping power approximation yields only a slightly lower value than the exact Bragg-curve (106.6 rad).

If the mean number of hits $\bar{\phi}$ in a spherical volume of tissue located in the plane of the section has to be determined by counting the mean number $\bar{\phi}_{cyl}$ of tracks crossing a cylinder located with its center 1.5 μm above the embedded tissue section (fig. 37), corrections have to be applied allowing for (i) the differences in the cross sections of cylinder and sphere, (ii) the underlap effect and (iii) the different particle ranges in embedding medium and tissue. These corrections lead to the relationship

$$\bar{\phi} = f \cdot \frac{R}{R_e} \cdot \frac{S_s}{S_{cyl}} \cdot \bar{\phi}_{cyl} .\qquad(4.12)$$

The underlap factor f was determined to 1.18 and represents an average for a distance up to the particle range from a planar source (fig. 37). The isotropic cross section of the cylinder is somewhat larger (S_{cyl} = 21.6 μm) than that of the sphere (S_s = 19.6 μm) and the particle range in embedded tissue R_e is smaller (30.7 μm) than in tissue (R = 37.4 μm). Consequently an overall correction factor of 1.30 results from eq. (4.12). Comparison of dose rate estimates calculated according to Mays [1958] from a measurement of the specific surface deposition by track counting (eq. 2.35) and those measured by the flux method depicted above showed reasonable agreement, but as James [1972] liked to emphasize, the flux method is insensitive with regard to the real thickness of the surface deposition of the nuclide, quite in contrast to the method of Mays.

One of the special track counting techniques applicable to SSNTDs exploits the property of etched tracks being holes: the spark counting technique developed first by Cross and Tommasino [1969, 1970] for fission fragment etch pits in polycarbonate foils (10 μm thick) and modified by Johnson et al. [1970] to work for α-tracks in cellulose nitrate as well. This method uses etch pits completely etched through the detectors. These serve as discharge channels if the detector is placed between electrodes with an appropriate voltage applied (400–700 V). One electrode is composed of a thin layer of aluminium evaporated on a Mylar backing. If a discharge occurs in one of the etched holes, the aluminium layer evaporates locally around the hole in a circular region of about 200 μm diameter, thus preventing a second discharge through the same hole. Consequently the discharge spark jumps from hole to hole and if the discharge circuit is coupled to a scaler, the events may be counted directly and the total number of tracks is obtained in a few seconds for relatively large detector areas of several cm^2. Moreover the perforated aluminium electrode provides a replica of the distribution pattern of the tracks visible by the unaided eye. It must be emphasized, however, that the spark counting technique has its inherent limitations in the fact, that relatively large perforations are produced and consequently the track density must not exceed 3,000/cm^2 to insure accurate counting. For the same reason the aluminium replicas only provide a qualitative impression of the distribution pattern in tissue [Becker, 1969; Becker and Johnson, 1970]. Nevertheless, the spark counting procedure may have its value even for mini- or microdosimetry because of its sensitivity in quantifying very minute amounts of activity, if its results can be correlated with precise informations about the distribution pattern on a true microscopic scale, derived from other methods.

A counting procedure capable to manage track densities up to 5×10^6 cm^{-2} is based on the effect of light scattering at the etch pits [Schultz, 1968; Khan, 1971]. As Schultz demonstrated, the amount of scattered light, measured under an angle ϑ with respect to the parallel collimated beam of incident light, is proportional to the track density over at least two decades. The lower limit depends on the background variation, the upper on the extent of overlapping of individual tracks. The signal to background ratio varies considerably for different scattering angles and was found to be optimum at $\vartheta = 30°$ for glass detectors and at $60°$ for polycarbonate detectors with fission fragment holes [Schultz, 1968]. Khan [1971] obtained an essentially non-linear response of the scattering signal, which might perhaps be attributed to a non-optimum setting of the scattering angle. No information regarding the limitations in geometrical resolution are available so far, as this parameter was not considered critical for the purposes of these authors. It may be suspected, that for light beam diameters below 100 μm the conventional light sources have to be replaced by a laser to provide the high light density necessary. The light scattering method seems to be a promising approach to track counting and still awaits a careful methodical investigation concerning its usefulness for high resolution scanning techniques.

A diffraction method with a coherent laser light source (He-Ne) for track density measurements was tested by Platzer et al. [1972]. The working principle of this coherent-optic integral method is that the diffraction pattern generated from the etch pits in a foil represents a Fourier transformation into spatial frequencies. The light intensity sensed in the Fourier domain by a photomultiplier should then be a linear function of the number of tracks present in the field of measurement. Such a linear relationship could indeed be verified for certain diffraction angles. But again the optical setup involved circular fields of 2 mm diameter in which the track density was evaluated, and thus the method is not yet ready for use in microscopic fields.

As with photographic autoradiographs, microphotometry holds promise of being a rapid and accurate technique for track density measurements in SSNTDs too. In preliminary experiments Kienzler and Polig [1975] showed that with both bright field and dark field transmitted light illumination the amount of light reaching the photomultiplier is almost linearly related to the density of α-tracks in a cellulose nitrate detector. The practical usefulness of the method was demonstrated by scanning detectors exposed on bone sections with ^{241}Am deposition. The square measurement aperture was 50 × 50 μm, and the scanning time necessary to cover a complete section was substantially reduced compared to the scanning procedure based on image analysis (to be discussed below). Detailed investigations concerning the various methodical aspects of microphotometric track counting with emphasis on fission fragment tracks in glass and polycarbonate detectors are presently under way [Kienzler, 1976]. These aspects include the optimal choice of the etching parameters, control of spatial resolution and registration efficiency by absorbers, self-absorption in hard and soft tissue sections, data handling procedures and spatial alignment of bone structure and radiation detectors. The latter is one of the key steps if a quantitative detailed autoradiography under computer control is aimed at. To date no method is conceivable by which it is possible to scan both tissue structure and detector without mutual disturbance. Consequently, the data obtained by scanning radiation detector and tissue sample separately has to be combined somehow. For the data stored on a mass storage device of a computer system this means, that one of the two images represented in a raster system, in general has to be subjected to a translation and a rotation to achieve alignment. The determination of the transformation equations for the image coordinates represents the basic problem which may be solved by:

1. subjective determination of the transformation equations by visual inspection of the track distribution and the tissue structure superimposed.

2. similarly as (1) but the computer does the alignment by means of correlating the two images according to an objective criterion.

3. landmarks on radiation detector and tissue images to identify corresponding points or areas on both images.

Especially the latter approach holds promise of providing an accuracy down to a few microns and is well suited for a computerized scanning system, if the determination of the transformation equations is carried out under software control employing a two dimensional correlational analysis of the environments of the landmarks in both images [Polig, 1976c]. Fig. 38 illustrates the result of such a computerized minidosimetry in the skeleton, displaying the combined information of bone structure and dose rate variation obtained by means of a microphotometer [Kienzler and Polig, 1976]. For this example none of the above mentioned methods of alignment was used, but the correspondence between the two images was insured by exposing and scanning the specimens in mechanical reference frames. The scans were generated with incident dark field illumination (Ultropak) and a measurement area of 20 × 20 μm. Dose rates were calculated from calibration curves, relating the photomultiplier signal to track density, approximated by third order polynomials.

With the progress in electronic image analysis and the advent of commer-

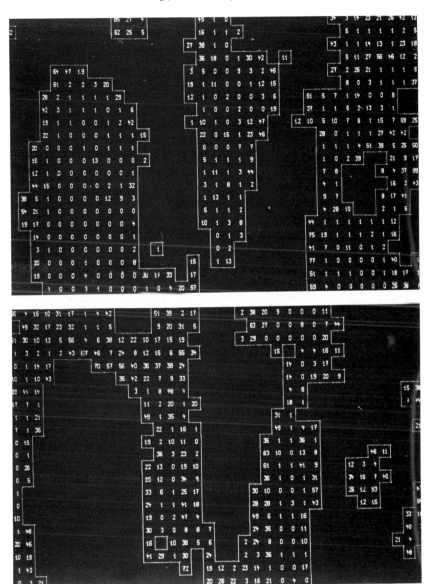

Fig. 38 Computer display of the dose rate pattern in the distal femur of a rat injected i.v. with 5 μCi/kg ^{239}Pu and killed 4 weeks after injection. The scans were generated by means of a microphotometer (MPV II, Leitz) controlled by a minicomputer (PDP 8E, Digital Equipment) in fields of 20 × 20 μm using incident light illumination (Ultropak). The soda lime glass detectors were exposed to a neutron flux of 7.26×10^{14} cm^{-2} (NIAR-technique). Top: dose rates (rad/day) in bone. Bottom: dose rates in marrow [Kienzler, 1967].

cially available image analyzers this method has gained increasing popularity for track counting too. It differs from the photometric method in the essential aspect, that the presence of tracks in a particular field is recognized by a logical analysis of a digitized image thereby identifying isolated objects which may be counted. The different instruments in use now can be classified into two categories: The flying spot scanner and the television image analyzer. Abmayr et al. [1969] described a flying spot system where the screen of a cathode ray tube projected on the object plane represents the scanning field. The light intensity of the spot is modulated by the sample and detected by a photomultiplier. After the conversion of the multiplier signal, with respect to a preset density level, into a sequence of pulses containing the information of the image, the logical analysis yielding the number of isolated objects in the field of view has to be carried out by a computer. The performance of the system was checked for the recognition of α-etch pits in a triacetate foil, but no further development of the system towards a routine application was reported. Most commercially built television image analyzers have a hardware counting logic as integral part of the system, which considerably speeds up the performance of these instruments. In a basic configuration they are equipped with three measurement functions, namely object count, area and intercept measurement. The structures to be analyzed are detected according to their brightness in relation to selected density discriminator levels. For measurements of relatively low densities where no track overlapping occurs, the image analyzer may display directly the number of tracks per field. The minimum track density measurable with sufficient statistical accuracy is then determined by the fluctuations of the background (not by the mean background!), and may be specified by choosing an appropriate criterion, e.g. that the minimum number of tracks per field should be at least three standard deviations of the background counts. Consequently the minimum track density countable in SSNTDs is largely influenced by the detector material, the etching procedure and the optical procedure to generate the image. For microdosimetric purposes, where highly non-uniform track densities are very common, it is essential, therefore, to have a track counting procedure that can cope with high track densities, for the sake of sufficient statistical accuracy in low density fields. Track overlapping becomes important for such high track densities. The probability of having no overlapping in a measurement field of size A_0 varies according to [Besant and Ipson, 1969],

$$p \text{ (no overlap)} = \exp(-2\pi r^2 \sigma_T^2 A_0) \,. \tag{4.13}$$

Eq. (4.13) holds under the assumption that tracks may be approximated by circles of radius r. Eq. (4.13) then gives for $r = 1$ μm and $\sigma_T = 10^6$ tracks cm^{-2} a probability p of 0.77 for finding an area $A_0 = 20 \times 20$ μm without overlapping tracks. In other words, for track densities exceeding 10^6 tracks cm^{-2} a considerable number of clusters is formed by overlapping of two and more tracks. The expectation value of the number of clusters C (including single isolated tracks) in field A_0 if N tracks and N_B background spots are present, approximated by circles of radius r and r_B, respectively, is then [Kendall and Moran, 1963; Polig, 1975b],

$$C = C_1 N \exp(-2pN) + C_2 \exp(-\tfrac{1}{2} N\hat{p}) \tag{4.14}$$

with

$$p = \pi r^2/A_0 \ ; \quad p_B = \pi r_B^2/A_0 \ ; \quad \hat{p} = p + p_B + 2(pp_B)^{1/2}$$

$$C_1 = \exp(-\tfrac{1}{2} N_B \hat{p}) \ ; \quad C_2 = N_B \exp(-2N_B p_B) \ .$$

If $N_B p_B \ll 1$, as is usually the case, then C_2 reduces to N_B. The second term in eq. (4.14) shows the variable background contribution to the number of clusters, which is decreasing for increasing N, as then the background spots become covered by the tracks. The same effect of covering tracks by background spots is displayed by C_1. It is important to notice, however, that N is a multivalued function of the measurable quantity C and thus eq. (4.14) is not applicable without additional information concerning the appropriate choice of N. Eq. (4.14) may be of use also for track counting by visual determination of C, as the human observer is able to decide whether C is small because only few tracks, or because many overlapping tracks are present. On the other hand, the total fraction G of the area A_0 not covered by tracks is a monotonically decreasing function of N and no problems of ambiguity arise. According to a derivation of Polig [1975b] but under somewhat more general conditions one obtains

$$N = \ln(G/\hat{C})/\overline{\ln(1-p)} \tag{4.15}$$

with $\overline{\ln(1-p)} = \int \ln(1-p) f(p) \, dp \ ; \quad \hat{C} - N_B \int \ln(1-p) f_B(p) \, dp$.

The above expression (4.15) applies for arbitrary shapes of the tracks, under the assumption that the distribution of the single track fraction p for tracks and background spots is characterized by the distribution functions $f(p)$ and $f_B(p)$, respectively. For small p one may approximate $\overline{\ln(1-p)}$ by $\ln(1-\bar{p})$, i.e. the distributions $f(p)$ and $f_B(p)$ need not be known, but it suffices to measure $\bar{p}$ directly by an image analyzer in fields of low track density. Another way of determining p is from a combined measurement of C and G employing eqs. (4.14) and (4.15) [Polig, 1975b]. The above relations for C and G were confirmed in their general behaviour by experimental verification with the image analyzer Quantimet 720. Two corrections had to be applied, however, imposed by technical features of image digitalization. For values of $G \leqslant 0.23$ N deviates from the value predicted by eq. (4.15), but the discrepancy can be

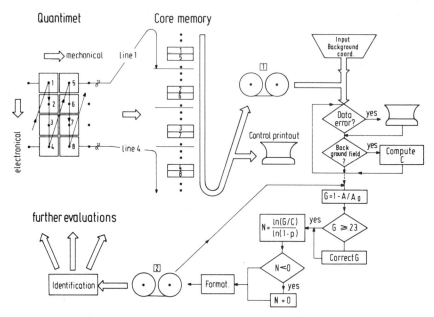

Fig. 39. Schematic representation of α-track scanning with an image analyzer (Quantimet, IMANCO) and an on line coupled minicomputer system (Wang 720C). Functions are explained in the text [Polig, 1975c].

accounted for by correcting G. C followed the expression (4.14) up to a track density of 10^7 cm^{-2} but with a parameter p larger than the true geometrical one. The procedure of counting α-tracks in SSNTDs outlined above was routinely applied to scan cellulose nitrate detectors exposed on 20 μm thick bone sections burdened with ^{241}Am. The measurement system used was a Quantimet 720 video image analyzer coupled on line to a Wang 720C minicomputer system, including external core memory and mass storage devices. In fig. 39 the principle of measurement and handling of data is depicted schematically [Polig, 1975c]. To speed up the performance of the scanner, a combined electronical and mechanical stepping of the measuring frame (80 × 80 μm or smaller) is provided. The resulting comb-like scan pattern requires a reorganization of the scan data (area measurements) by means of the fast external core memory. After the raw data have been stored on magnetic tape, the areas of the scan in which the background parameter C (eq. 4.15) has to be computed are determined by visual inspection of the scan plots, and the coordinates fed into the computer. Then in a first run, computation of C is carried out checking simultaneously for data errors. During a second pass the number N of tracks per field is calculated according to eq. (4.15), with provisions made for corrections in the high track density range by adjustment of G and in the low density range where a negative N may occur (fig. 39). After storing the data on tape again, changing the data format, they are available for a sequence of evaluation programmes, to obtain such informations as dose rate distributions in two dimensions or profiles, distribution functions, iso-dose plots etc.

By and large it must be stated, that the various measurement procedures in localized dosimetry of internal α-emitters have not yet reached a really satisfactory state, and the whole field is open for further improvements. Visual track counting is tedious and time consuming and has turned out to be a major obstacle for continuing microdosimetric investigations in a systematic way and on a large scale. The shortcomings of James' method are its limited applicability for a certain range of specific activities only, which must be high enough to register a sufficient number of target hits in the biologically relevant time. The densitometric method of track counting in emulsions is also confined to relatively high specific activities to avoid very long exposure times. With SSNTDs the situation is somewhat more favourable for, because of the larger track size, smaller track densities and consequently shorter exposure times are required for quantitative α-autoradiography. For nuclides fissionable with thermal neutrons the NIAR technique provides sufficient sensitivity and must be regarded as the method of choice. A direct recording of dose rates via track distributions is difficult because fission fragments have a smaller mean range than α-particles. To master the large amount of data generated in high resolution quantitative α-autoradiography and facilitate an appropriate data reduction and extraction of information wanted, sophisticated and expensive computerized measurement systems are an indispensable tool which, preferably, should also have the facilities of interactive man-machine communication.

5. Localized dosimetry of radionuclides in the skeleton

5.1. Radium-226

The radium isotopes are bone-seeking elements, due to their chemical similarity with the other alkaline earths like Ca and Ba. From the viewpoint of radiation dosimetry the deposition pattern of these radionuclides in the skeleton is of interest because of the largely inhomogeneous deposition reflecting the physiological and anatomical characteristics of this organ, and the consequent non-uniform impartment of radiation energy on a cellular or supercellular scale. Our present knowledge of the toxicity of Ra in humans stems essentially from three groups of people, namely the dial painters ingesting radium from luminous paints, chemists inhaling or ingesting radium and people which were treated with radium orally or intravenously for a variety of medical conditions. Among these individuals the principal effects of clinical significance from their Ra body-burden were osteoporosis and bone necrosis, osteosarcomas distributed throughout the entire skeleton and head carcinomas of epithelial or mesenchymal origin. From the vast amount of literature on this subject it may suffice to mention only a few publications, dealing with the description of the clinical findings, frequency of occurrence of osteosarcomas and carcinomas and its relation to cumulative average skeletal doses and other parameters in man [Evans et al., 1969; Evans, 1974; Finkel et al., 1969; Spiess, 1969], dogs [Dougherty, 1962; Mays et al., 1969], and mice [Finkel et al., 1969]. As a result of extensive experimental work the conclusion was drawn that osteosarcomas which are the primary radiogenic malignancy from Ra and other bone-seeking radionuclides are arising on endosteal and periosteal surfaces, that the osteogenic cells have to be considered as cells at risk, and that this risk is connected with the proliferating potential of the cells [ICRP, 1967]. In the light of these observations it is obvious that "dose to certain identifiable tissues within or associated with bone is the important parameter determining maximum permissible levels, rather than dose to bone regarded as a single uniform medium", as the ICRP stated. The situation is illustrated in fig. 40 displaying graphically a piece of trabecular bone and the osteogenic cell populations involved. The continuous remodelling of bone, i.e. the resorption of old and accretion of new bone material which is even going on to a certain extent in adults, is maintained by osteoblasts (formation) and osteoclasts (resorption) (fig. 40). Osteoblasts are initiating the formation of bone by secreting the organic matrix (osteoid) which subsequently mineralizes [Vaughan, 1970]. Remodelling rates reported in the literature differ somewhat but figures of 0–5% for human cortical bone and 10%/year for cancellous (trabecular) bone seem to lie within the correct range [Rowland, 1960; Schlenker and Farnham, 1975]. These figures give the fraction of total bone volume replaced by the apposition of new bone within the time period specified. Only a small percentage of the bone surfaces is growing or resorbing at any time. By measuring the length of the osteoid seams which are the indicators of appositional bone growth a range of 2–14% was determined in the human and in the dog vertebra [Jee et al., 1965; Bromley et al., 1966]. Thus, most of the bone surface are quiescent with osteoblasts of the typical flat and oblong form which makes them easy to distinguish from active osteoblasts (fig. 40). During their secretory work active osteoblasts may eventually bury themselves and become osteocytes. The proliferating pool of the osteoblast population is on the average located some cell diameters away from the surface. These so-called osteoprogenitor cells are however still within the range of the α-particles arising from surface

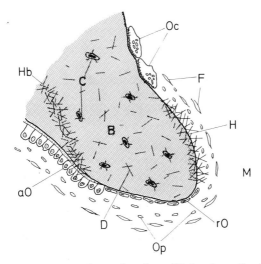

Fig. 40. Typical portion of a piece of cancellous bone (B) showing active (aO) and resting osteoblasts (rO), osteoprogenitor cells (Op), osteoclasts (Oc) and fibrocytes (F) in the marrow (M) near the bone surfaces. Osteocytes (C) are located in bone lacunae. The α-particle tracks from ^{226}Ra and its α-emitting daughters are found as superficial (H) or buried hotspots (hot lines) (Hb) and as a low level diffuse label throughout bone volume (D).

deposits or from atoms located up to a certain depth within the mineralized tissue. Besides the processes of apposition and resorption affecting growth and architecture of bone, there are two other processes affecting bone density. These processes called augmentation and diminution govern the long term exchange of calcium to and from bone volume via the system of canaliculi interconnecting the osteocytic lacunae, and play an important role in the mechanism of radium acquisition in the skeleton [Marshall, 1969].

For most of the human cases mentioned above, the longlived isotope ^{226}Ra and its α-emitting daughter products have been of widespread concern and are responsible for causing the observed deleterious effects in the skeleton. Its distribution pattern in bone is characterized by two components reflecting the very existence of the surface and volume processes as indicated above. A rather uniform labeling of pre-existing bone at a relatively low specific activity is often referred to as the "diffuse" component and must be attributed to augmentation and diminution, while the so-called "hotspot" component, i.e. small areas of bone or bone-tissue interface with a higher specific activity originates from those bone surfaces (periosteal, endosteal, trabecular, haversian and volkmann) laid down at the time ^{226}Ra is present in the blood stream (fig. 40). Specific activities were measured autoradiographically in bone sections obtained from patients in areas of the diffuse and the hotspot component [Spiers, 1953; Hindmarsh et al., 1958, 1959; Rowland and Marshall, 1959; Rowland, 1959, 1960; Lloyd, 1961; Schlenker and Farnham, 1975]. These investigations were complemented by animal studies with mice and dogs [Marshall and Finkel, 1959; Rowland, 1961; Sears et al., 1963]. Table 9 summarizes some of the results. If the specific activity of the diffuse component is related to the specific activity of a uniform label, i.e. a hypothetical distribution of the total skeletal burden throughout bone volume, the ratio diffuse to uniform specific activity is approximately 0.5 [Rowland and Marshall, 1959; Rowland, 1959; Lloyd, 1961]. From these results obtained in samples of human cortical bone it was concluded that the diffuse distribution makes up a significant fraction of the total ^{226}Ra burden. Taking into account the experimental results of Sears et al. [1963] who measured the diffuse to uniform ratio in both cortical and spongy bones of beagles, the ratio of 0.5 was confirmed in cortical bone but in trabecular bone occasionally was as high as 2 with an average of 1, indicating that the diffuse component may even constitute a higher fraction of the total amount of activity than might be suspected from analyzing cortical bone alone (table 9). Schlenker and Farnham [1975] demonstrated that, in spite of some variations from site to site in the skeleton, the specific activity of the diffuse component is rather uniform with a value of 2.8 for the ratio of the highest to the lowest concentration.

The pattern of hotspot distribution is determined by the location of the growing surface during the period of ^{226}Ra uptake into the blood, and by the topography of remodelling during the times thereafter. The association of hotspots to identifiable anatomical entities differs considerably from individual to individual and may depend on the manner of acquisition of the isotope. In cortical bone Lloyd [1961] showed that in 2 out of 3 cases the hotspots were associated with haversian systems. Between 1 to 6% of the haversian systems contained hotspots. If the period of uptake exceeds the time required for the formation of an osteon, complete labeling of the haversian systems from the outer cement line to the inner canal may be found [Rowland, 1960].

^{226}Ra set free into the circulation by the processes of bone resorption or diminution is again redeposited on growing surfaces. This recirculation, however, proceeds at a much lower level of the blood specific activity and consequently these hotspots are not so "hot" as those formed by initial deposition of the nuclide. Typical values of the maximum hotspot concentrations to the diffuse concentration lie between 80 and 100, but may range up to 200 (table 9) [Rowland, 1960; Lloyd, 1961; Rowland and Marshall, 1959]. In dogs this ratio seems to be generally lower by a factor of 4 to 5 [Rowland, 1961; Sears et al., 1963]. The frequency distribution of the hotspot specific activity (fig.

Table 9
Summary of ^{226}Ra results

Author(s)	Source	Maximum hotspot/Diffuse		Diffuse/Uniform		Average dose rates [a] (rad kg/µCi day)	
		Average	Range	Average	Range	D	Location
Rowland [1960]	19 patients	73	18–218	0.47	0.23–0.81	221	10 µm lacuna in hotspots
						3.6	10 µm lacuna in diffuse
Lloyd [1961]	3 patients	92		0.41		184	10 µm lacuna in hotspots
						158	20 µm lining in Haversian canal (hotspot)
						62	10 µm lining on bone surface (hotspot)
						26.2	40 µm lining on bone surface (hotspot)
Rowland [1961]	mongrel dog	23	19.4–28.4				
Sears et al. [1963]	beagle dogs	19	13.4–21.6	0.59 (1.0) [b]		39 [c]	10 µm lining on vertebral surface with hotspot
						2.7 [c]	10 µm lining on vertebral surface (diffuse)

[a] Normalized to 1 µCi/kg terminal body burden.
[b] First figure for cortical, second for spongy bone.
[c] Recalculated to terminal body burden using retention equation $0.79\, t^{-0.2}$.

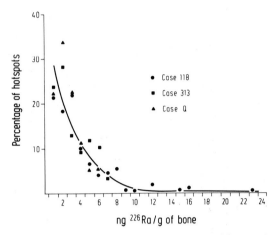

Fig. 41. Frequency of hotspot specific activity in sections of human bone. The percentage values refer to class intervals as indicated on the abscissa. Case 118: Female patient, 10 μg ^{226}Ra terminal body burden, 36 years after acquisition. Case 313: Female dial painter, 1.3 μg ^{226}Ra terminal body burden, 29 years after acquisition. Case Q: Male patient, 1.2 μg ^{226}Ra terminal body burden, 26 years after acquisition [Lloyd, 1961].

41) reveals that hotspots are becoming more frequent as their specific activity decreases, and a considerable fraction may have a specific activity of only 3–4 times the average diffuse activity (case 118 in fig. 41). In experiments with canine bone Rowland [1961] was able to demonstrate that there is no discrimination of ^{226}Ra against Ca in the transfer from plasma to bone at growing sites, i.e. hotspots are formed at the same specific activity as existent in blood.

Calculations of radiation dose have to take into account the complete chain of α-emitting daughters resulting from the decay of the longlived parent ^{226}Ra. Two of these daugher α-particles (RaA-^{218}Po, RaC'-^{214}Po) follow promptly after the decay of ^{222}Rn. The concentration of the third, RaF (^{210}Po), builds up slowly because of the 22 years halflife of RaD (^{210}Pb). An additional complication arises from the fact that in vivo radon escapes from bone via the blood stream, becoming exhaled in the breath. The fractional radon retention (f_{Rn}) increases with time after uptake of the ^{226}Ra-burden, amounting to 15–30% between 1 and 10 years [Mays et al., 1958; Mays et al., 1975]. If, however, cut or sawed bone sections are exposed, the "in vitro" fractional radon retention is usually higher than in vivo, depending on the conditions prevailing during exposure. Thus the number of α-particles produced per decay of ^{226}Ra is

$$1 + f_{Rn}(3 + f_{RaF}), \tag{5.1}$$

where f_{RaF} is the fraction of equilibrium activity of RaF (at equilibrium the activities of parent and all daughters are equal). The above equation provides conversion factors for specific activities in embedded bone sections to bone in vivo and from terminal dose rates to those at earlier times (neglecting loss of radium) [Spiers, 1953; Rowland and Marshall, 1959; Schlenker and Farnham, 1975]. Table 9 gives some dose rate figures for soft tissue with geometrical conditions as specified. To ease intercomparison the values were normalized to a terminal body burden of 1 µCi/kg ^{226}Ra. Rowland [1960] used the geometry factor derived by Spiers from parallel plate geometry which tends to underestimate the dose rate (fig. 13). Lloyd applied the more exact factors of Kononenko [1957] (fig. 8) but nevertheless arrived at somewhat lower values for a 10 µm lacuna in hotspots than Rowland. The dose estimates of Sears et al. [1963] are based on calculations of Mays and Sears [1962]. The lower dose rate (39 rad kg/µCi day) in a 10 µm surface lining of a dog vertebra near a hotspot as obtained by Sears et al., compared to a corresponding location in human bone (62 rad kg/µCi day) is in accordance with the lower hotspot to diffuse ratio in dogs. A comparison between the values of Rowland for the 10 µm lacuna in the diffuse component (3.6 rad kg/µCi day) with the 10 µm surface lining in dogs displays a remarkable agreement if one bears in mind that the geometry factor for the lacuna is larger than that for the surface layer. Mays and Tueller [1964] and Harley and Pasternack [1976] carried out theoretical calculations yielding the dose rate near a plane surface of bone uniformly labeled with ^{226}Ra and daughters (fig. 42). Mays' and Tueller's curve was constructed by means of the geometry factors in section 2.4 for the constant LET approximation (eq. 2.32), whereas Harley's and Pasternack's curves were based on experimental stopping power determinations and must be considered more accurate. Fig. 42 clearly underlines the importance of an exact knowledge of the fractional radon retention. In the figure the dose rate is plotted as a function of the distance from the plane surface of a piece of bone contaminated uniformly with 1 µCi/g ^{226}Ra. Even at low distances the dose contribution from the daughters excluding RaF (^{210}Po) is well above 50% (for 30% ^{222}Rn retention) increasing to 100% beyond the particle range of ^{226}Ra, due to the long range alphas from the daughters (table 1). The build-up of ^{210}Po does not much alter the curve in the proximity of the surface, and has no influence on dose rates beyond 39.3 µm. It must be stressed, however, that such dose rate distributions are not static, even at quiescent surfaces, but changing with time. This comes about because the fractional radon retention increases with time and, on the other hand, the specific activity of ^{226}Ra and its daughters decreases. An additional complication is introduced by allowing for different retention kinetics for ^{226}Ra and RaD. In this case a complex relationship between decay kinetics and retention kinetics results. Such calculations including the dynamic behaviour of all the factors

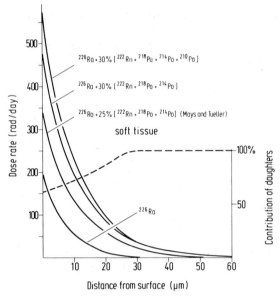

Fig. 42. Dose rates at the surface of bone uniformly contaminated with 1 μCi/g ^{226}Ra. The curves represent different contributions from daughters or ^{226}Ra alone. The dashed curve indicating the percentage of total dose originating from the daughters refers to ^{226}Ra + 30% [^{222}Rn + ^{218}Po + ^{214}Po] [Mays and Tueller, 1964; Harley and Pasternack, 1976].

involved have not yet been done. Depth-dose calculations for the frontal sinuses of humans were done by Kolenkow [1967]. The two frontal sinus cavities in the head are overlaid by a thin layer of epithelial tissue subject to α-irradiation from ^{226}Ra and its daughters in bone and from the radioactive atmosphere of ^{222}Rn in the cavities. The computations were performed under the assumption that the bone enclosing the cavity is labeled uniformly in a thick layer adjacent to tissue, that the stopping power S can be described by the expression $1/S = K_1 E + K_2$ with appropriate choice of parameters K_i for bone and tissue (see section 2.3) and that the retention of ^{226}Ra and ^{222}Rn follows (different) power functions while neither RaD nor RaF is excreted. Slightly modified calculations yielded depth-dose relationships for radioactive air layers of varying thickness (0.5, 1, 2 cm) corresponding to characteristic dimensions of the cavities. Mays and Sears [1962] derived a formula by which the average dose rate to a 10 μm layer of soft tissue at a bone volume with a uniform specific activity of V μCi/g bone of ^{226}Ra may be evaluated,

$$\bar{D}_{10\mu} = V[100 + 477 F_1 + 119 F_2] \text{ (rad/day)}. \tag{5.2}$$

F_1 is the fractional radon retention and F_2 the ^{210}Po/^{226}Ra activity ratio, which can be neglected for the first years after ^{226}Ra uptake. According to Lloyd et al. [1976] the activity of RaD in bone which decays via RaE(^{210}Bi) to the α-emitting RaF is on the average less than 1% of the deposited ^{226}Ra activity during the first 10 years. For F_1 the recently reported expression of Mays et al. [1975] should be inserted:

$$F_1 = 0.075\, t^{0.158}(1 - \exp(-0.181\, t)) \qquad (t \text{ in days}). \qquad (5.3)$$

F_1 has been shown to depend almost exclusively on time following ^{226}Ra deposition and is independent of age or species.

If accumulated doses are calculated from dose rates obtained at time of sampling, possible loss of ^{226}Ra from the skeleton has to be taken into account. Rowland [1960] estimated doses received by osteocyte lacunae from terminal dose rates in human bone samples by multiplying with the time the activity was carried. He termed the enormous doses ranging in hotspots between 10,000 and 430,000 rad as minimum lifetime dose, indicating that they may underestimate the dose due to loss of activity from the skeleton. Measurement of the gross retention of ^{226}Ra in the skeleton of man [Norris et al., 1955], beagle dogs [Van Dilla et al., 1958] and mice [Marshall and Finkel, 1959] (in femur only) as a function of time could be described by a power function $R = at^b$, where b was found to be -0.52, -0.2 and -0.3, respectively. Loss of radioactivity in human bone was demonstrated autoradiographically by Rowland and Marshall [1959]. They found the specific activity in hotspots to be smaller by a factor of about 10 than would be predicted assuming that no discrimination of ^{226}Ra against Ca takes place and that the hotspots are laid down at the same ^{226}Ra/Ca ratio as existent in blood. Rowland [1961] showed that in the bone of a mongrel dog and within a time interval of 353 days the hotspot activity decreased by 34% whereas the diffuse activity was found to have diminished by 25%. The bone samples were obtained by amputation four weeks after multiple ^{226}Ra injections and at time of sacrifice. From all the experimental evidence it was concluded that diminution is the main process responsible for the reduction of the skeletal burden and that this same process is removing both the activity from hotspots and from the diffuse component by exchanging ^{226}Ra with Ca in the blood-bone equilibrium at the same rates. The more rapid reduction of hotspots compared to the diffuse component in the experiment was explained conclusively, as the diminution of hotspots starts immediately if the blood specific activity drops after the last ^{226}Ra injection, whereas ^{226}Ra inflow into the diffuse component still continues until eventually the blood specific activity falls below the diffuse specific activity too. The results of Sears et al. [1963] supported the view that hotspot and diffuse deposits are reduced by identical mechanisms. By measuring the rate of loss in different bones of the beagle and fitting the power function to the data of individual bones or subunits of bones, the coefficients b turned out to be similar for the maximum specific activity in hotspots and the average diffuse activity, with exception of the femoral epiphysis. But the variation from bone to bone was considerable. Some bones like the tibia and the femoral metaphysis released their activity much more slowly than the skeleton as a whole. According to Lloyd [1969] similar differences in retention kinetics exist in human bones. She found the hotspot concentrations to decrease more slowly than the diffuse or uniform label in the midcortex of the tibia. Groer et al. [1972] examined the time dependence of the hotspot to diffuse ratio which remained constant in cancellous bones but decreased in cortical bone. Anyhow, the retention data in different bones of the skeleton of man and animals are too fragmentary and inconsistent at present, to provide a sound basis for exact dosimetric calculations.

Comparing the exponential factor b for the human skeleton with those of dogs and mice it might be suspected that the rate of loss in human bones is relatively high and consequently the reduction of the body burden too during some decades. This is not necessarily true for other species and even an increase of dose rates may be anticipated if the diminishing ^{226}Ra content in the skeleton is overcompensated by the increase of the fractional radon retention. This was borne out in autoradiographic studies conducted by Marshall and Finkel [1959]. Female CF1 mice were injected intravenously with 30 μCi/kg body weight ^{226}Ra, a dose which turned out to be most effective in producing osteosarcomas, and killed from 15 min. to 18 months. The dose rates were measured densitometrically. The time dependence of the dose to a 10 μm lacuna in bone at specific sites in the femur is displayed in fig. 43. The rapid drop of the dose rates in the distal plate is caused by appositional growth in this very active portion of the femur, thereby pushing the initial zone of high specific activity out of this area. Both the diffuse and the hotspot dose rates in the cortex are increasing in time. In the same figure calculated curves for the Utah beagles are drawn, based on the regression equations of Sears et al. [1963] and eqs. (5.2) and (5.3). The calculations for hotspots do not allow for remodelling and burial and therefore hold for hotspots remaining fixed at the surfaces. As the calculated curves represent the average dose rates to a 10 μm soft tissue layer, adjacent to a bone surface, they must be multiplied by a factor of 3.8 to compare with the measurements of Marshall and Finkel. In the shaft the dose rates fall off very slowly, at maximum metaphyseal hotspots they remain practically constant. The calculations yield a slow increase of dose rates from the diffuse deposits in the epiphysis (not drawn).

Among the two other processes envisaged as cause for the changing of dose rates in time, namely short term exchange and bone remodelling, the former

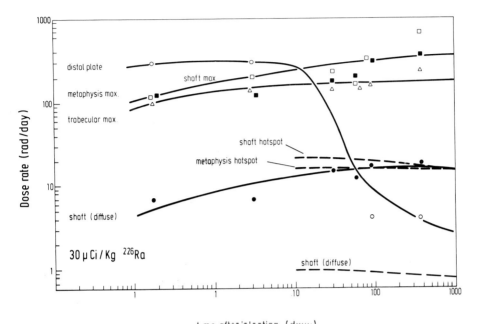

Fig. 43. Variation of dose rates with time after injection. Solid curves for an injection of 30 μCi/kg ^{226}Ra in mice [Marshall and Finkel, 1959] and calculated for a 10 μm lacuna in bone. Dashed curves represent the dose rates to a 10 μm lining on endosteal surfaces of dogs assuming 100% ^{226}Ra retention in the skeleton and using the retention equations of Sears et al. [1963]. To compare with the mouse data the dog curves have to be multiplied by a factor of 3.8.

is of less importance from a purely dosimetric point of view. In rabbits and dogs an uptake of ^{45}Ca on bone surfaces immediately after injection was observed [Rowland, 1966]. A similar phenomenon has to be expected for ^{226}Ra, but the surface labeling disappears too quickly to allow an appreciable dose accumulation. While the effect of bone remodelling plays a minor role in gross skeletal retention, it has considerable consequences on radiation doses delivered to osteogenic cells near hotspot locations. As previously mentioned, hotspots are formed at growing bone surfaces thereby delivering rather high doses to the osteoblasts just engaged in the production of the bone matrix. If the dose rates are not high enough to kill or inactivate a substantial fraction of the cells, these hotspots become buried within the mineralized tissue, provided ^{226}Ra intake into the blood ceases and the blood specific activity drops rapidly (fig. 40). Continuing growth subsequently generates a shielding of newly formed bone, protecting the sensitive cells on the surface from the

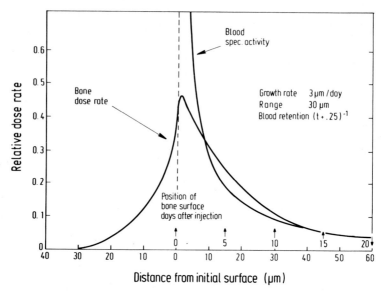

Fig. 44. Calculated dose rate profile for the burial of a ^{226}Ra hotspot at a growing surface of bone in a uniform bone medium (relative scale). The skewed shape of the profile results from the relatively slow decrease of the blood specific activity. Arrows indicate the position of the surface at the times specified [Marshall, 1960].

intense hotspot bombardment. To estimate the surface dose from hotspots and the relative importance of hotspot and diffuse irradiation Marshall [1960, 1962] presented a lucid mathematical analysis based on the three fundamental assumptions:

1. The radioactivity from ^{226}Ra is deposited on bone surfaces at the specific activity of the blood.
2. The particle range is short ($\simeq$30 µm).
3. Surface growth is maintained for a few days or weeks after injection.

Mathematically an infinite growth at constant growth rate (r) is assumed. Fig. 44 shows the calculated dose rate profile across a surface that is to be expected following a single injection of ^{226}Ra. The curve was constructed by averaging over 10 µm. From the arrows indicating the position of the bone surface at intervals of 5 days it is obvious, that the dose rates decrease sharply if growth proceeds at a rate of 3 µm/day, and continues at least 10 days. The reason for the skewed shape of the profile is that, in spite of the rapid decrease of the blood level, the specific activity remains significant over a few days (fig. 44) and accordingly ^{226}Ra is buried with decreasing concentration in a volume extending from the plane representing the surface at the time of injection to the actual surface. Assuming a uniform medium, i.e. neglecting the different stopping powers in bone and soft tissue, it is obvious from fig. 44 that the hotspot dose D_h accumulated at the growing surface is

$$D_h = \int_0^\infty \dot{D} \, dt = \frac{1}{r} \int_0^\infty \dot{D} \, dx \cong \frac{1}{2r} \int_{-\infty}^\infty \dot{D} \, dx, \qquad (5.4)$$

i.e. the cumulative dose is proportional to the area under the curve in fig. 44. The profile is assumed to be approximately symmetrical. Bearing in mind postulate 1, the total amount A of activity per unit surface deposited in the surface is

$$A = r\rho c \int_0^\infty B \, dt \, ; \qquad (5.5)$$

where ρ is the density of bone, c is the calcium content of bone (g_{Ca}/g_{bone}), B is the blood specific activity (μCi ^{226}Ra/g_{Ca} in blood). In the diffuse component ^{226}Ra is deposited by augmentation at an augmentation rate f. f is the fraction of blood calcium delivered to the bone volume per unit time, and assumed to be identical for ^{226}Ra [Marshall et al., 1959]. Then it follows that the dose rate on a bone surface $\dot{D}_d$ due to the diffuse component is

$$\dot{D}_d = \tfrac{1}{2} fKc \int_0^\infty B \, dt \, . \qquad (5.6)$$

The factor K converts specific activity to dose rate (rad/day $\times$ g_{bone}/μCi). Application of the principle of energy conservation and appropriate elimination of parameters then yields

$$D_h \cong \dot{D}_d / f. \qquad (5.7)$$

The apposition rate r has cancelled out. The reason for the surprising simplicity of the formula is that the ratio of hotspot to diffuse activity is independent of the amount of activity injected. Surfaces which are growing more rapidly take up more activity, but burial and consequently the reduction of surface dose rates is also more rapid. Consequently eq. (5.7) does not contain r. As an example Marshall [1960, 1962] used the figure of 0.0025/day for f in young mice. This would mean that, according to eq. (5.7), the surface dose accumulated from the diffuse deposit equals that of the hotspot dose after 400 days. After all, the analysis shows that burial of hotspots might be important in reducing the dose rates to osteogenic cells. But as Marshall stated, the relative importance of the hotspot and diffuse component is species dependent through the age dependent augmentation rate f and may change during lifetime. It must be recalled that one essential point of the above analysis is the assumption of infinitely continuing growth, which in practice means growth continuing at least for 50 μm. If at a particular site apposition ceases before complete burial is accomplished, the cumulative dose could well be an order of magnitude higher than in eq. (5.7). Nevertheless Marshall's formula has its value in setting up a lower limit for the cumulative dose to cells on the bone surfaces. It may be combined with the previously discussed procedure to calculate hotspot doses by means of eq. (5.2) and the power function retention in a particular bone, which on the other hand yields an upper limit for the case of a permanent hotspot at a surface.

5.2. Plutonium-239

Plutonium is expected to come into use in the foreseeable future in amounts of thousands of kilograms [Stannard, 1973] because of the central role ^{239}Pu plays in the fuel cycles of nuclear power plants and the importance ^{238}Pu gains as an efficient source of thermoelectric power in biomedical and space technology applications. The safe handling of such sizeable amounts of "the most toxic substance known to man", as ^{239}Pu was characterized sometimes, clearly necessitates the accumulation of biomedical data concerning its toxicity. In the course of experiments conducted during the last three decades ^{239}Pu was recognized as an osteotrope element, the high toxicity of which must be exclusively attributed to its decay via the emission of an α-particle and the radiation energy expended in tissue thereby. The brief review of Bair and Thomson [1974] represents for the interested reader a valuable source of background information concerning the various aspects of physico-chemical properties, routes of uptake, organ distribution and biological effects. A comprehensive survey of the literature is given in the Handbook of Experimental Pharmacology (Vol. XXXVI) where in connection with the current topic the chapter of Vaughan [1973] is of particular interest. The major long-term biological change in several animal species after incorporation of ^{239}Pu is the formation of bone tumours, chiefly of osteogenic origin [Finkel and Biskin, 1962; Jee et al., 1962; Bensted et al., 1965; Dougherty and Mays, 1969; Mays et al., 1969; Stevens et al., 1975]. On the basis of average cumulative skeletal radiation doses the RBE of ^{239}Pu to ^{226}Ra was 6 in beagle dogs, i.e. Pu is considerably more effective in producing osteosarcomas than ^{226}Ra [Dougherty and Mays, 1969]. Such differences point directly to the significance of dissimilarities in the distribution of the nuclides on a microscopic level, and confirm the notion of cells near bone surfaces as being responsible for tumour formation, rather than cells in bone volume. Besides bone tumours a multitude of other histopathological changes, as peritrabecular fibrosis, atypical bone formation, altered vascularity, osteocyte death, haversian canal plugs, abnormal resorption etc. are manifesting particularly at the higher dose levels [Jee et al., 1962; Jee and Arnold, 1961; Jee, 1972]. But the induction of osteogenic sarcomas is by far the most sensitive index of skeletal toxicity of ^{239}Pu. Plutonium is a typical "surface-seeker", a term coined by Marshall [1969] to characterize its pronounced affinity to bone surfaces, which is in contrast to the "volume-seeker" radium. Initially ^{239}Pu is laid down in a relatively uniform layer on all calcified osseous surfaces which have a blood supply: trabecular, endosteal, periosteal and haversian [Arnold and Jee, 1957, 1959]. The affinity of ^{239}Pu to the various classes of surfaces is different and varies with the character of the surfaces too, depending on whether it is quiescent, active or resorbing [Herring et al., 1962]. At this point it is imperative to recall that ^{226}Ra is concentrated in hotspots on active surfaces only. Indications are that binding of ^{239}Pu occurs preferably to the organic matrix of bone [Taylor and Chipperfield, 1971] and that among the various organic constituents of the matrix, the glycoproteins may play a dominant role [Chipperfield and Taylor, 1968]. The different avidity of bone surfaces for ^{239}Pu may be expressed by the mean specific surface deposition (activity per unit surface). Twente et al. [1960] determined a ratio of approximately 1 : 3 for the periosteal to endosteal surface concentration of ^{239}Pu in the lumbar vertebral centra of dogs injected with 2.7 μCi/kg. This is the same ratio as in the cortex of mouse tibiae after 0.1 μCi/kg ^{239}Pu [Rosenthal et al., 1968]. In the distal femur of dogs receiving 0.015 μCi/kg Dockum et al. [1964] reported ratios of 1 : 1.2 : 1.5 : 2.6 : 3 for periosteal, haversian, epiphyseal trabecular, epiphyseal endosteal and metaphyseal trabecular surfaces, indicating that the surface deposition is dissimilar even within the same bone at various anatomical sites. A much higher ratio of the periosteal to endosteal concentration was obtained by Kimmel et al. [1975] in the lumbar vertebra of dogs (1 : 78), whereas in the same animals almost equal deposition on both surfaces in the ulna was found (1 : 0.9). For both bones the surface activity in haversian canals was equal to or smaller than on periosteal deposits. The drastically increased periosteal to endosteal ratio compared to the results of Twente et al. [1960] was attributed to possible saturation of binding sites at endosteal surfaces which should come into play at injected doses as high as 2.7 μCi/kg, or to radiation injury to blood vessels and cells. It should be borne in mind, however, that such distinct dose effects are not likely to occur at short times after injection and that the experiment of Dockum et al. [1964] was carried out at the same low dose level as Kimmel's. Probably an extensive morphometric investigation of the vascular system could contribute to a better understanding of the disparities observed in both the microanatomical deposition and gross retention in individual bones of the same animal. With due precaution regarding the incomplete information available at present the affinity of ^{239}Pu might be ranked as haversian $\leqslant$ periosteal $\leqslant$ endosteal $\leqslant$ trabecular. The ratios discussed above

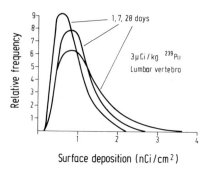

Fig. 45. Frequency distribution of the specific surface deposition of ^{239}Pu in the lumbar vertebra of beagle dogs at various time intervals after i.v. injection [Twente et al., 1959].

refer to average surface concentrations. Appreciable variation of the surface deposition exists within a particular type of surface. Fig. 45 illustrates the unevenness of the surface density of ^{239}Pu in adult beagles by frequency distributions [Twente et al., 1959]. The surface density of ^{239}Pu becomes increasingly non-uniform with time after injection and the average concentration still rises between one and seven days. It is not quite obvious from the reported details of the experiment of Twente et al., whether the distribution functions in fig. 45 obtained by photometric density measurements [Twente et al., 1958; Twente and Jee, 1961] really reflect the non-uniformity of the source of α-particles. Only for high lineal track densities the Poisson fluctuation becomes negligible. In general the true surface distribution has to be constructed by an unfolding procedure. This was done by Bleaney [1967] for the femurs and vertebrae of young rabbits. Below the epiphyseal cartilage plate where ^{239}Pu appears in a highly radioactive band, the dose rate distribution and correspondingly the surface density was rather uniform, displaying only minor deviations from a Poission curve. The concentration on trabecular bone could be described by the superposition of up to four Poisson curves, thereby defining a minimum and a maximum concentration and corresponding fractions of the total surface involved. According to Bleaney [1967] the ratio of maximum to minimum intensity of the deposits is not more than 3 : 1 at 24 h after i.v. injection. The subsequent alteration of the initial distribution function differs markedly from those shown in fig. 45, due to remodelling and burial, particularly prominent in growing animals (to be discussed below). In femur shafts of young adult rabbits Bleaney and Vaughan [1971] measured a rather uniform surface intensity over large regions, if the marrow was extruded previously from the shafts. The plutonium remaining was localized below the osteoid border and thus represented the actual surface deposit.

Another question arising in conjunction with discussing the laying down of plutonium on bone surfaces is: What is the thickness of the plutonium deposits? Can they really be regarded as superficial deposits? This question has an immediate bearing on the localized dosimetry close to bone surfaces, which critically depends on the geometrical conditions of the source (see section 2.4). The thickness of the deposits of ^{239}Pu and ^{226}Ra in the tibial and femoral shaft of dogs was measured by Schlenker and Marshall [1975] by means of α-spectrometry. From the distortions of the measured spectra compared with a genuine planar source the depth of the source could be inferred using the energy loss relationship of Kolenkow and Manly [1967]. For an animal sacrificed 5 h after injection, the thickness of the endosteal deposits of ^{226}Ra was found to be ⩽6.5 μm, whereas the layer of ^{239}Pu activity was ⩽0.2 μm. In a subsequent calculation Schlenker and Marshall determined the ratio of energy loss from a plane source of variable thickness to the energy loss from a source of infinitesimal thickness. This ratio is 0.98 for a source thickness of ⩽0.2 μm and amounts to 0.733 for 5 μm thickness. In the calculations a uniform dispersion of the activity throughout the depth of the source was assumed. The degree of self-absorption for the source thickness, as measured by Schlenker and Marshall, is very small and may be neglected without introducing any appreciable error. These results partially confirm the earlier findings of Bleaney and Vaughan [1971] who concluded that the narrowest spreading of ^{239}Pu on periosteal and endosteal surfaces of the femur shaft of rabbits must be very small, if inferred from the width of fision track bands in NIAR samples, and if marrow was extruded previously. For marrow remaining in situ the spreading was at least 6 μm wide and occasionally amounted to 35–40 μm. The interpretation of this observation, namely that ^{239}Pu is concentrated throughout an osteoid layer of varying thickness has to be judged in the light of observations made by Arnold and Jee [1957], that ^{239}Pu is present in mineralized tissue only, not in osteoid. A certain fraction of surface ^{239}Pu is certainly in endosteal cells closely applied to bone. Bleaney and Vaughan [1971] reported this fraction to range between 30 and 50% in femur shafts of rabbits, largely located as aggregates in a layer of spindle-shaped cells of 2–5 μm width. These findings point towards the question about the role bone cells play in determining the fate of the deposits after ^{239}Pu is laid down on all surfaces in the phase of initial uptake. The pattern of ^{239}Pu deposition in the skeleton is a dynamic one, even in the adult skeleton, and a precise knowledge of its change, not only in space but also in time, is of paramount importance if informations about the dose to the cells at risk have to be extracted from available data. The translocation of Pu from the site of early deposition may be summarized as follows:

1. Plutonium on growing surfaces becomes buried by the apposition of new bone, which thereby subsequently builds up a radiation shielding with respect to the sensitive cells near surfaces. This process is analogous to the

burial of ^{226}Ra hotspots (section 5.1). Fig. 36 showing a NIAR detector [Fellows et al., 1975] exposed on a piece of trabecular bone from the distal femur of a dog gives an a impressive illustration, displaying intense hot lines located within bone.

2. Buried hot lines may evolve at bone surfaces again, either by bone resorption from the same side they were buried, or from the opposite side. Such a passage of ^{239}Pu deposits through bone may be characterized by a mean transit time or bural time, which is dependent on species, age and probably varies from bone to bone and even within a bone. This reappearance of buried activity and its movement within bone is still a matter of dark ignorance. On the basis of the present knowledge ^{239}Pu located deeply in bone must be considered as unaccessible to any of the cellular processes or other possible agents, instrumental in removing the nuclide from the skeleton. There are no indications of something like a long-term exchange process known to remove ^{226}Ra from bone volume.

3. On resorbing surfaces plutonium is progressively concentrated in osteoclasts, which upon digesting the organic matrix of bone take up the nuclide into their cytoplasm [Twente et al., 1960; Arnold and Jee, 1957, 1959, 1962; Jee, 1971]. Three types of osteoclasts were identified in the skeleton of rats, with different capability to accumulate plutonium. Occasionally the activity of individual osteoclasts corresponded to a volume of resorbed bone 5 or 10 times the cell volume itself [Arnold and Jee, 1957]. While labeled osteoclasts appeared already 24 h after injection in young adult rats [Arnold and Jee, 1957], the first osteoclasts in adult dog bones were not detected until 2 weeks after injection [Twente et al., 1960; Jee, 1971]. So the amount of plutonium associated with osteoclasts, which in the aforementioned experiment varied between 4 and 16.5% of the total activity present in a bone section and showed spurts of increased osteoclastic activity, largely depends on the remodelling rate and thus on both species and age.

4. At about the same time with labeled osteoclasts plutonium can be found in macrophages. This activity is believed to originate from osteoclasts, either killed by the radiation from their accumulated ^{239}Pu burden or ending their normal life-span. By phagocytizing the cell debris of the decaying osteoclasts some of the activity enters the cytoplasm of the macrophages and is often found in association with haemosiderin [Twente et al., 1960; Vaughan et al., 1967; Bleaney and Vaughan, 1971; Jee, 1971], an iron-protein complex. The number of labeled macrophages increases with decreasing injected dose, but in dogs injected with 0.015 µCi/kg no labeled macrophages were detected [Jee, 1971, 1972]. The reduced number of labeled macrophages for increasing doses of ^{239}Pu is the consequence of a depressed bone resorption and a corresponding diminution of osteoclastic activity. Fig. 36 gives indirect evidence of plutonium-laden macrophages in the marrow, manifesting as fis-

sion track clusters on NIAR detectors. Plutonium may reach the circulation immediately after liberation from bone by osteoclastic resorption, or by the efflux of macrophages from the marrow.

5. Recirculating ^{239}Pu is supposed to behave much alike ^{239}Pu entering the blood stream initially. A large fraction of it is redeposited on bone surfaces again at a low level of blood specific activity and appears either as a diffuse distribution in newly formed bone, or as a more or less intense surface deposition at resting surfaces, depending on the rate of skeletal turnover [Arnold and Jee, 1962]. The combined effects of firm fixation in bone, burial, and recirculation are the cause for the relatively slow clearance from the skeleton [Stover et al., 1972; Taylor et al., 1961].

6. Depending on the colloidal state, a varying amount of plutonium appears far from bone surfaces in the marrow as a low level diffuse distribution with a superimposed pattern of particulate plutonium, identifiable as stars of particle tracks in autoradiographs [Vaughan et al., 1967; Bleaney, 1967]. The star pattern becomes increasingly pronounced if the degree of polymerization is enhanced [Rosenthal et al., 1968].

With regard to the dose rate to endosteal cells one may summarize the above qualitative description as follows: There are five different kinds of plutonium deposits capable to irradiate the sensitive cell populations, namely (1) the surface deposits (static or newly formed), (2) activity buried in bone, (3) a diffuse labeling in post-injection bone, (4) a diffuse distribution in marrow, and (5) aggregates of plutonium contained in cells of marrow. The relative importance of these five components in inducing the known harmful effects in the skeleton has not yet been worked out so as to give detailed quantitative information.

Table 10 presents an abbreviated summary of the local dose determinations in several species and under various experimental conditions. To facilitate intercomparison between different experiments, absolute values of dose rates were normalized to unit amount injected and recalculated (last figures in dose rate column) to average dose rates over the entire particle range. According to Twente et al. [1960] the average surface dose in the lumbar vertebra of beagles injected with 3 μCi/kg rises from 28 rad/day to a maximum of 53.5 rad/day two weeks after injection and then declines. The percentage of ^{239}Pu associated with cortical bone and periosteal surfaces did not change with time and was less than 5% of the total activity found in the section. The reduction of the surface burden was attributed to osteoclastic activity, which was constantly going on during lifetime showing up to 5% of the total activity associated with osteoclasts. The percentage of ^{239}Pu deposited in post injection bone amounted to at least 10% after 3 years. These diffuse deposits had a concentration ranging between 0.05–0.1 μCi/cm^3 and delivered a dose rate of about 1–2 rad/day to adjacent tissue. The changes in the pattern of ^{239}Pu

Table 10
Localized dosimetry of ^{239}Pu in the skeleton

Author(s)	Species	μCi/kg injected	Bone	Time after injection (days)	Dose rate rad/day (rad kg/day μCi)			Location [c]
Twente et al. [1960]	dogs	3 i.v.	lumbar vertebra centra	1	28	(9.3)	9.3 [a]	Trabecular surfaces, average over entire particle range
				28	47	(16.5)	16.5	
				400	32	(10.8)	10.8	
				1	8	(2.7)	2.7	Periosteal surfaces, average over entire particle range
				28	16	(4.7)	4.7	
				400	11	(3.7)	3.7	
				400	1.3	(0.43)	0.43	^{239}Pu in newly formed bone over entire particle range
Bleaney [1967]	young rabbits	1.25 i.v.	metaphysis of femur	1	20	(16)	13.1	
				112	6.8	(5.4)	4.4	
			lumbar vertebra	1	13.5	(10.8)	8.8	
				112	2.6	(2.1)	1.7	
Bleaney [1969]	young adult rabbits	1.07–2.30 i.v.	epiphysis of femur	1		(8)	6.5	10 μm from trabecular surfaces
				365		(9)	7.4	
			lumbar vertebra	1		(10.2)	8.3	
				365		(10.8)	8.8	

Reference	Subject	Dose	Location	Time (days)	Values	Notes
		2.61–2.68 i.m.	epiphysis of femur	8	(2) 1.6	
				63	(9.5) 7.8	
				365	(4) 3.3	
			lumbar vertebra	8	(3) 2.5	
				63	(8) 6.5	
				365	(3) 2.5	
James [1972]	young rats	4.5 i.v.	epiphysis of femur	1	33 (7.3) 3.8	5 μm from trabecular surfaces
					10 (2.2) 4.2	20 μm surfaces
				336	35 (7.8) 4.1	5 μm
					15 (3.3) 6.2	20 μm
Schlenker and Oltman [1974]	young woman	0.006 [d] i.v. (0.3 μCi total)	lumbar vertebra	initial	(31) [b,d] 26	10 μm from trabecular surfaces
			humerus		(20) 16	
			femur		(18) 14	
			humerus		(14) 11	10 μm from endosteal surfaces
			femur		(3.1) 2.5	

[a] First value as measured, in parentheses normalized to 1 μCi/kg injected ^{239}Pu, third value recalculated to average dose over entire particle range.
[b] Calculated by means of eq. (2.35) from surface deposition reported.
[c] "Location" refers to the dose rate figures in the first and second column.
[d] For 50 kg body weight.

distribution revealed, that remodelling was going on to a very low extent only. It is known, however, that at this rather high dose an inhibition of bone resorption may occur [Arnold and Jee, 1959]. The experiments of Bleaney [1967, 1969] show clearly the effect of age and different routes of entry on local dose rates (table 10). In weanling rabbits extensive remodelling had lowered the surface dose rates considerably after 112 days, both in the femur metaphysis and in the lumbar vertebra. The initial uptake seems to have been less affected by age and species. Comparisons of the normalized dose rate values of Twente [1960] for the lumbar vertebra of dogs with those for weanling [Bleaney, 1967] and young adult rabbits [Bleaney, 1969] at the first day shows remarkable agreement (9.3, 8.8 and 8.3 rad kg/μCi day, respectively). In weanling rabbits the dose rates from volume deposits of ^{239}Pu were negligible at 16 weeks compared with those from stable surface deposits, ranging from 0.04 to 0.39 rad/day. The ratio of maximum to minimum surface dose in the femur and vertebra was between 2.4 and 2.9. At certain sites on periosteal surfaces of the femurs of weanling rabbits the dose rates increased with time and eventually were higher than at endosteal locations by nearly the same factor (2.7), as the endosteal doses were higher than periosteal ones after 24 h (2.5). This phenomenon has to be explained by the transversal of former endosteal deposits through the volume of cortical bone and reappearance as activity located at the periosteum. In young adult rabbits i.m. and i.v. injections resulted in a similar pattern of distribution [Bleaney, 1968, 1969] with again a factor of 3–5 variation in dose rates at trabecular surfaces. After i.v. injection the average and maximum dose rates were increasing in the epiphysis of the femur until 16 weeks and then remained virtually constant. In the vertebra they were constant from the first day after injection. The surface concentrations of ^{239}Pu were rising much slower after an i.m. than after i.v. injection, though a significant amount of the activity injected appeared in the skeleton [Bleaney, 1968, 1969]. The ratio of the maximum average dose rates after i.v. and i.m. injection was nearly 4, with respect to unit intake. The slower increase of the endosteal doses is obviously due to the slow resorption of ^{239}Pu from the intramuscular deposit into the blood. James [1972] applied the α-particle flux method (section 4.4) to determine dose rates typical for a 5 μm diam. spherical target. The highest doses were measured in the distal part of the femur of rats in the zone of endochondral ossification (35 rad/day) at the first day. Already after 4 days this high dose level had dropped to zero. The change of dose rates in time in the femoral epiphysis represents a lucid demonstration of the influence of apposition and resorption on the irradiation of cells near surfaces. Between the first and the fourth day the dose rates were still increasing due to deposition of ^{239}Pu from the initial injection and from ^{239}Pu recirculating early from bones with higher remodelling rates than the epiphysis. During the following time interval up to

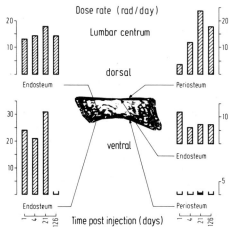

Fig. 46. Average dose rates at endosteal and periosteal surfaces in the lumbar centrum of rats injected i.v. with 4.5 µCi/kg ^{239}Pu. The values given are for a distance of 20 µm from the surfaces and were determined by an α-particle flux method [James, 1972].

the 252th day the average dose rate declined to about 50% of the maximum and then rose again. This holds for all three distances measured (5, 12.5, 20 µm). According to the interpretation of James the dose at trabecular surfaces after 4 days is contributed both by ^{239}Pu in bone and in cellular elements, and the reduction of the dose rate between 4 and 252 days represents the clearance of cell-bound plutonium. The observed terminal increase in dose rates demonstrates that buried plutonium has reappeared at a resorbing surface again. This "liberated" activity appeared as punctate deposits on the surface opposite to that where the activity was initially laid down. A summary of average surface dose rates in the lumbar centrum is shown in fig. 46. The changes in dose rates at the dorsal periosteum and endosteum again reflect the influence of appositional bone growth, whereas the trabecular surface deposits are relatively stable over the time period considered. This is impressively demonstrated by the distinct rise of the periosteal doses, which correlates with a corresponding lowering of the irradiation level at the dorsal endosteum. Stripping film autoradiographs showed that the hot line of initial surface deposition at the endosteum traversed the neural canal wall leaving behind a diffuse labeling of the cortex. James made additional calculations relating the measured particle fluxes to hit frequencies over a typical cell cycle time of 100 h. At the neural wall 11–12% of the cell nuclei (5 µm diam.) received just one hit and 4–5% two or more hits. For the cell nuclei 20 µm from trabecular surfaces the corresponding figures are 19–29% and 10–14% respectively. These figures suggest a considerable extent of cell killing at the given ^{239}Pu dose (4.5 µCi/kg), bearing in mind that at locations closer to the surfaces the hit probability rises steeply. Applying the same method, a similar experiment was carried out quantitating the effect of DTPA on local radiation doses [James and Taylor, 1971]. DTPA (Diethylenetriaminepentaacetate) is a decorporating agent acting by the formation of chelate complexes with a potentially hazardous metal atom and subsequent elimination from the organ by the natural excretion routes [Catsch, 1964]. From the deviation of the flux distribution measured at epiphyseal surfaces in the femurs of rats from a single Poisson distribution, the authors concluded that a significant fraction of the radiation dose is contributed by ^{239}Pu in cellular elements near the surfaces. DTPA therapy commenced 7 days after injection of the nuclide resulted in a reduction of dose rates at the epiphyseal bone plate and trabecular surfaces. Dose rates at growing travebular surfaces were not significantly altered by DTPA treatment. This is consistent with the notion that DTPA affects principally the ^{239}Pu content in osteoclasts and macrophages, thereby reducing that fraction of radiation dose to osteogenic cells which is of cellular origin.

At present the autoradiographic analysis (NIAR) of only one human case, namely the i.v. injection of a young woman with 0.3 μCi ^{239}Pu, is available [Schlenker and Oltman, 1974; Schlenker et al., 1975]. Most of the endosteal surface concentrations were extremely low (patient died 17 months after injection), but buried hot lines indicated the position of the surfaces at the time of acquisition of ^{239}Pu. In the lumbar vertebra, for instance, the burial depth ranged between 14 and 120 μm (57 μm average) and was similar in the femur [Schlenker et al., 1975]. In table 10 the dose rates calculated from the reported initial surface concentrations [Schlenker and Oltman, 1974] are presented. In the lumbar vertebra, where direct comparison with the dog data is possible, a higher specific surface deposition is indicated.

At this point it is of interest to discuss the specific surface deposition n_A (decays per unit area) on a more general basis, to gain some insight into the dependence of this quantity on the morphological and physiological parameters. If one considers the skeleton as a whole, a single bone or even a piece of bone with weight fraction f referring to the total body weight and a fractional retention r of the injected amount, then the average surface concentration at early times after injection of the amount a per unit body weight (uptake completed) may be calculated from

$$n_A = \frac{a \cdot r(\rho_B + \rho_M \cdot q)}{f \cdot S_v} \; ; \tag{5.8}$$

S_v is the surface volume ratio (e.g. cm^2/cm^3) of that particular part of the skeleton or an average of the total skeleton, ρ_B and ρ_M are the densities of bone and marrow, respectively, and q is the ratio of marrow to bone volume. Eq. (5.8) elucidates the significance of S_v in determining the average local dose rates at bone surfaces, whereas variations in q normally have a smaller influence on n_A (for $\rho_B = 2$ gcm^{-3} and $\rho_M = 1$ gcm^{-3}). The fractional skeletal volume occupied by bone appears to be 15–17% ($q = 4.7$–7.7) in man and 15–45% ($q = 3.2$–7.7) in dogs [Jee et al., 1973; Lloyd and Hodges, 1971]. The surface volume ratio S_v of cancellous bone is distinctly higher in dogs than in man. Jee et al. [1973] reported values of 300–380 cm^{-1} in the lumbar vertebra of dogs, Lloyd and Hodges [1971] 225 cm^{-1} for dogs and 120 cm^{-1} for man. Consequently, the surface concentration and the corresponding dose rates in the lumbar vertebra of man are expected to be about twice as high as in the dog vertebra, provided all the other factors are nearly identical. According to table 10 this factor is somewhat smaller than two but as estimated by Schlenker and Oltman the skeletal retention of this individual which suffered from Cushing's syndrome was below normal. The differences between the dose rates on trabecular surfaces in the femur and lumbar vertebra are most probably affected by differences in f and r for the two bones. The ratio r/f which is proportional to the percentage of the total ^{239}Pu dose per g of organ weight is in the femur nearly one half of that in the vertebra [Durbin, 1972] which could explain the ratio of the trabecular dose rates in table 10.

The ratio of the local dose rates averaged over a distance x from the bone surfaces to the average skeletal dose calculated under the assumption of uniform deposition is a measure of non-uniformity, commonly used to characterize a distribution. Clearly this ratio is affected by the translocation of activity from the surface to the volume of bone in the sense, that under the condition of a slowly declining total skeletal burden this ratio decreases in time because of remodelling. Moreover such a definition neglects the variations in local dose rates on the surfaces itself. E.g., the ratio of maximum local to uniform dose rates in general is considerably higher than average local to uniform. In the lumbar vertebra of beagles the former was found to vary within 60–86 for injected amounts of 0.015–2.7 μCi/kg [Jee, 1964]. Average local to uniform ratios are listed in table 11 for dogs, rabbits and rats [James and Kember, 1972]. The low ratio in the 0.3 μCi/kg dogs compared to the other dose levels is still unexplained. The conclusion that might be drawn from table 11 is, that the non-uniformity ratio in rabbits is comparable or even higher than in dogs, and in rats it is distinctly lower than in rabbits and dogs. By means of eq. (5.8) it is possible to make some theoretical generalizations displaying the functional dependence on relevant parameters. Such an expression for the non uniformity ratio holds, like eq. (5.8), for any part of the skeleton that can be characterized by an average surface volume ratio S_v and becomes particularly simple if it refers to the "local" uniform dose rate in the specific part of the skeleton contemplated, rather than to the total skeleton. Then the parameters r and f (eq. 5.8) cancel and one has for

Table 11
^{239}Pu non-uniformity factors [a,b]

Author(s)	Species	μCi/kg injected	Site	Time	10 μm surf. dose: av. skel. dose	
					measured	Q calc. [c]
Twente et al. [1960]	dogs	2.7	lumb. vert. trab.	over 1st year	21	
Dockum et al. [1964]	dogs	0.3	lumb. vert. trab.	over 1st year	10	3.5–6.3
		0.015	lumb. vert. trab.	over 1st year	23	(6.6–11.8) [d]
		0.015	metaphyseal trab.	over 1st year	14	
		0.015	diaphyseal endost.	over 1st year	13	
		0.015	epiphyseal trab.	over 1st year	8	
Bleaney [1969]	rabbits	1.2	lumb. vert. trab.	over 1st year	27	
		1.2	metaphyseal trab.	over 1st year	26	
James and Kember [1972]	rats	5.1	lumb. vert. trab.	1 day	4.1	
		4.8	diaphyseal endost.		3.4	
		4.8	epiphyseal trab.		2.7	0.9–3.9

[a] James and Kember [1972].
[b] Local dose averaged over first 10 μm from surface.
[c] Calculated according to eq. (5.9) for $x = 10$ μm.
[d] Recalculated allowing for non-uniformity of skeletal retention.

the non-uniformity factor Q:

$$Q = \frac{\text{average surface dose to distance } x}{\text{uniform skeletal dose}} \quad (5.9)$$

$$Q = \frac{1 + \ln(R/x)}{2R} \cdot \frac{1}{S_v} \cdot \left(\frac{\rho_B}{\rho_M} + q\right).$$

In deriving eq. (5.9) use was made of Mays' [1958] formula giving the average surface dose to distance x. The first quotient of eq. (5.9) exhibits the energy dependence. For $x = R$ it becomes simply $\sim E^{-1.43}$ (see section 2.3). Spiers and Whitwell [1975] determined the energy dependence of non-uniformity factors by means of a Monte Carlo procedure, developed earlier in order to calculate mean endosteal and marrow doses from α-emitting bone-seekers [Whitwell and Spiers, 1971]. The method is based on pathlength distributions obtained by bone mensuration data of humans [Darley, 1967]. Q of eq. (5.9) declines more rapidly with particle energy than in Spiers and Whitwell's curves. Moreover they showed that the energy dependence of Q is different for cortical and trabecular bone. The non uniformity factor Q is inversely proportional to the surface volume ratio S_v, as it was for n_A (eq 5.8). Available morphometric data of the dog lumbar vertebra are inconsistent but the average surface to volume ratio of trabecular bone may lie somewhere between 225–380 cm^{-1} and varies with age [Lloyd and Hodges, 1971; Jee et al., 1973]. This would correspond to the range of Q listed in the last column of table 11 (3.5–6.3). If, however, one takes into account that the relative concentration in the beagle lumbar vertebra is 1.88 times the average skeletal concentration [Atherton et al., 1972], then the higher values (in parenthesis) result and are consistent with the measurement at the 0.3 µCi/kg dose level. Surface to volume ratios in the rat are higher than in the dog and range in the femur epiphysis between 350 and 500 cm^{-1} [Polig, 1976a]. Accordingly, Q lies within 1 and 3.9 (table 11), in agreement with the result of James and Kember [1972]. An average of Q over the whole skeleton of different mammalian species can be calculated by means of eq. (5.9), employing the figures of Lloyd and Hodges for the average S_v in man and dog (50 cm^{-1} and 77 cm^{-1}, respectively). Assuming 25, 30, 35% bone content in the skeletal volume of man, dog and rat, respectively, and an average trabecular S_v of 400 cm^{-1} for the rat and 30 cm^{-1} for cortical bone (80% cortical, 20% trabecular) one arrives at 24.8, 17.4 and 11.5 for Q (first 10 µm) in man, dog and rat, respectively. It should be pointed out that these figures do not represent a relative hazard index in the three species. The calculated Q holds for early times after injection only and neglects burial and uptake by cells. Moreover if the hazard is proportional to the number of cells subject to a certain radiation dose, then the two effects of dose and size of cell populations cancel for a linear dose response.

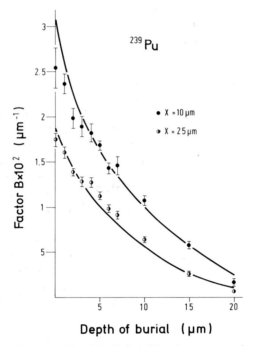

Fig. 47. Variation of the quantity $B(x, d)$ describing the average endosteal dose rate to a tissue layer of thickness x due to a plane α-particle source of infinite thickness with depth of burial d. Points calculated by a Monte Carlo procedure [Thorne, 1976], and curves by means of eq. (5.11) in the text.

In spite of the paramount and generally recognized importance of the modelling and remodelling processes in determining the radiation doses to the cells at risk, there is only relatively little quantitative information available at present. A theoretical analysis concerning the reduction of dose rates with respect to burial depth was done by Thorne [1976]. The geometric factor $B(x, d)$ describing the dependence of the marrow dose rates averaged over the first x µm, if the depth of burial is d

$$\bar{D}(x, d) = \frac{n_A E_0}{\rho} B(x, d) , \qquad (5.10)$$

(ρ is the marrow density) was calculated by means of a Monte Carlo procedure. The Bragg equation was employed in the form as given by Harley and Pasternack [1972], with an additional correction making the stopping power fit the values of Walsh [1970] below 2 MeV. Fig. 47 shows the variation of $B(x, d)$ with the depth of burial d for two thicknesses of the endosteal surface layer. The error bars indicate the statistical errors arising from the limited number of Monte Carlo iterations. Another approach is to derive an analytical expression in the constant stopping power approximation by appropriate integration of eq. (2.44). This yields

$$B(x, d) = \begin{cases} \frac{1}{2x}\left[u(1 - \ln u) - \frac{d}{R_B}\left(1 - \ln \frac{d}{R_B}\right)\right] & \text{for } \frac{d}{R_B} \leq 1 \\ 0 & \text{for } \frac{d}{R_B} > 1 \end{cases} \quad (5.11)$$

with

$$u = \begin{cases} \frac{x}{R} + \frac{d}{R_B} & \text{if } d \leq R_B - \frac{R_B}{R}x \\ 1 & \text{else} \end{cases}$$

(R, R_B is the particle range in marrow and bone, resp.). The curves computed according to eq. (5.11) are in good agreement with Thorne's results (fig. 47). This is surprising in view of the simple stopping power approximation on which the derivation is based. The information contained in fig. 47 thus provides the starting point for a determination of the dose rate distribution from all ^{239}Pu deposits, buried or superficial, in a particular bone. It requires, however, the accurate measurement of the burial depth d along with the relative size of the surface associated with a particular value of d. Though such information is not available it is possible to obtain an impression of the rate of burial from existing data. Jee et al. [1969] investigated ^{239}Pu burial in 1.5 years old beagles injected with 0.3 μCi/kg and 0.015 μCi/kg. In these adult dogs remodelling still proceeded at a relatively high rate so that within three months the bone surfaces were almost free of ^{239}Pu. In the distal femur metaphysis of animals injected with 0.3 μCi/kg 25% of the bone surfaces were laden with ^{239}Pu at 40 days and 10% at 187 days. The percentage of buried activity attained a maximum of 67% at 187 days. The amount of surface activity dropped more rapidly in the lumbar vertebral bodies where only 2.7% labeled surfaces were found at 187 days. This is consistent with the fact, that the lumbar vertebra has a higher remodelling rate than the femoral metaphysis. At the 0.015 μCi/kg dose level the percentage of buried activity was nearly twice as high in the femur metaphysis as in the lumbar vertebra. At all dose levels the fraction of buried ^{239}Pu decreased with time after passing through a maximum (21–67%) at two to three months after injection [Jee et al., 1969, 1972]. The higher fraction of buried ^{239}Pu in 0.3 μCi/kg animals was taken as evidence for a partial inhibition of resorption processes. It should be noted, however, that these data might be subject to methodical uncertainties. In a later study using the NIAR technique, Kimmel et al. [1975] found the surface labeling in the lumbar vertebra to drop more slowly than in the previous experiments [Jee et al., 1969, 1972], and pointed out that in bone sections not heavily exposed to detect even very minute surface concentrations the fraction of ^{239}Pu laden surfaces might be underestimated. The percentage of labeled surfaces in the ulna and lumbar vertebra in Kimmel's et al. experiment could be fitted to a single exponential term. The half-life of the vertebral trabecular surface deposits was 78 days, much shorter than in the ulna (1353 days). In summarizing it may be said that the rate of transfer of ^{239}Pu from the surfaces to bone volume may vary in different bones depending on the rate of remodelling. In bones of high remodelling activity an almost complete liberation of bone surfaces from ^{239}Pu may be achieved in a few months even in young adult dogs. The extent of burial is affected by irradiation damage and becomes higher for increasing doses injected. Determining the effect of burial on radiation doses involves the determination of the location of buried activity relative to the bone surfaces. The NIAR study of Jee et al. [1975] in the metaphysis of proximal ulnae of beagles (0.015 μCi/kg) provides a partial answer. They divided the bone area in the sample section into the two regions $\leqslant$25 μm and $>$25 μm from surfaces. Thus, by counting the lineal track density of buried hot lines in the two portions of bone, a relative measure of the surface dose rate and the wasted dose was obtained, assuming that the α-particle range in bone is 25 μm. Shortly after injection (5 days) the ratio of the number of tracks $\leqslant$25 μm to $>$25 μm was 21.4. At 746 days the ratio was nearly one. A high and approximately constant number of tracks remained within 25 μm beneath the surface over the whole time of observation. The measured track density corresponds to about 0.8 pCi/cm^2 on the bone surface. In deep bone many hot lines were present at 2 years post injection with an average intensity of 5 pCi/cm^2 on bone surfaces at the time of deposition. A more detailed analysis with regard to the geometrical resolution was published by Storr et al. [1975]. They divided the region on both sides of the trabecular surfaces of rats into 6 adjacent bands each of 15 μm width (3 bands in bone, 3 in marrow). The adult rats were injected with 0.1 μCi/kg monomeric ^{239}Pu. Spatial correspondence between the bone samples and the NIAR detectors was achieved by means of alignment with a comparison microscope. The results are shown in fig. 48, displaying the track densities in each of the 6 bands for one hour up to one year after injection. A remarkable phenomenon is the similarity of the relative distribution in all six bands within the first three weeks, though the absolute number of tracks still increases during that time. Only after one year the relative distribution was found to have changed substantially. At this time the trabeculae were labeled throughout the whole volume with the highest concentration of ^{239}Pu in the centre. From fig. 48 a distinct reduction in dose rates near surfaces between 3

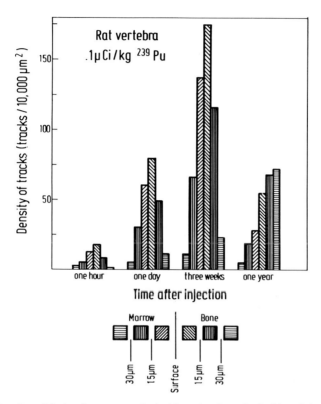

Fig. 48. Density of fission fragment tracks in 15 μm bands on both sides of the bone surfaces [Storr et al., 1975].

weeks and one year may be inferred from the diminished track densities in the two bands centering around the surfaces. Obviously, the contribution of the two bands in the marrow nearest to the surface is significant and even more important at earlier times.

These findings together with the aforementioned autoradiographic studies of other authors point towards the possible relevance of the radiation dose arising from ^{239}Pu in cells in the marrow in evoking the known lesions. The marrow distribution of ^{239}Pu(NO$_3$)$_4$ (1.25 μCi/kg) was studied quantitatively by Vaughan et al. [1967], particularly with respect to the potential leukaemic risk. The colloidal state of the injection solution, which is known to affect the distribution pattern and the relative partition between marrow and bone surfaces [Rosenthal et al., 1968; Vaughan, 1967] was not specified but it may be suspected that a polymeric fraction was present. A topographical analysis showed that 7 weeks after injection particulate deposits were concen-

Table 12
^{239}Pu [a] in bone marrow of rabbits [b]. Dose rates from stars and diffuse component

Time (days)	Density of stars (mm^{-3})	No. of Pu atoms in stars	Dose rates (rad/day)		
			stars + diffuse	diffuse	10 μm from star centre
1	54,000	9 × 10^6	5	~1.2	1.5
8	40,000	8 × 10^6	3.8	~1.2	1.5
49	13,000	8 × 10^6	1.4	0.3	1.5
210	35,000	2 × 10^6	0.16	~0	0.4

[a] 1.25 μCi/kg ^{239}Pu(NO$_3$)$_4$ i.v.
[b] Vaughan et al. [1967].

trated close to the endosteal surface of the femur shaft and in bands across the marrow cavity. In the femur epiphysis patchy agglomerates of stars appeared in the autoradiographs. The density of stars in central marrow decreased with time, as can be seen from table 12, but the number of ^{239}Pu atoms producing individual stars remained remarkably constant. In central marrow far from the heavy star concentrations in the endosteal bands, the average dose rates from both the particulate sources and the diffuse label was determined as listed in table 12. The diffuse label contributed about one third to one fourth of the total dose rate until 7 weeks and was virtually absent at 7 months. The dose rates given do not reflect the local variations within the marrow and the non-uniformity of the distribution, but, as the authors pointed out, may nevertheless be meaningful in view of the mobility of the marrow cells. Cell movement possibly averages out local variations and the given values may well represent the average dose an individual moving cell experiences during its life-time.

One of the objectives of localized dosimetry in the skeleton is to establish a measure of toxicity of a particular radionuclide, for which the skeleton is the critical organ. If this measure has to be an absolute one, such a toxicity estimation would involve the detailed quantitative knowledge of the sequence of steps from the primary physical events to the pathological endpoints. Such a knowledge is not available for any of the histopathological manifestations. It has becomes customary, therefore, to base toxicity comparisons on the concept of the relative biological effectiveness (RBE) by determining the ratio of skeletal burdens of two different isotopes known to produce the same effect. In the case of ^{239}Pu the endpoint of main interest is the occurrence of osteosarcomas. This approach is based on the philosophy that it is the energy deposition in a sensitive structure that matters in evoking harmful effects, not the properties of the source and its temporal and spatial arrangement, and that equal radiation doses delivered to that target produce the same effect. By forming the ratios of average skeletal burdens producing the same radiation dose to the sensitive target structures the intricacies of the biological mechanisms are cancelled out. The justification of the above fundamental principle may then be tested against experimental animal data and, moreover, this procedure allows to establish standards for the exposure of man where toxicity data are available only for ^{226}Ra. Thus, the accumulation of local doses in the case of ^{226}Ra may serve as baseline. The translation of RBE values derived from animal species to man, however, requires some precautions and indeed may be completely misleading. Lloyd and Marshall [1972] emphasized that because of differences in relevant morphometric parameters and remodelling rates the RBE ^{239}Pu/^{226}Ra in dogs cannot be regarded as being equal to that of man. This was confirmed in a theoretical study conducted by the same authors [Marshall and Lloyd, 1972]. The mathematical model was based on the assumptions that all ^{239}Pu is deposited initially on bone surfaces and that all resorbed activity is redeposited throughout the skeleton. The final plutonium distribution is considered to be a volume source, whereas the radium distribution is a volume source at all times. For the surface dose rate $\dot{D}_s$ due to the surface-seeker ^{239}Pu Marshall and Lloyd then postulated the expression

$$\dot{D}_s = Q_s \dot{D}_{av} \exp(-\lambda_v t) + Q_v \dot{D}_{av}(1 - \exp(-\lambda_v t)) \quad (5.12)$$

to hold. Q_s and Q_v are the non-uniformity factors representing the ratio of the local doses at the surfaces to the average skeletal dose rate $\dot{D}_{av}$ for a surface distribution and a volume distribution, respectively. λ_v is the rate of volumization of the activity initially laid down on surfaces. According to Marshall and Lloyd this parameter is proportional to the rate of bone turnover and S_v. For 16-month-old dogs and adult man they arrived at estimates of 100%/year and 10%/year, respectively. If then one defines

$$\text{RBE (Pu/Ra)} = \frac{\text{average skeletal dose radium}}{\text{average skeletal dose plutonium}}$$

for equal doses to bone surfaces, i.e. assuming that equal doses to the cells at risk signify an equal probability of osteogenic sarcomas, one obtains

$$\text{RBE (Pu/Ra)} = (\rho - 1)\left[\frac{1 - \exp(-\lambda_v t_m)}{\lambda_v t_m}\right] + 1, \quad (5.13)$$

with $\rho = Q_s/Q_v$. This equation implies the notion that the accumulated dose up to the time t_m of tumour appearance is the relevant parameter for toxicity comparisons. For $\lambda_v t_m$ the values of 10 in man and 3 in dogs were taken, based on a median appearance time of 10 years for dogs [Mays et al., 1969]

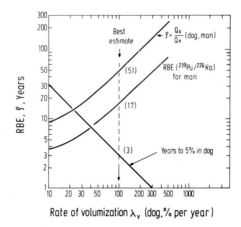

Fig. 49. Dependence of the estimated RBE and factor ρ (see text) on the rate of volumization λ_v in canine trabeculae. The curve "years to 5% in dog" gives the corresponding time necessary to transfer all but 5% of the initial surface deposits to bone volume [Marshall and Lloyd, 1972].

and 30 years for man [Rowland et al., 1972]. Inserting the known RBE for dogs (= 6) into eq. (5.13) and solving for ρ one obtains $\rho = 51$. Postulating this ρ to be the same as in man, eq. (5.13) then yields RBE (Pu/Ra) in man = 17 [Marshall and Lloyd, 1972], i.e. the RBE for man is about 3 times higher for man than for dogs. The authors emphasized the uncertainty of the value of λ_v for dogs, which is fundamental for the calculation. Therefore they calculated curves for ρ and the resulting RBE (man) for different values of the rate of volumization (fig. 49). If λ_v for dogs is actually lower than 100%/year a smaller RBE (man) would result. The model of Marshall and Lloyd is certainly a rigorous simplification of the real situation. At present there is no experimental verification of eq. (5.12). It is possible to make an independent determination of ρ by means of eq. (5.9) for a surface-seeker. With respect to the dose rates in a 10 μm surface layer Q_s, and accordingly ρ, is distinctly smaller for the skeleton of dog and man ($Q_v \approx 1$ for dog and man) than calculated by means of eq. (5.13). Other calculations of Marshall et al. [1974] showed that in man the ratio of the endosteal doses for equal average doses to bone for ^{239}Pu relative to ^{226}Ra is about 28. Moreover ρ needs not necessarily be equal in dog and man and indeed is very different according to eq. (5.9) in cancellous bone, where the dog is known to have a surface volume ratio twice as high as that of man. Nevertheless the model is a useful approximation, bearing in mind the present lack of more detailed data and elucidates the manner in which the concepts of localized dosimetry could contribute to the confirmation or revision of existing exposure standards. Another risk estimation for ^{239}Pu and ^{226}Ra in man was set up by Spiers and Vaughan [1976]. Proceeding from calculated dose rates to trabecular marrow, endosteum and air-sinus epithelium for a 1 µCi skeletal burden of ^{226}Ra or ^{239}Pu they calculated risk rates using risk coefficients (cases per 10^3 per year for 1 rad/year). The dose rates were determined by the Monte Carlo procedure (see above) [Whitwell and Spiers, 1971] taking into account the physical dimensions of the trabeculae and marrow spaces and amounted to 19 rad/year for ^{226}Ra and daughters and 129 rad/year for ^{239}Pu, averaged over a trabecular layer with depth equal to the particle range. Though it was concluded from the calculations that the current permissible body burdens for radium and plutonium are justified it should be emphasized that no allowance was made for burial.

In attempting to improve the present knowledge on the distribution of plutonium and the consequent irradiation to sensitive cells it seems promising to continue research along the following lines:

1. Quantitative description of burial with geometrical resolution as high as possible and over a considerable fraction of the total life-span. The results could be presented in the form of time-dependent frequency distributions of the depth to bone surfaces.

2. Examination of the liberation of buried hot lines and diffuse activity by resorption. Contribution of this process to the endosteal dose.

3. Contribution of particulate sources of plutonium in osteoclasts and macrophages to the endosteal dose. Significance of the movement of these cells. Is there a "hot particle problem" in the skeleton?

4. More precise information on the location of the cells at risk and variations of the density of these cells along surfaces and in different parts of the skeleton. Possibly, it would be a more appropriate description of the location to specify a probability distribution of distances weighted with the corresponding cell densities than assuming a fixed distance.

5.3. Americium-241

^{241}Am like ^{239}Pu exhibits a pronounced affinity to the skeleton, though the initial partition of the nuclide between liver and skeleton, the main organs of deposition, is in favour of the liver and, consequently, the skeletal retention in rats and dogs is somewhat lower than for ^{239}Pu [Taylor, 1962; Belyayev, 1969; Lloyd et al., 1970; Durbin, 1973]. The effectiveness of americium to induce osteosarcomas in the rat is evidently lower than for plutonium [Bensted et al., 1965; Taylor and Bensted, 1969]. In spite of the close analogy between the two isotopes with respect to the gross skeletal retention in different bones [Lloyd et al., 1972] and the slow release from the skeleton [Taylor et al., 1961], there are apparent dissimilarities on a microscopic level. Autoradiographic studies revealed that ^{241}Am is a surface-seeker too [Taylor et al., 1961; Herring et al., 1962; Williamson and Vaughan, 1963; Durbin et al., 1969; Lloyd et al., 1972; Nenot et al., 1972] but with the endosteal and periosteal surface concentrations being more equal. In the distal femoral metaphysis of dogs endosteal concentrations were 1.5 times higher than on the periosteum; with ^{239}Pu the ratio was 3.4 [Lloyd et al., 1972]. Taylor et al. [1961] even found periosteal deposits in rat femurs slightly more intense than on the endosteum. No evidence of americium uptake in the marrow could be found, except in cells at sites of bone resorption [Herring et al., 1962]. The ^{241}Am labeling of the walls of blood vessels in the cortical bone of rats [Taylor et al., 1961; Durbin et al., 1969] is absent in the case of plutonium.

Until now only little work has been done to determine local radiation doses in the skeleton. Taylor et al. [1961] measured average dose rates in the femur of rats by means of emulsion autoradiography. The animals received a single injection of 3.3 µCi/kg. From local measurements in fields of 100 square microns the authors deduced dose rates between 21 and 362 rad/day in the metaphysis (average: 138 rad/day) in the highly radioactive band formed initially at the zone of endochondral ossification beneath the cartilage plate (see also fig. 50). Outside this band the dose rates ranged between 27 and 169 rad/day (average: 71 rad/day) two days after injection. Periosteal dose rates at the diaphyseal cortex were always higher on the average than endosteal ones but were nearly equal at 21 days (21–24 rad/day). The dosimetric study of Polig [1976b] was directed towards a topographical description of the dose distribution in the femur of young rats injected with 30 µCi/kg ^{241}Am. In this experiment solid state track detectors (cellulose nitrate) were used as autoradiographic medium and the measurements were performed by scanning the detectors in 80 × 80 µm fields with a computer-controlled image analyzer (fig. 39). Fig. 50 displays the localization and dose rate distribution in the form of print symbols representing the dose rate levels as specified in the legend. The aforementioned highly radioactive band in the metaphysis is to be seen clearly (fig. 50 left) 7 days after administration of ^{241}Am [Taylor, 1961; Williamson and Vaughan, 1964; Durbin et al., 1969]. Other zones of relatively high concentrations are the ossification zones in the epiphysis near the articular and the epiphyseal cartilage. The dose rate profile along the bone axis measured 56 days after injection still shows the three peaks corresponding to these deposits (fig. 51 and also fig. 50). The metaphyseal band has a maximum dose rate of 170 rad/day, which is higher than at the 7th day (110 rad/day). Microradiographic investigations demonstrated unresorbed bone remnants in the metaphysis pointing towards severe radiation damage and inhibition of resorption in the metaphyseal band. Thus the increase of dose rates in this domain can be explained by redeposition of ^{241}Am translocated

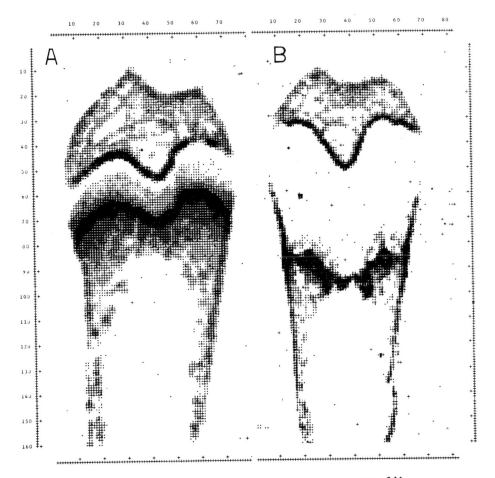

Fig. 50. Scan of the rat femur (distal end) after i.v. injection of 30 μCi/kg ^{241}Am 7 days (A) and 56 days (B) after injection. Plot code (rad/day): blank, <5; •, 5–10; +, 10–20; *, 20–40; *+, 40–80; *W, >80. The last two ranges are represented by the two symbols superimposed on one another [Polig, 1976b].

from other parts of the skeleton, which is particularly pronounced in growing rats. The phenomenon of translocation and recirculation is also illustrated in fig. 50 (left) by the lighter concentrations of ^{241}Am above the metaphyseal band in post injection bone. In comparing dose measurements with the results of other investigations it should be borne in mind that relatively large scan fields were used involving considerable averaging. Normalized to unit intake the values reported by Taylor et al. are considerably higher than those of Polig [1976b] and the ^{239}Pu results of James [1972]. If average dose rates at

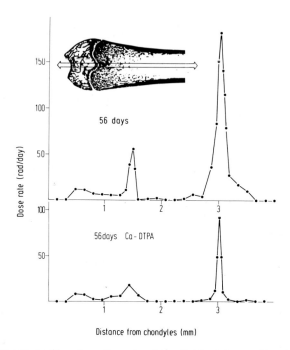

Fig. 51. Dose rate profiles along the axis of the distal rat femur after the injection of 30 μCi/kg ^{241}Am. The measurements were made in a stripe of 10 scan fields width (800 μm) as indicated in the microradiograph on the top. The points are the average of 3 sections in one femur. DTPA therapy was commenced 1.5 h after injection of the nuclide and continued with weekly injections of 30 μmole/kg Na$_3$Ca-DTPA.

trabecular surfaces in the epiphysis are calculated from the total number of tracks and the surface volume ratios measured in cross sections [Polig, 1976a] values between 0.93 and 1.4 rad kg/μCi day are obtained. If these results are to be compared with James' dose rates of approx. 4 rad kg/μCi day (table 10) methodical differences, a possible lower gross skeletal retention of ^{241}Am compared to ^{239}Pu and the well known lower avidity of trabecular surfaces for ^{241}Am should be kept in mind. At the endosteal and periosteal sites of the diaphyseal cortex the dose rates were nearly equal at 7 days after injection, varying between zero and 40 rad/day. Profile scans through the cortex revealed that labeling throughout the bone volume was present. The endosteal dose rate profile occasionally was interrupted by peaks indicating cellular deposition with dose rates exceeding the local environment by two to three times [Polig, 1976b].

The present status of the local ^{241}Am-dosimetry does not allow far reaching conclusions concerning the long-term behaviour of this nuclide in the skeleton nor is it possible to characterize precisely and quantitatively the essential differences compared with ^{239}Pu. Apart from the growing importance of ^{241}Am in the nuclear industries, necessitating more detailed dosimetric information, just the disparities between ^{241}Am and ^{239}Pu deposition, which obviously are the cause for the observed differences in toxicity deserve attention and should motivate continuation of ^{241}Am work. Maybe that going the roundabout way over ^{241}Am contributes to a better understanding of radiation hazards from ^{239}Pu in the skeleton also.

6. Conclusions and future outlook

The preceding paragraphs were intended to give an idea of the scope and significance of the achievements brought about by a detailed consideration of the radiation dose distribution on a microscopic or submicroscopic scale. Though comprising a good deal of physics the aims of localized α-dosimetry go far beyond being merely a branch of applied physics theory but are directed towards establishing a unified physico-biological description of at least the initial stages of biological effects evoked by irradiation of tissue. Such a statement might perhaps sound somewhat pretentious but the progress made during the last decade and with a very limited number of experiments encourages an optimistic outlook to future developments.

As far as the physical basis is concerned some satisfaction with the present state of knowledge is justified. Qualitative and quantitative information on the fundamental processes of energy dissipation already appear accurate enough to provide a reliable tool at least for nonstochastic microdosimetry, particularly in view of the considerable gaps in the understanding of the biological substrate involved. A break-through towards elucidating some of the deleterious effects caused by radiation in terms of a quantifiable sequence of molecular events might well necessitate further refinements of the physical theory. But this is not a problem of pressing importance at the present time.

The author's opinion that stochastic microdosimetry should more frequently be applied to the treatment of problems with internally deposited α-emitters, although mentioned several times in the sections, should be reemphasized here. One of the basic objectives should be to expand the conceptual fundamentals of the current approach to localized dosimetry, which so far rest almost exclusively on the absorbed dose concept.

However, they should include such notions as hit frequency and specific energy density together with its distribution functions. This then would provide a natural transition towards an extension by taking into account the stochastic variables of physiological origin, e.g. probability of location of cells, branching process of a dividing cell population, probability of promotion and expression of an initial lesion etc. With the extension of information and understanding of the mechanisms of radiation effects in living organisms the absorbed dose notion may prove to be inadequate for the description of certain phenomena urging replacement by other quantities of which the investigator should be aware in time.

Another point urgently needs intensification of future efforts: the improvement of the methodical tools, particularly of instrumental and technical nature. Cumbersome visual measurement techniques and the large body of data to be handled often have turned out a barrier for systematic continuation of experiments initiated. Progress in improving the technical facilities could render the whole field more attractive to a wider range of investigators now still showing some reluctance because of the expenditure such experiments require. A principal shortcoming seems to be the insufficient sensitivity of methods to detect very minute amounts of nonfissionable α-emitting nuclides deposited in an organ if a precise topographical mapping of its concentration is asked for. Consequently, the possibility to transfer results and conclusions obtained from experiments with relatively large amounts of incorporated radioactivity to concentrations relevant for man in his environment is always hampered by side-effects peculiar to irradiation at high dose rates. A way out of this dilemma is not to be seen at present.

In future experiments detailed consideration should be given to a complete description of changes in time and comparison between different nuclides and species. One of the most promising aspects in the developing area of localized α-dosimetry is the linking of metabolic and dosimetric models to the different mathematical theories of tumour induction presently under discussion [e.g. Marshall and Groer, 1975a, 1975b; Mayneord and Clarke, 1975; Walsh, 1975; Baum, 1976]. The induction of tumours as a late effect is of primary concern particularly for the uptake of radium and plutonium in the skeleton. It would mean a clearcut justification of efforts and a major step towards the aforementioned physico-biological description of radiation effects in vivo if — in the course of time — a general theory of radiation carcinogenesis in the skeleton would evolve, valid for surface- and volume-seekers and different species including man. Such a unification of dosimetric and tumour models also has some bearing on another area of current interest, namely the deposition of particulate sources of α-radiation in the lungs and the so-called "hot particle problem" which could not be discussed here because of lack of space.

Acknowledgements

The author owes much to his teacher Prof. Dr. A. Catsch who died in spring 1976 while this article was in preparation. It is a testimony for the impressive scope of interests and the intellectual abilities of this man, known as the father of chelation therapy throughout the world, that he initiated localized dosimetry as a new line of research in our institute and gave the author every possible encouragement and support in his scientific activities.

I am greatly indebted to Prof. Dr. K.G. Zimmer whose continued interest, valuable suggestions and criticism was of great help in preparing the manuscript.

I am particularly grateful to Miss M. Schneider, librarian, whose dedicated work was invaluable in providing the necessary literature and correcting the references.

Last not least I want to express my thanks to Mrs. S.I. Wibowo, Mrs. B. Heinold and Mrs. Ch. Heinold for preparing the figures and typing the manuscript and to the editors for their patience and appreciation.

References

Abmayr, W., Grünauer, F. and Burger, G. 1969. Proc. Int. Topical Conf. on Nuclear Track Registration, Clermont-Ferrand, Vol. 1. Eds. Isabelle and Monnin, p. III 46.
Al-Bedri, M.B. and Harris, S.J. 1975. Health Phys. 28, 816.
Armstrong, T.W. and Chandler, K.C. 1973. Nucl. Instrum. Methods 113, 313.
Arnold, J.S. and Jee, W.S.S. 1957. Am. J. Anat. 101, 367.
Arnold, J.S. and Jee, W.S.S. 1959. Lab. Invest. 8, 194.
Arnold, J.S. and Jee, W.S.S. 1962. Health Phys. 8, 705.
Aspin, N. and Johns, H.E. 1963. Br. J. Radiol. 36, 350.
Atherton, D.R., Stover, B.J., Jee, W.S.S., Stevens, W. and Bruenger, F.W. 1972. Rep. COO-119-246, p. 126.
Auxier, J.A., Beach, J.L., Becker, K., Gammage, R.D., Henley, L.C. and Parkinson, W.W. 1975. In: The Health Effects of Plutonium and Radium. Ed. Jee (J.W. Press, Salt Lake City) p. 553.
Baily, N.A. and Steigerwalt, J.E. 1975. In: Adv. Radiat. Biol. Vol. 5. (Academic Press, New York) p. 1.
Bair, W.J. and Thompson, R.C. 1974. Science 183, 715.
Barkas, W.H. 1963. Nuclear Research Emulsions, Vol. I (Academic Press, New York).
Barkas, W.H. and Berger, M.J. 1964. In: Studies in Penetration of Charged Particles in Matter. Publ. 1133 (Natl. Acad. Sci.-Natl. Res. Council, Washington D.C.) p. 103.
Baum, J.W. 1976. Health Phys. 30, 85.
Becker, K. 1969. Proc. Int. Topical Conf. on Nuclear Track Registration, Clermont-Ferrand, Vol. 2. Eds. Isabelle and Monnin, p. V 2.
Becker, K. and Johnson, D.R. 1970. Science 167, 1370.
Belyayev, Yu.A. 1969. In: Radioaktivnye izotopy i organizm. Ed. Moskalev, (Izdat. Med., Moscow). English translation AEC-tr-7195 p. 168.
Bensted, J.P.M., Taylor, D.M. and Sowby, F.D. 1965. Br. J. Radiol. 38, 920.
Benton, E.V. 1968. Rep. USNRDL-TR-68-14.
Benton, E.V. and Nix, W.D. 1969. Nucl. Instrum. Methods 67, 343.
Besant, C.B. and Ipson, S.S. 1969. Rep. AEEW-M881.
Bethe, H. 1930. Ann. Phys. (Leipzig) 5, 325.

Biavati, B.J. 1966a. Rep. NYO-2740-3, p. 119.
Biavati, B.J. 1966b. Rep. NYO-2740-3, p. 131.
Biavati, B.J., Gross, W., Rossi, H.H. and Kellerer, A.M. 1968. Rep. NYO-2740-5, p. 61.
Bichsel, H. 1968. In: Radiation Dosimetry, Vol. I. Eds. Attix and Roesch (Academic Press, New York), p. 157.
Bichsel, H. 1969. 2nd Sympos. Microdosimetry, Stresa, EUR 4452 d-f-e (Euratom) p. 511.
Birkhoff, R.D., Turner, J.E., Anderson, V.E., Feola, J.M. and Hamm, R.N. 1970. Health Phys. 18, 1.
Blanford, G.E., Walker, R.M. and Wefel, J.P. 1970. Radiat. Eff. 3, 267.
Bleaney, B. 1967. Phys. Med. Biol. 12, 145.
Bleaney, B. 1968. In: Delayed Effects of Bone-Seeking Radionuclides. Eds. Mays et al. (University of Utah Press, Salt Lake City) p. 125.
Bleaney, B. 1969. Br. J. Radiol. 42, 51.
Bleaney, B. and Vaughan, D.M. 1971. Br. J. Radiol. 44, 67.
Bloch, F. 1933. Ann. Phys. (Leipzig) 16, 285.
Bohr, N. 1948. Mat. Fys. Medd. 18 (8).
Booz, J., Giglio, C., Waker, A. and Gaggero, G. 1971. 3rd Sympos. Microdosimetry, Stresa, EUR 4810 d-f-e (Euratom) p. 833.
Booz, J., Smit, Th. and Waker, A. 1972. Phys. Med. Biol. 17, 477.
Bourland, P.D., Chu, W.K. and Powers, D. 1971. Phys. Rev. 38, 3625.
Bourland, P.D. and Powers, D. 1971. Phys. Rev. 38, 3635.
Bradt, H.L. and Peters, B. 1948. Phys. Rev. 74, 1828.
Brendle, M., Gugel, F. and Steidle, G. 1975. Nucl. Instrum. Methods 130, 253.
Bromley, R.G., Dockum, N.L., Arnold, J.S. and Jee, W.S.S. 1966. J. Gerontol. 21, 537.
Butts, J.J. and Katz, R. 1967. Radiat. Res. 30, 855.
Caswell, R.S. 1966. Radiat. Res. 27, 92.
Catsch, A. 1964. Radioactive Metal Mobilization in Medicine (Ch.C. Thomas, Springfield).
Charlton, D.E. and Cormack, D.V. 1962a. Radiat. Res. 17, 34.
Charlton, D.E. and Cormack, D.V. 1962b. Br. J. Radiol. 35, 473.
Chipperfield, A.R. and Taylor, D.M. 1968. Nature 209, 609.
Cole, A., Simmons, D.J., Cummins, H., Congel, F.J. and Kastner, J. 1970. Health Phys. 19, 55.
Costa-Ribeiro, C. and Lobão, N. 1975. Health Phys. 28, 162.
Cross, W.G. and Tommasino, L. 1969. Proc. Int. Topical Conf. on Nuclear Track Registration, Clermont-Ferrand, Vol. 1. Eds. Isabelle and Monnin, p. III 73.
Cross, W.G. and Tommasino, L. 1970. Radiat. Eff. 5, 85.
Czuber, E. 1884. Sber. Akad. Wiss. Wien, Abt. 2, 90, 719.
Darley, P.J. 1967. 1st Sympos. Microdosimetry, Ispra, EUR 3747 d-f-e (Euratom) p. 509.
Dockum, N.L., Mical, R.S., Tegge, R., Lowe, M., Mays, Ch.W. and Jee, W.S.S. 1964. Rep. COO-119-229, p. 127.
Doniach, I. and Pelc, S.R. 1950. Br. J. Radiol. 23, 267.
Dörmer, P. 1967. Leitz-Mitt. Wiss. Tech. 4, 74.
Dougherty, Th.F. 1962. In: Some Aspects of Internal Irradiation. Eds. Dougherty et al. (Pergamon Press, Oxford) p. 47.
Dougherty, Th.F. and Mays, Ch.W. 1969. Rep. COO-119-240, p. 86.
Durbin, P.W., Jeung, N. and Williams, M.H. 1969. In: Delayed Effects of Bone-Seeking Radionuclides. Eds. Mays et al. (University of Utah Press, Salt Lake City) p. 137.

Durbin, P.W. 1972. In: Radiobiology of Plutonium. Eds. Stover and Jee (J.W. Press, Salt Lake City) p. 469.
Durbin, P.W. 1973. In: Handbook of Experimental Pharmacology, Vol. XXXVI. Eds. Eichler et al. (Springer, Berlin) p. 739.
Elkind, M.M. and Whitmore, G.F. 1967. The Radiobiology of Cultured Mammalian Cells (Gordon and Breach, New York).
Evans, R.D. 1962. In: Some Aspects of Internal Irradiation. Eds. Dougherty et al. (Pergamon Press, Oxford) p. 381.
Evans, R.D., Keane, A.T., Kolenkow, R.J., Neal, W.R. and Shanahan, M.M. 1969. In: Delayed Effects of Bone-Seeking Radionuclides. Eds. Mays et al. (University of Utah Press, Salt Lake City) p. 157.
Evans, R.D. 1974. Health Phys. 27, 497.
Fellows, M.H., Clark, L., O'Toole, J., Kimmel, D.B. and Jee, W.S.S. 1975. Health Phys. 29, 97.
Fink, D. 1973. Nucl. Instr. Methods 107, 615.
Finkel, A.J. Miller, Ch.E. and Hasterlik, R.J. 1969. In: Delayed Effects of Bone-Seeking Radionuclides. Eds. Mays et al. (University of Utah Press, Salt Lake City) p. 195.
Finkel, M.P. and Biskis, B.O. 1962. Health Phys. 8, 565.
Finkel, M.P., Biskis, B.O. and Jinkins, P.B. 1969. In: Radiation-Induced Cancer Proc. IAEA Sympos., Athens, p. 369.
Fischer, H.A. and Werner, G. 1971. Autoradiographie (Walter De Gruyter and Co., Berlin).
Fleischer, R.L., Price, P.B. and Walker, R.M. 1965a. Ann. Rev. Nucl. Sci. 15, 1.
Fleischer, R.L., Price, P.B. and Walker, R.M. 1965b. J. Appl. Phys. 36, 3645.
Fleischer, R.L., Price, P.B. and Walker, R.M. 1967. Phys. Rev. 156, 353.
Fleischer, R.L., Price, P.B. and Walker, R.M. 1969. Sci. Am. 220 (6) 30.
Fleischer, R.L., Alter, H.W., Furman, S.C., Price, P.B. and Walker, R.M. 1972. Science 178, 255.
Fleischer, R.L., Price, P.B. and Walker, R.M. 1975. Nuclear Tracks in Solids. Principles and Applications (University of California Press, Berkeley).
Gahan, P.B. (Ed.) 1972. Autoradiography for Biologists (Academic Press, London and New York).
Glass, W.A. and Braby, L.A. 1969. Radiat. Res. 39, 230.
Green, D., Howells, G.R. and Humphreys, E.R. 1975. Health Phys. 29, 798.
Groer, P.G., Marshall, J.H., Simmons, D.J., Rabinowitz, A. and Goldman, M. 1972. Rep. ANL-7960 (II) p 48
Gross, W., Biavati, B.J. and Rossi, H.H. 1969a. 2nd Sympos. Microdosimetry, Stresa, EUR 4452 d-f-e (Euratom) p. 249.
Gross, W., Biavati, B.J. and Rossi, H.H. 1969b. Rep. NYO-2740-6, p. 113.
Gross, W., Rodgers, R., Rossi, H.H. and Kitzman, J. 1970. Rep. NYO-2740-7, p. 53.
Gude, W.D. 1968. Autoradiographic Techniques (Prentice-Hall, New York).
Gullberg, J.E. 1957. Exp. Cell Res. Suppl. 4, 222.
Hamilton, E.I. 1968. Int. J. Appl. Radiat. Isot. 19, 159.
Harley, N.H. and Pasternack, B.S. 1972. Health Phys. 23, 771.
Harley, N.H. and Pasternack, B.S. 1976. Health Phys. 30, 35.
Harvey, J.R. 1971. Health Phys. 21, 866.
Harvey, J.R. and Townsend, S. 1971. Rep. RD/B/N 2127.
Henke, R.P. and Benton, E.V. 1971. Nucl. Instrum. Methods 97, 483.
Herring, G.M., Vaughan, J. and Williamson, M. 1962. Health Phys. 8, 717.
Herz, R.H. 1969. The Photographic Action of Ionizing Radiations. (Wiley–Interscience, New York).

Hindmarsh, M., Owen, M., Vaughan, J., Lamerton, L.F. and Spiers, F.W. 1958. Br. J. Radiol. 31, 518.
Hindmarsh, M., Owen, M. and Vaughan, J. 1959. Br. J. Radiol. 32, 183.
Hogeweg, B. and Barendsen, G.W. 1971. 3rd Sympos. Microdosimetry, Stresa, EUR 4810 d-f-e (Euratom) p. 857.
Howarth, J.L. 1965a. Radiat. Res. 24, 158.
Howarth, J.L. 1965b. Br. J. Radiol. 38, 51.
Hsieh, J.J.C., Hungate, F.P. and Wilson, S.A. 1965. Science 150, 1821.
Hug, O. and Kellerer, A.M. 1966. Stochastik der Strahlenwirkung. (Springer, Berlin).
ICRP 1959. Publication 2. Report of Committee II on Permissible Dose for Internal Radiation. (Pergamon Press, Oxford).
ICRP 1967. Publication 11. A Review of the Radiosensitivity of the Tissues in Bone (Pergamon Press, Oxford).
ICRP 1975. Publication 23. Report of the Task Group on Reference Man. (Pergamon Press, Oxford).
ICRU 1970. Linear Energy Transfer. Report 16.
ICRU 1971. Radiation Quantities and Units. Report 19.
James, A.C. 1969. Ph.D. Thesis, University of London.
James, A.C. and Kember, N.F. 1970. Phys. Med. Biol. 15, 39.
James, A.C. and Taylor, D.M. 1971. Health Phys. 21, 31.
James, A.C. and Kember, N.F. 1972. In: Radiobiology of Plutonium. Eds. Stover and Jee (J.W. Press, Salt Lake City) p. 281.
James, A.C. 1972. Radiat. Res. 51, 654.
Jee, W.S.S. and Arnold, J.S. 1961. Lab. Invest. 10, 797.
Jee, W.S.S., Stover, B.J., Taylor, G.N. and Christensen, W.R. 1962. Health Phys. 8, 599.
Jee, W.S.S., Arnold, J.S., Cochran, T.H., Twente, J.A. and Mical, R.S. 1962. In: Some Aspects of Internal Irradiation. Eds. Dougherty et al. (Pergamon Press, Oxford) p. 27.
Jee, W.S.S. 1964. In: Assessment of Radioactivity in Man. (IAEA, Vienna) p. 369.
Jee, W.S.S., Bromley, R.G., Dockum, N.L., Lowe, M., Burggraaf, R., Mical, R., Dedekind, K. and Arnold, J.S. 1965. Rep. COO-119-232, p. 99.
Jee, W.S.S., Park, H.Z. and Burggraaf, R. 1969. Rep. COO-119-240, p. 188.
Jee, W.S.S. 1970. Rep. COO-119-242, p. 268.
Jee, W.S.S. 1971. Rep. COO-119-244, p. 228.
Jee, W.S.S. 1972. Health Phys. 22, 583.
Jee, W.S.S., Dell, R.B. and Miller, L.G. 1972. Health Phys. 22, 761.
Jee, W.S.S. 1972. In: Radiobiology of Plutonium. Eds. Stover and Jee. (J.W. Press, Salt Lake City) p. 171.
Jee, W.S.S., Kimmel, D.B., Hashimoto, E.G., Dell, R.B. and Woodbury, L.A. 1973. Rep. COO-119-248, p. 255.
Jee, W.S.S., Smith, J.M., Kimmel, D.B., Wronski, T.J., Dell, R.B. and Schlenker, R.A. 1975. Rep. COO-119-250, p. 166.
Jee, W.S.S., Kimmel, D.B., Wronski, T.J., Gotcher, J.E. and Dell, R.B. 1976. Rep. COO-119-251, p. 253.
Johnson, D.R., Boyett, R.H. and Becker, K. 1970. Health Phys. 18, 424.
Jones, W.D. and Neidigh, R.V. 1967. Appl. Phys. Letters 10, 18.
Kappos, A.D. 1967a. Biophysik 4, 137.
Kappos, A.D. 1967b. 1st Sympos. Microdosimetry, Ispra, EUR 3747 d-f-e (Euratom) p. 569.
Katz, R. and Butts, J.J. 1965. Phys. Rev. 137, B 198.
Katz, R. and Kobetich, E.J. 1968. Phys. Rev. 170, 401.

Katz, R. and Kobetich, E.J. 1969. Phys. Rev. 186, 344.
Katz, R., Ackerson, B., Homayoonfar, M. and Sharma, S.C. 1971. Radiat. Res. 47, 402.
Katz, R., Sharma, S.C. and Homayoonfar, M. 1972. In: Radiation Dosimetry Suppl. 1. Eds. Attix and Roesch. (Academic Press, New York) p. 317.
Kellerer, A.M. 1966. Panel Rep. No. 58. Biophysical Aspects of Radiation Quality. (IAEA, Vienna) p. 95.
Kellerer, A.M. 1967. 2nd Panel Rep. Biophysical Aspects of Radiation Quality (IAEA, Vienna) p. 89.
Kellerer, A.M. 1968. Report B1, Gesellschaft für Strahlenforschung, Neuherberg, München.
Kellerer, A.M. 1969. 2nd Sympos. Microdosimetry, Stresa, EUR 4452 d-f-e (Euratom) p. 107.
Kellerer, A.M. and Rossi, H.H. 1969. 2nd Sympos. Microdosimetry, Strasa, EUR 4452 d-f-e (Euratom) p. 843.
Kellerer, A.M. and Rossi, H.H. 1971. Radiat. Res. 47, 15.
Kellerer, A.M. 1971. Rad. Res. 47, 359.
Kellerer, A.M. and Rossi, H.H. 1972. Curr. Top. Radiat. Res. Q. 8, 85.
Kellerer, A.M. and Rossi, H.H. 1973. 4th Sympos. Microdosimetry, Verbania Pallanza, EUR 5122 d-e-f (Euratom) p. 331.
Kellerer, A.M. 1975. 5th Sympos. Microdosimetry, Verbania Pallanza, EUR 5452 d-e-f (Euratom) p. 409.
Kellerer, A.M. and Chmelevsky, D. 1975a. Radiat. Environ. Biophys. 12, 61.
Kellerer, A.M. and Chmelevsky, D. 1975b. Radiat. Environ. Biophys. 12, 205.
Kellerer, A.M. and Chmelevsky, D. 1975c. Radiat. Environ. Biophys. 12, 321.
Kellerer, A.M. and Chmelevsky, D. 1975d. Radiat. Res. 63, 226.
Kendall, M.G. and Moran, P.A.P. 1963. Geometrical Probability. (Charles Griffin, London) p. 116.
Kerr, G.D., Hairr, L.M., Underwood, N. and Waltner, A.W. 1966. Health Phys. 12, 1475.
Khan, H.A. 1971. Radiat. Eff. 8, 135.
Kienzler, B. and Polig, E. 1975. Radiat. Environ. Biophys. 12, 77.
Kienzler, B. 1976. Thesis, University of Karlsruhe.
Kienzler, B. and Polig, E. 1976. Unpublished.
Kimmel, D.B., Wronski, T.J., Taylor, G.N., Dell, R.B. and Jee, W.S.S. 1975. Rep. COO-119-250, p. 153.
Kimmel, D.B., Jee, W.S.S., Wronski, T.J., Atherton, D.R., Schlenker, R.A. and Stover, B.J. 1975. In: The Health Effects of Plutonium and Radium. Ed. Jee (J.W. Press, Salt Lake City) p. 105.
Kleeman, J.D. and Lovering, J.F. 1967. Science 156, 512.
Kleeman, J.D. and Lovering, J.F. 1970. Radiat. Eff. 5, 21.
Kolenkow, R.J. and Manly, P.J. 1967. Rep. MIT-952-4, p. 132.
Kolenkow, R.J. 1967. Rep. MIT-952-4, p. 163.
Kononenko, A.M. 1957. Biofizika 2, 98.
Lamerton, L.F. and Harriss, E.B. 1954. J. Photogr. Sci. 2, 135.
Lea, D.E. 1956. Actions of Radiations on Living Cells, 2nd ed. (Cambridge University Press).
Lederer, C.M., Hollander, J.M. and Perlman, I. 1967. Table of Isotopes, 6th Ed. (John Wiley & Sons, Inc., New York).
Lindenbaum, A., Rosenthal, M.W. and Smoler, M. 1967. In: Diagnosis and Treatment of Deposited Radionuclides. Proc. Sympos. Richland, Washington. (Excerpta Medica Foundation) p. 65.

Lindenbaum, A. and Lund, C.J. 1969. Radiat. Res. 37, 131.
Lindenbaum, A. and Russel, J.J. 1972. Health Phys. 22, 617.
Lindhard, J. and Scharff, M. 1961. Phys. Rev. 124, 128.
Lloyd, E. 1961. Br. J. Radiol. 34, 521.
Lloyd, E., Marshall, J.H., Butler, J.W. and Rowland, R.E. 1966. Nature 211, 661.
Lloyd, E. 1969. Rep. ANL-7615, p. 49.
Lloyd, E. and Hodges, D. 1971. Clin. Orthop. Relat. Res. 78, 230.
Lloyd, E. and Marshall, J.H. 1972. In: Radiobiology of Plutonium. Eds. Stover and Jee. (J.W. Press, Salt Lake City) p. 377.
Lloyd, R.D., Taylor, G.N. and Atherton, D.R. 1970. Health Phys. 18, 149.
Lloyd, R.D., Jee W.S.S. Atherton, D.R., Taylor, G.N. and Mays, Ch.W. 1972. In: Radiobiology of Plutonium. Eds. Stover and Jee. (J.W. Press, Salt Lake City) p. 141.
Lloyd, R.D., Mays, Ch.W., Atherton, D.R., Taylor, G.N. and Van Dilla, M.A. 1976. Radiat. Res. 66, 274.
Lück, H.B. 1974. Nucl. Instrum. Methods 116, 613.
Madhvanath, U., Murthy, M.S.S., Vishwakarma, R.R., Subrahmanyam, P. and Das, G.C. 1974. Health Phys. 27, 469.
Marshall, J.H., Rowland, R.E. and Jowsey, J. 1959. Radiat. Res. 10, 258.
Marshall, J.H. and Finkel, M.P. 1959. Rep. ANL-6104, p. 48.
Marshall, J.H. 1960. Rep. ANL-6297, p. 16.
Marshall, J.H. 1962. In: Radioisotopes and Bone. Eds. Lacroix and Budy. (Blackwell, Oxford) p. 35.
Marshall, J.H. 1969. In: Delayed Effects of Bone-Seeking Radionuclides. Eds. Mays et al. (University of Utah Press, Salt Lake City) p. 7.
Marshall, J.H. and Lloyd, E. 1972. In: Radionuclide Carcinogenesis. Eds. Sanders et al., Proc. 12th Ann. Hanford Biol. Sympos., Richland. (USAEC) p. 421.
Marshall, J.H., Keefe, D.J., Groer, P.G. and Selman, R.F. 1973. Rep. ANL-8060 (II) p. 242.
Marshall, J.H., Groer, P.G. and Schlenker, R.A. 1974. Rep. ANL-75-3(II) p. 71.
Marshall, J.H. and Groer, P.G. 1975a. Rep. ANL-75-60(II) p. 1.
Marshall, J.H. and Groer, P.G. 1975b. In: The Health Effects of Plutonium and Radium. Ed. Jee, (J.W. Press, Salt Lake City) p. 717.
Matsuoka, O., Yoshikawa, K. and Fukumoto, T. 1967. J. Japan Health Phys. Soc. 2, 121. Translation: NSJ-Tr 136, Sept. 1968.
Mayneord, W.V. and Clarke, R.H. 1975. Br. J. Radiol. Suppl. No. 12.
Mays, Ch.W., Floyd, R.L. and Arnold, J.S. 1958. Radiat. Res. 8, 480.
Mays, Ch.W. 1958. Rep. COO-217, p. 161.
Mays, Ch.W. 1960. Rep. COO-220, p. 200.
Mays, Ch.W. and Sears, K.A. 1962. Rep. COO-226, p. 78.
Mays, Ch.W. and Tueller, A.B. 1964. Rep. COO-119-229, p. 199.
Mays, Ch.W., Dougherty, Th.F., Taylor, G.N., Lloyd, R.D., Stover, B.J., Jee, W.S.S., Christensen, W.R., Dougherty, J.H. and Atherton, D.R. 1969. In: Delayed Effects of Bone-Seeking Radionuclides. Eds. Mays et al. (University of Utah Press, Salt Lake City) p. 387.
Mays, Ch.W., Lloyd, R.D. and Van Dilla, M.A. 1975. Health Phys. 29, 761.
Munson, R.J. 1950. Br. J. Radiol. 23, 505.
Nenot, J.C., Masse, R., Morin, M. and Lafuma, J. 1972. Health Phys. 22, 657.
Neufeld, J. and Snyder, W.S. 1960. In: Selected Topics in Radiation Dosimetry (IAEA, Vienna) p. 35.
Norris, W.P., Speckman, T.W. and Gustafson, P.F. 1955. Am. J. Roentgenol. Radiat. Therap. Nucl. Med. 73, 785.

Northcliffe, L.C. and Schilling, R.F. 1970. Nuclear Data Tables, A 7. (Academic Press, New York) p. 233.
Numakunai, T. 1974. Radioisotopes 23, 474.
Paretzke, H.G., Benton, E.V. and Henke, R.P. 1973. Nucl. Instrum. Methods 108, 73.
Piesch, E. 1969. Proc. Int. Topical Conf. on Nuclear Track Registration, Clermont-Ferrand, Vol. 1. Eds. Isabelle and Monnin, p. III 66.
Piesch, E. and Sayed, A.M. 1974. Nucl. Instrum. Methods 119, 367.
Platzer, H., Abmayr, W. and Paretzke, H.G. 1972. Atomkernenergie 20, 162.
Polig, E. 1975a. Int. J. Appl. Radiat. Isot. 26, 471.
Polig, E. 1975b. Int. J. Appl. Radiat. Isot. 26, 519.
Polig, E. 1975c. IMANCO Sympos. Fortschritte der quantitativen Bildanalyse, Frankfurt/M., p. 285.
Polig, E. 1976a. Radiat. Environ. Biophys. 13, 27.
Polig, E. 1976b. Radiat. Res. 67, 128.
Polig, E. 1976c, unpublished.
Rauth, A.M. and Simpson, J.A. 1964. Radiat. Res. 22, 643.
Reynolds, H.K., Dunbar, D.N.F., Wenzel, W.A. and Whaling, W. 1953. Phys. Rev. 92, 742.
Roesch, W.C. and Attix, F.H. 1968. In: Radiation Dosimetry, Vol. I. Eds. Attix and Roesch. (Academic Press, New York) p. 1.
Roesch, W.C. and Glass, W.A. 1971. Radiat. Res. 45, 1.
Roesch, W.C. 1975. Rep. BNWL-SA-5550.
Rogers, A.W. 1969. Techniques of Autoradiography. (Elsevier Publ. Co., Amsterdam).
Rosenthal, M.W., Marshall, J.H. and Lindenbaum, A. 1968. In: Diagnosis and Treatment of Deposited Radionuclides. Proc. Sympos. Richland, Wash. (Excerpta Medica Foundation) p. 73.
Rossi, H.H. and Rosenzweig, W. 1955. Radiology 64 (3), 404.
Rossi, H.H. and Failla, G. 1956. Nucleonics 14 (2), 32.
Rossi, H.H. 1960. Radiat. Res. Suppl. 2, 290.
Rossi, H.H. 1961. Radiat. Res. 15, 431.
Rossi, H.H. 1966. Biophysical Aspects of Radiation Quality. IAEA Tech. Rep. 58, p. 81.
Rossi, H.H. 1967a. Adv. Biol. Med. Phys. 11, 27.
Rossi, H.H. 1967b. ist Sympos. Microdosimetry, Ispra, EUR 3747 d-f-e (Euratom) p. 27.
Rossi, H.H. 1968. Radiation Dosimetry, Vol. I. Eds. Attix and Roesch. (Academic Press, New York) p. 43.
Rossi, H.H. 1971. In: Biophysical Aspects of Radiation Quality. Proc. IAEA, Vienna, p. 333.
Rossi, H.H. and Kellerer, A.M. 1973. 4th Sympos. Microdosimetry, Verbania Pallanza, EUR 5122 d-e-f (Euratom) p. 315.
Rotblat, J. and Ward, G. 1956. Phys. Med. Biol. 1, 57.
Rotondi, E. 1968. Radiat. Res. 33, 1.
Rowland, R.E. and Marshall, J.H. 1959. Radiat. Res. 11, 299.
Rowland, R.E. 1959. Rep. ANL-6104, p. 16.
Rowland, R.E. 1960. In: Radioisotopes in the Biosphere. Eds. Caldecott and Snyder. (University of Minnesota) p. 339.
Rowland, R.E. 1961. Radiat. Res. 15, 126.
Rowland, R.E. 1966. Clin. Orthop. 49, 233.
Rowland, R.E., Keane, A.T. and Lucas, H.F., Jr. 1972. In: Radionuclide Carcinogenesis. Eds. Sanders et al. Proc. 12th Ann. Hanford Biol. Sympos., Richland. (USAEC) p. 406.

Schaefer, H. 1948. Strahlentherapie 77, 613.
Schlenker, R.A. and Oltman, B.G. 1973. Rep. ANL-8060 (II) p. 163.
Schlenker, R.A. and Oltman, B.G. 1974. Rep. ANL-75-3 (II) p. 82.
Schlenker, R.A., Oltman, B.G. and Cummins, H.T. 1975. In: The Health Effects of Plutonium and Radium. Ed. Jee. (J.W. Press, Salt Lake City) p. 321.
Schlenker, R.A. and Farnham, J.E. 1975. In: The Health Effects of Plutonium and Radium. Ed. Jee. (J.W. Press, Salt Lake City) p. 437.
Schlenker, R.A. and Marshall, J.H. 1975. Health Phys. 29, 649.
Schubert, J., Fried, J.F., Rosenthal, M.W. and Lindenbaum, A. 1961. Radiat. Res. 15, 220.
Schultz, W.W. 1968. Rev. Sci. Instrum. 39, 1893.
Sears, K.A., Jee, W.S.S., Haslam, R.K. and Mays, Ch.W. 1963. Rep. COO-227, p. 90.
Seelmann-Eggebert, W., Pfennig, G. and Münzel, H. 1974. Nuklidkarte (Gersbach u. Sohn, München).
Seidel, A. 1975. Strahlentherapie 149, 442.
Simmons, D.J. and Fitzgerald, K.T. 1970. Rep. ANL-7760 (II) p. 208.
Somogyi, G. 1966. Nucl. Instr. Methods 42, 312.
Somogyi, G., Várnagy, M. and Petö, G. 1968. Nucl. Instr. Methods 59, 299.
Somogyi, G. and Szalay, S.A. 1973. Nucl. Instrum. Methods 109, 211.
Spiers, F.W. 1949. Br. J. Radiol. 22, 521.
Spiers, F.W. 1953. Br. J. Radiol. 26, 296.
Spiers, F.W. 1968. In: Radioisotopes in the Human Body: Physical and Biological Aspects. (Academic Press, New York).
Spiers, F.W. 1970. In: Radiation Dosimetry, Vol. I. Proc. Int. Summer School Radiat. Prot., Cavtat, Ed. Mirić, p. 199.
Spiers, F.W., Zanelli, G.D., Darley, P.J., Whitwell, J.R. and Goldman, M. 1971. In: Biomedical Implications of Radiostrontium Exposure. Eds. Goldman and Bustad. (USAEC Conf., 710201) p. 130.
Spiers, F.W. and Whitwell, J.R. 1975. In: The Health Effects of Plutonium and Radium. Ed. Jee. (J.W. Press, Salt Lake City) p. 537.
Spiers, F.W. and Vaughan, J. 1976. Nature 259, 531.
Spiess, H. 1969. In: Delayed Effects of Bone-Seeking Radionuclides. Eds. Mays et al. (University of Utah Press, Salt Lake City) p. 227.
Srdoč, D. 1970. Radiat. Res. 43, 302.
Stannard, J.N. 1973. In: Handbook of Experimental Pharmacology, Vol. XXXVI. (Springer, Berlin) p. 309.
Stevens, W., Atherton, D.R., Buster, D.S., Grube, B.J., Bruenger, F.W. and Lindenbaum, A. 1975. Rep. COO-119-250, p. 128.
Stolz, W. and Dörschel, B. 1968. Kernenergie 11, 137.
Storr, M.C., Hollins, J.G. and Clarke, R.L. 1975. Int. J. Appl. Radiat. Isot. 26, 708.
Stover, B.J., Atherton, D.R. and Buster, D.S. 1972. In: Radiobiology of Plutonium. Eds. Stover and Jee. (J.W. Press, Salt Lake City) p. 149.
Taylor, D.M., Sowby, F.D. and Kember, N.F. 1961. Phys. Med. Biol. 6, 73.
Taylor, D.M. 1962. Health Phys. 8, 673.
Taylor, D.M. and Bensted, J.P.M. 1969. In: Delayed Effects of Bone-Seeking Radionuclides. Eds. Mays et al. (University of Utah Press, Salt Lake City) p. 357.
Taylor, D.M. and Chipperfield, A.R. 1971. Seminar on Radiation Protection Problems Relating to Transuranium Elements. EUR 4612 d-f-e (Euratom) p. 187.
Thomas, J., Machek, J., Hanzlik, J. and Sedlák, A. 1973. Proc. 5th & 6th Conf. Radiation Hygiene (Med. Res. & Postgr. Inst. Press, Hradec Králové).

Thorne, M.C. 1976. Nature 259, 539.
Tisljar-Lentulis, G., Feinendegen, L.E. and Walther, H. 1975. Sympos. Adv. Biomed. Dosimetry. (IAEA, Vienna) p. 645.
Tisljar-Lentulis, G., Walther, H. and Feinendegen, L.E. 1976. Radiat. Environ. Biophys. 13, 197.
Turner, J.E. 1964. In: Studies in Penetration of Charged Particles in Matter. Publ. 1133. (Natl. Acad. Sci.-Natl. Res. Council, Washington D.C.) p. 99.
Twente, J.A., Arnold, J.S., Mays, Ch.W., Taysum, D.H. and Jee, W.S.S. 1958. Rep. COO-215, p. 98.
Twente, J.A. and Jee, W.S.S. 1958. Rep. COO-217, p. 147.
Twente, J.A., Butler, E.G., Freudenberger, O. and Jee, W.S.S. 1959. Rep. COO-218, p. 190.
Twente, J.A. and Atherton, D.R. 1959. Rep. COO-218, p. 207.
Twente, J.A., Butler, E.G. and Jee, W.S.S. 1960. Rep. COO-220, p. 168.
Twente, J.A. and Jee, W.S.S. 1961. Health Phys. 5, 142.
Van Dilla, M.A., Stover, B.J., Floyd, R.L., Atherton, D.R. and Taysum, D. 1958. Radiat. Res. 8, 417.
Vaughan, J., Bleaney, B. and Williamson, M. 1967. Br. J. Haematol. 13, 492.
Vaughan, J.M. 1970. The Physiology of Bone (Clarendon-Press, Oxford).
Vaughan, J.M. 1973. In: Handbook of Experimental Pharmacology, Vol. XXXVI. Eds. Eichler et al. (Springer, Berlin).
Vavilov, P.V. 1957. Sov. Phys. JETP 5, 749.
Venkataraman, G., Murthy, M.S.S. and Viswakarma, R.R. 1975. Health Phys. 28, 461.
VonSeggen, W.W., Farnham, J.E., Rybicki, K.A. and Sonza-Novera, J. 1973. Rep. ANL-8060 (II) p. 268.
Walsh, P.J. 1970. Health Phys. 19, 312.
Walsh, P.J. and McRee, D.I. 1971. Health Phys. 20, 352.
Walsh, P.J. and Pendergrass, F. 1972. Health Phys. 23, 701.
Walsh, P.J. 1975. In: The Health Effects of Plutonium and Radium. Ed. Jee. (J.W. Press, Salt Lake City) p. 657.
Wenger, F., Gardner, R.P. and Verghese, K. 1973. Health Phys. 25, 67.
Whaling, W. 1958. In: Handbuch der Physik. Vol. XXXIV. Ed. Flügge. (Springer, Berlin) p. 193.
Whitwell, J.R. and Spiers, F.W. 1971. Proc. 5th Int. Meeting French Soc. Radiat. Prot. (Grenoble) p. 401.
Williamson, C.F., Boujot, J.P. and Picard, J. 1966. Rapport CEA-R 3042. (Centre D'Etudes Nucléaires De Saclay).
Williamson, M. and Vaughan, J. 1963. Bone and Tooth. Proc. 1st Europ. Sympos. Ed. Blackwood. (Pergamon, Oxford) p. 71.
Wilson, K.S.J. and Emery, E.W. 1967. 1st Sympos. Microdosimetry, Ispra, EUR 3747 d-f-e (Euratom) p. 79.
Wilson, K.S.J. 1969. 2nd Sympos. Microdosimetry, Stresa, EUR 4452 d-f-e (Euratom) p. 235.
Woodard, H.Q. 1962. Health Phys. 8, 513.
Zaidins, C.S. 1974. Nucl. Instrum. Methods 120, 125.
Zimmer, K.G. 1961. Studies on Quantitative Radiation Biology. (Oliver and Boyd, Edinburgh).

SUBJECT INDEX

Adipose cells in skin, 4
AET and double strand breaks, 76, 78
After effects of ^{32}P and ^{33}P decays in DNA, 70
α-autoradiography and SSNTD, 260
α emitters, dose distribution, 206
— —, dosimetry, 189
— —, physical characteristics, 193
α particles, energy dissipation, 194
— — range in tissue bone and water, 205
^{241}Am, 313
Anagen, 12
Apoptic cells in epidermis, 31
Atomic stopping cross-section in tissue, 198
Autoionization, 67
Autoradiography, emulsion, 249
—, stripping film, 250

Bacteria, ^{32}P and ^{33}P, 82
Basal cell density, 34
— cells, labelling index, 43
Biological target, 223
Bone, elementary composition, 196
— and ^{239}Pn α dose, 247

Cancellous bone, 281
Cell cycle time, ear epidermis, 8
— migration and epithelialization, 23
— — and sensitivity, 160
Clonogenic basal cells, 25

D_0, endothelial cells, 15
—, epidermal cells, 22
—, epithelial cells, 29
δ rays, 224
— ray dose in water, 203
Deferred death, 92
Depigmented hairs, 15
Dermis, 4
— and irradiation, 16

Developmental stages of the weevil, 103
Dielectric track detectors, 258
DNA breaks and temperature, 74
— — — transmutation, 77
— double strand breaks after ^{32}P decay, 73
— and transmutation, 67
Dose average LET, 228
— distribution for α emitters, 206
Dosimetry localized skeleton, 279
Dose-time relationships, weevils, 167
Double strand breaks and AET, 76
Dual radiation action theory, 232

Ear epidermis, cycle time, 8
Elementary composition of tissues, 196
Emulsion autoradiography, 249
Energy dissipation, α, 191
— — near boundaries, 209
— straggling, 239
Endothelial cells, D_0, 15
Epidermal cells, D_0, 22
— cell kinetics, 7
— — response to irradiation, 17
— — survival, 17
— organisation, 6
Epidermis, 4
—, apoptic cells, 31
—, pycnotic cells, 31
—, ultrastructural abnormalities, 41
Epilation, 11
Epithelialization and cell migration, 23
Epithelial cells, D_0, 29
— cell numbers, 23
Erythema and skin reaction, 10

Fission cross section, thermal neutrons, 264
Follicles and skin tumours, 13
Fractionation and iso-effect dose, 163
— — mitotic delay, 133
Fraction number and LD50, 140

329

330 Subject Index

Grain weevil culture, 104
Grid counter, 233

Hair follicle, 4
— — and irradiation, 13
— — stimulation, 11
Homeostasis in repopulation, weevils, 160
Hydroperoxide radical, 69
Hyperpigmentation, 15

I decay and inactivation probability, 72
Inactivation probabilities of P and I decays, 72
Internal α-emitters microdosimetry, 224
Iso-effect dose and fractionation, weevils, 163
— — — and overall time, 128
— — relationships and repair, 100

Labelling index, irradiated basal cells, 43
Langerhans cells and irradiation, 45
— —, function, 51
LD50 and fraction number weevils, 140
— in weevils and temperature, 104
LET and δ rays, 200
— dose average, 228
— track average, 228
Lethal effects of ^{32}P, 71
— events in phages, 80
Linear energy transfer, see LET
Localized dosimetry, radionucleotides in skeleton, 279

Mammalian cells, ^{32}P, 85
Melanocytes and irradiation, 15
Methyl cholanthrene and sebaceous glands, 14
Microdosimetry, internal α emitters, 244
—, stochastic, 191, 223
Mitotic delay and fractionation weevils, 133
— — — radiation, weevils, 160
Mouse epidermal proliferative units, 7
Multifraction irradiations, weevils, 102
Muscle cells in skin, 4

Neutron induced autoradiography, 263
NIAR see neutron induced autoradiography

Overall time and fractionation, 128

P decay and inactivation probability, 72
^{32}P and double strand breaks, 73
— in yeast and mammalian cells, 85
— in transmutation, lethal effect, 71
^{32}P and ^{33}P decays, after effects in DNA, 70
^{32}P and ^{33}P in bacteria, 82
Phosphorous tracers, 63, 65
Phages, lethality, 80
Proportional counter, wall-less, 234
Protons, stopping power, 189
Protracted versus acute irradiation, weevils, 155
^{239}Pu, 292
— α particle dose in bone, 247
Pycnotic cells in epidermis, 31

^{226}Ra, 279
— hot spots in human bone, 284
— incorporation, 280
— recirculation, 282
Radiobiology of skin, 3
Radionucleides in skeleton dosimetry, 279
Recoil energy, 65
— spectrum of ^{32}S and ^{33}S, 66
Recovery and repopulation, weevils, 164
Repair processes and isoeffect relationships, 100
Repopulation, 160
— and damage, weevils, 142, 148
— — homeostasis, 160
— — recovery, weevils, 164
— in skin, 183

^{32}S and ^{33}S recoil spectrum, 66
Sebaceous gland, 4
— glands and irradiation, 14
— — — methylcholanthrene, 14
Sensitivity and cell cycle, weevils, 160
Shake off energy, 65
Skin histology, 3
— and radiation, 1
— repopulation, 183
— response to irradiation, history, 9
— tumour, follicle, 13
Soft tissue, elementary composition, 196
Solid state nuclear track detectors (SSNTD), 258
Spark counting technique, 272

Subject Index

Spherical counters, 233
Split dose recovery, weevils, 125
SSNTD see also solid state nuclear track detectors
− and α autoradiography, 260
Stimulation, hair follicle, 11
Stopping cross section of tissue, 197
− power, 194
Stripping film autoradiograph, 250
Sweat glands, 4
Sub-lethal damage repair, weevils, 160

Telogen, 12
Temperature and DNA breaks, 74
− − LD50, 104
Thermal neutrons, fission cross section, 264
Tissue, elementary composition, 196
Tongue filiform papilla, 7
Track average LET, 228
− counting, 260, 269
− etching SSNTD, 259
− geometry, 260
− segments, 226
− viewing, 268
Transmutation, atomic, 65
− effects, 61
− − in DNA, 67
− and DNA breaks in phages, 77
−, electronic effects, 65
Two dose responses of weevils, 148

Ultrastructural abnormalities, irradiated epidermis, 41

Vasculature and irradiation, 14
Vascular supply to skin, 4

Wall-less proportional counter, 234
Weevil, developmental stages, 103
−, irradiated, 97

Yeast, ^{32}P, 85

AUTHOR INDEX

Adloff, J.P., see Apelgot, 61

Apelgot, S. and Adloff, J.P., Transmutation effects of ^{32}P and ^{33}P incorporated in DNA, 61

Dale, R.G., see Liversage, 97

Liversage, W.E. and Dale, R.G., Dose-time relationships in irradiated weevils and their relevance to mammalian systems, 97

Polig, E., The localized dosimetry of internally deposited alpha-emitters, 189

Potten, C.S., The cellular and tissue response of skin to single doses of ionising radiation, 1